9th Annual International Wafer-Level Packaging Conference 2012

(IWLPC 2012)

San Jose, California, USA
5-8 November 2012

ISBN: 978-1-62276-897-4

Printed from e-media with permission by:

Curran Associates, Inc.
57 Morehouse Lane
Red Hook, NY 12571

Some format issues inherent in the e-media version may also appear in this print version.

Copyright© (2012) by Surface Mount Technology Association (SMTA)
All rights reserved.

Printed by Curran Associates, Inc. (2013)

For permission requests, please contact Surface Mount Technology Association (SMTA)
at the address below.

Surface Mount Technology Association (SMTA)
5200 Wilson Road
Suite 215
Edina, MN 55424

Phone: (952) 920-4682
Fax: (952) 926-1819

www.smta.org

Additional copies of this publication are available from:

Curran Associates, Inc.
57 Morehouse Lane
Red Hook, NY 12571 USA
Phone: 845-758-0400
Fax: 845-758-2634
Email: curran@proceedings.com
Web: www.proceedings.com

IWLPC 2012 CONFERENCE PROGRAM

Tutorials

Monday, November 5th

T1: TSV and Other Key Enabling Technologies for 3D IC/Si Integration
John H. Lau, Ph.D., Industrial Technology Research Institute (ITRI)
8:30am-12:00pm
Monterey

T2: IC Package Cost Reduction Using Supply Chain Modeling
Chet Palesko, SavanSys Solutions LLC and Jan Vardamann, TechSearch International, Inc.
8:30am-12:00pm
Carmel

T3: Embedded Components Design and Process Implementation
Vern Solberg, Invensas Corporation
1:30pm-5:00pm
Monterey

T4: Failure Analysis Techniques for a 3D World
Chris Henderson, Semitracks, Inc.
1:30pm-5:00pm
Carmel

Tuesday, November 6th

T5: Wafer-Level Packaging
Luu Nguyen, Ph.D., Texas Instruments, Inc.
8:30am-12:00pm
Monterey

T6: PiezoMEMS from Design to Packaging
Dag Wang, Ph.D., SINTEF ICT
8:30am-12:00pm
Carmel

T7: Packaging for MEMS
Andy Oliver, Ph.D., Wireless Integrated MicroSensing and Systems Research Center (WIMS2)
University of Michigan
1:30pm-5:00pm
Monterey

T8: Failure Mode Analysis of Flip Chip and Advanced Package and Board Assemblies
Brian Lewis, Engent, Inc.
1:30pm-5:00pm
Carmel

IWLPC 2012 CONFERENCE PROGRAM

Wednesday, November 7th
Opening Comments
Andrew Strandjord, Ph.D., PacTech-USA, Conference General Chair
9:00am – 9:10am, Cedar Ballroom

Morning Plenary
Silicon Interposer: Much More than a "Piece of Silicon"
Nicolas Sillon, Ph.D., CEA-Leti
9:10am – 10:00am, Cedar Ballroom

Coffee Break
10:00am – 10:30am
Exhibit Hall, Pine/Fir/Oak Ballroom

WLP TRACK

Session 1 – Wafer-Level Testing: Challenges and Solutions
Chair: Ted Tessier, FlipChip International
10:30am – 12:00pm, Monterey

10:30am
Embedded Barrel Spring Probe – Solution for WLCSP Testing   1
Frank Zhou, Ph.D., IDI, Smith Group

11:00am
Wafer-Level Testing Challenge for Flip Chip and Wafer-Level Packages   6
Muru Yogathasan, STATS ChipPAC, Ltd. speaking on behalf of Lim Kok Hwa, STATS ChipPAC, Ltd.

11:30am
Processing, Bumping and Assembly of Single Chip Plated Ni/Pd Over ALCAP Bond Pads for Flip Chip Applications and Prototyping   11
Brian Lewis, Engent, Inc.

3D TRACK

Session 2 – Process and Materials
Chair: Keith Cooper, SET North America
10:30am – 12:00pm, San Carlos

10:30am
Understanding the Stacked Dies Interface Temperature and its Influence During the 3D IC Thermocompression Stacking Process   12
Robert Daily, IMEC

11:00am
Evaluating Methods of Shipping Thin Silicon Wafers For 3D Stacked Applications   17
Richard Allen, SEMATECH

11:30am
3D Packaging- Synthetic Quartz Substrate and Interposers for High Frequency Applications   23
Vern Stygar, Asahi Glass Corporation

MEMS TRACK

Session 3 –Wafer Bonding MEMS and Hermeticity Standards
Chair: Russell Shumway, Amkor Technology
10:30am – 12:00pm, Santa Clara

10:30am
Co-Design Strategies for MEMS Packaging   28
Mary Ann Maher, SoftMEMS

11:00am
Yield and Strength of Metal Wafer-Level MEMS Device Sealing Using Al, Au, or Ti   32
Kari Schjølberg-Henriksen, Ph.D., SINTEF ICT

11:30am
Sealing Dispensing for MEMS Wafer Capping   39
Heakyoung Park, Nordson ASYMTEK

Lunch Break
12:00pm to 1:30pm
Exhibit Hall, Pine/Fir/Oak Ballroom

PANEL DISSCUSION

MEMS Integration Strategies: From A Packaging Perspective
1:30pm – 3:00pm, Cedar Ballroom
Moderator: Roger Grace, Roger Grace Associates & Russell Shumway, Amkor Technology
Panelists:
- Matthew Apanius, Desich SMART Center
- Mary Ann Maher, SoftMEMS
- Sean Ding, Ph.D., MEMSIC
- Thava Thavarajah, Fairchild Semiconductor

Coffee Break
3:00pm – 3:30pm
Exhibit Hall, Pine/Fir/Oak Ballroom

WLP TRACK

Session 4 – Wafer-Level Packaging Materials & Process
Chair: Steven Xu, Qualcomm
3:30pm – 5:00pm, Monterey

3:30pm
Low Stress Thick Film Photopatternable Thick Film Silicones for Large Die Wafer-Level Applications   45
Herman Meynen, Dow Corning Europe S.A.

4:00pm
A New Single Wafer Cleaning Technology for Advanced Packaging Applications   51
Richard Peters, Ph.D., Dynaloy, LLC

4:30pm
Silicone and Cleaning Solvent Compatibility   55
Michelle Velderrain, NuSil Technology, LLC

3D TRACK

Session 5 - TSV's and Wafer Thinning
Chair: Peter Ramm, Fraunhofer EMFT
3:30pm – 5:00pm, San Carlos

3:30pm
TSV Process Variations for 2.5D and 3D Semiconductor Packaging   62
Vern Solberg, Invensas Corporation

4:00pm
Single Sided Wet Etching for Thinning, Packaging, and Texturing Applications   69
Ricardo Fuentes, Ph.D., Matech

4:30pm
Deposition Processes for Competitive Through Silicon Via Interposer For 3D   74
Cyprian Uzoh, Invensas, Corporation

MEMS TRACK

Session 6 —MEMS WLP, 3D Integration, and Reliability
Chair: Maaike M.V. Taklo Ph. D., SINTEF ICT
3:30pm – 5:00pm, Santa Clara

3:30pm
Bonding and Contacting of Vertically Integrated 3-D Microscanners   80
Maik Wiemer, Ph.D., Fraunhofer Institute for Electronic Nanosytems (ENAS)

4:00pm
MEMS Hermeticity and Reliability Testing Today   87
Mike Shillinger, Innovative Micro Technology

4:30pm
Reliability of TSV and Wafer-Level Bonding for a 3D Integrable SOI Based MEMS Application 93
Maaike M.V. Taklo, Ph.D., SINTEF, ICT

EXHIBITS & WELCOME RECEPTION
5:00pm – 6:00pm
Exhibit Hall, Pine/Fir/Oak Ballroom

KEYNOTE DINNER AND ADDRESS
A Trojan Chip in Your Smartphone? It's Coming...
John Ellis, *bestselling author of 'Dormant Curse'*
Wednesday, November 7, 2012 | 6:00pm - 8:00pm, Cedar Ballroom

IWLPC 2012 CONFERENCE PROGRAM

Thursday, November 8th

Morning Plenary
3D Integration – A Corner Technology for Heterogeneous Integration
Paul Marchal, Ph.D., IMEC
Cedar Ballroom
9:10am – 10:00am

Coffee Break
10:00 – 10:30am
Exhibit Hall, Pine/Fir/Oak Ballroom

WLP TRACK

Session 7 – Wafer-Level Packaging Reliability
Chair: Janet Love, Interconnect Devices, Inc.
10:30am – 12:00pm, Monterey

10:30am
Characterization of eWLB PoP Structures   109
Tom Strothmann, STATS ChipPAC, Ltd.

11:00am
Marked Reliability Increase of Plastic-Cored Solder Ball for Large Size Wafer-Level CSP   114
Hiroya Ishida, Sekisui Chemical Co., Ltd.

11:30am
Pad Lift Failure Mode Investigation for Wafer-Level Package   100
Laurent Gay, STMicroelectronics is not available to speak due to travel restrictions
* Paper and presentation will be available on proceedings

3D TRACK

Session 8 – TSVs and Lithography
Chair: Laurette Nacamulli, Dow Chemical Company
10:30am – 12:00pm, San Carlos

10:30am
3D TSV Micro Cu Pillar Chip-To-Substrate/Chip Assembly/Packaging Technology   123
Tom Strothman, Ph.D., STATS ChipPAC, Ltd

11:00am
Verification of Back-to-Front Side Alignment for Advanced Packaging   127
Robert Hsieh, Ph.D., Ultratech, Inc.

11:30am
A Study of a Development Lithography Processes for 3Di Plating Applications   134
Patrick Kearney, P.E., Tokyo Electron Europe, Ltd.

Lunch Break
12:00pm to 1:00pm
Exhibit Hall, Pine/Fir/Oak Ballroom

PANEL DISSCUSION

3D Integration: How Did We Get Here? Where Do We Need To Go Now?
1:00pm – 2:30pm, Cedar Ballroom
Moderator: Keith Cooper, SET North America

Panelists:
- Jeff Calvert, Ph.D., Dow Chemical
- John Lau, Ph.D., Industrial Technology Research Institute (ITRI)
- David Love, Rambus
- Garret Oakes, EV Group
- Peter Ramm, Ph.D., Fraunhofer EMFT
- Tom Strothmann, STATS ChipPAC

Coffee Break
2:30pm – 3:00pm
Conference Foyer

WLP TRACK

Session 9 - Fan-Out Wafer-Level Packaging Technologies
Chair: Curtis Zwenger, Amkor Technology
3:00pm – 5:00pm, Monterey

3:00pm
Innovative 2.5D Solution: Extended/Flip Chip eWLB (Embedded Wafer Level Ball Grid Array) Technology   139
Tom Strothman, STATS ChipPAC, Ltd.

3:30pm
Developments of Fan-Out Wafer-Level Packaging Technology for System-in-Package on Wafer-Level (WLSiP)   145
Jose Campos, NANIUM, S.A.

4:00pm
Adaptive Patterning for Panelized Packaging   153
Chris Scanlan, Deca Technologies

3D TRACK

Session 10 – 3D Materials and Debonding
Chair: Laurette Nacamulli, Dow Chemical Company
3:00pm – 5:00pm, San Carlos

3:00pm
Optical Profilometry of Substrate Bow Reduction Using Temporary Adhesives   159
John Moore, Daetec LCC

3:30pm
Wafer Spray Coating for Pre-Applied Underfill   165
Akira Morita, Nordson ASYMTEK

4:00pm
Equipment and Process Solutions for Low Cost High Volume Manufacturing of 3D Integrated Devices   171
Garrett Oakes, EV Group, Inc.

Additional Papers

Heterogeneous Packaging for MEMS   175
Matt Apanius

OMEMS Integration Issues   178
Mary Ann Maher

Package Modeling of MEMS Devices 181
AManickam Thavarajah, John Bloomsburgh, Fairchild Semiconductor

EMBEDDED BARREL SPRING PROBE – SOLUTION FOR WLCSP TESTING

Jiachun Zhou (Frank)*, Jon Diller**
IDI, Smith Group
*Gilbert, AZ, ** Kansas City, KS
fzhou@idinet.com

ABSTRACT

Interconnect Devices, Inc. (IDI) has developed a new technology, the Embedded Barrel Spring Probe (EBSP), for very fine wafer level chip scale package test. IDI uses a special process to embed a spring contact probe barrel into a socket body. When assembled with other probe components, this creates a rugged contact solution capable of extraordinarily fine pitch. Benefits include:

1. Reliable contact to device through greater compliance (up to 500um), more contact force (16gf and above), and a four point crown contact tip.
2. Improved accuracy by eliminating the gap between the barrel and the corresponding cavity in the socket.
3. Improved signal integrity with shorter total connection length (~ 4mm) or signal path from device to test board.
4. Field maintainable; the contact element can be replaced at the production site.

Commercial embodiments of this technology are referred to by IDI as "Monet Sockets". A patent of this technology is in process.

Key words: WLCSP testing socket, Wafer testing

INTRODUCTION

The package technologies for smaller semiconductor devices, such as wafer level chip scale package (WLCSP) and micro chip scale package (MicroCSP), have been accelerated to grow for high volume production and to be used in consumer products. While these small devices offer functional and cost advantages, they present significant manufacturing and test demands. One of these challenges comes from their smaller pitch among device leads. Traditional IC packages usually have pad pitch of >0.4mm. Product miniaturization and cost per die have driven WLCSP pitches to 250 μm and below. This change to smaller pitch requires the final package test industry to find a new type of testing set up, such as test socket and contactor, to meet mechanical and electrical performance of IC device in testing.

The typical approach to WLCSP test borrows from conventional front-end test techniques, and employs probe cards based either on cantilever needles or 'vertical probe cards' based primarily on buckling beam technology. Each of these, beyond the attraction of their familiarity, has significant limitations.

Cantilever cards are incapable of RF testing as is often required with WLCSP final test. They are difficult to maintain and require extensive training and capital to repair. They can only be designed to probe an array of limited depth; high pin count devices cannot be addressed. They are easily damaged in handling.

Vertical cards based on buckling beams are more reliable and provide better RF performance; their array depth is unlimited. However, many embodiments cannot be repaired in the field at all. They represent a significant capital outlay, especially since their long repair cycle means that more spares must be kept on hand.

Both of these technologies are most significantly handicapped by their lack of compliance. While this can be overcome in contacting very flat wafer bond pads, it is a significant liability when attempting to make contact to the more diverse topography of WLCSP bumps.

In response to this, the industry has been using technology borrowed from package final test – interposers based on spring contact probes. WLCSP pitches had been limited to 400 μm until recently, and spring contact probes of that pitch are relatively common. Spring probe-based interposers provide exceptional reliability in part due to their outstanding compliance-to-length ratio. They are relatively inexpensive and extremely easy to maintain in the field with minimal training and tooling. However, its reliability and performance of contact are significantly degraded when its diameter decreases, especially at pitches below 400 μm, due to mechanical limitations.

We may review as an example a typical 250 μm probe as is available from several vendors, shown in Figure 1.

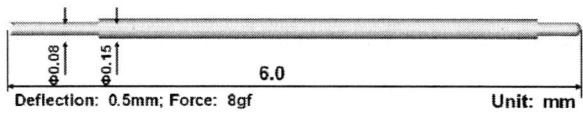

Figure 1. Traditional Spring Probe, 0.25mm Pitch

This probe is fragile due to its very low ratio of diameter (0.15mm) and length (6mm), which brings difficulties in handling process and actual usage in testing. Contact force is extremely limited as well, making reliable performance over cycles difficult to achieve.

Consequently, there is a strong desire in the marketplace for an interposer solution which provides the benefits of spring contact probe technology at pitches finer than 250 μm. To achieve this, it is necessary to depart from the normal paradigm of spring contact probe structure.

EMBEDDED BARREL SPRING PROBE INTERPOSER STRUCTURE

The primary obstacle in creating a reliable <250 μm spring contact probe is the barrel, a housing into which the spring and plungers are normally crimped. At the necessary diameters for such fine pitches, the barrel wall must be extremely thin. A thinner barrel permits a larger diameter plunger and spring, increasing the robustness and usability of the system.

However, making the barrel too thin poses serious risks. While techniques such as electroforming can produce very thin barrels, they are very fragile; damage to the outside creates performance problems that may be difficult to detect in process, and poses serious challenges for field replaceability.

Various companies produce spring contact probes that function without a barrel, using the support of the interposer cavity to guide the spring and plunger. There are two serious limitations with this technique. Electrical resistance will be relatively high because the current is passing through the entire length of the spring; this can limit the measurement capability and current carrying capacity of the interposer. Secondly, the machined plastic interposer cavity often provides less than ideal guidance and support for the necessarily precise travel of the plungers and spring.

It is possible through special processes to embed a very thin barrel in the plastic interposer body. This allows the plunger and spring to be as large as possible, enhancing the function and reliability of the system, while retaining the benefits of a barrel with respect to both mechanical guiding and electrical resistance. Figures 2 and 3 show the major structure of test probe tower with an embedded spring probe, including the contactor cartridges, the fan out PCB, and the bottom side contactor. Contactor cartridges use embedded spring probes, shown in Figure 3, as an interposer to drive contact between the device lead and the related pads on a fan out PCB. The fan out PCB has the function of converting small pitch pads (such as <0.25mm) to large pitch pads (>0.8mm pitch). The bottom side contactor connects the fan out PCB to the test load board at a manageable pitch, easing routing and simplifying the load board; this not only reduces capital expenditure, but also increases the reliability of supply of a functional load board. The bottom side contactor uses spring probes with diameters greater than 0.6mm to provide very reliable electromechanical connections between fan out PCB pads to test load board.

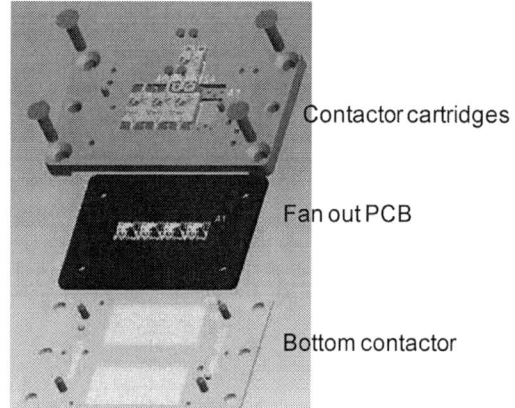

Figure 2. EBSP Contactor

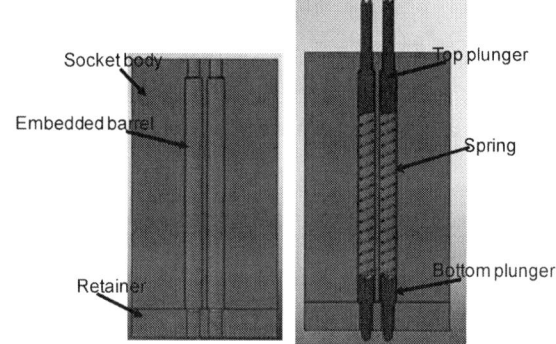

Figure 3. EBSP Internal Structure

The contactor cartridge is a key sub-assembly in this test probe tower. Usually, the contactor cartridge body uses high strength composite dielectric materials, such as ceramic PEEK, to provide very solid support to all contactor barrels and handle the compression force during testing. There are cavities holes in the body that are lined with barrels. These barrels are formed from high strength alloys and are lined with gold on the internal surfaces of the barrel. The contactor body provides dielectric insulation between all contacts. The contact of the device lead to the fan out PCB pad relies on the balance of probe components, being the top plunger, spring, and bottom plunger. The spring, contained within and guided by the barrel, provides the mechanical compression force to keep good connection among all pads and tips. WLCSP testing is usually multi-site; quad site and octal site applications are common, with sixteen sites being the current maximum. Thus, one test probe tower may have 4, 8, and even more contactor cartridge, depending on the desired throughput of the system, the available routing space in the load board, and the available resources in the tester.

Advantages of this embedded barrel spring probe contactor can be summarized as:
1) Pitches below 200 μm, providing needed contact density
2) Electrical paths as short as 3.2 mm, which enables good RF performance
3) Contact force as high as 15 grams, which permits reliable contact to solder-coated bumps

4) Compliance as extensive as 250 micron, providing ample overdrive for easy setup and repeatable contact.
5) Excellent pointing accuracy

PERFORMANCE EXCELLENCE

The embedded barrel spring probe preserves the key advantages of traditional spring contact probes – excellent compliance-to-length ratios, high contact force, and RF performance -- while avoiding their limitation of pitch. At same time, EBSPs offer several performance advantages over traditional wafer probe cards such as more reliable contact, better compliance, easier repair and maintenance, and better signal integrity for higher frequency.

Reliable Contacts

The key function of the contactor in both wafer and package testing is to provide reliable electrical contact between each pad or ball on the device under test and the corresponding pads on the tester interface board. Compliant interconnects such as spring contact probes accomplish this by compensating for all z-axis dimensional tolerances of the device under test and all components of the test interface. These tolerance ranges increase in proportion to the physical size of the device and numbers of sites in each test interface unit; thus, as package complexity and the number of available test resources increases, the compliance of the interconnect must proportionally increase.

Cantilever test probes provide compliance by bending; they move both laterally (or 'scrub') and in the Z axis. This lateral movement and the elastic range of the contact limit the usable compression of the cantilever probe, and most cantilever test probes provide at most 100 microns of travel – less as pitch decreases and small device lead size creates potential for the probe to scrub off of the lead.

Vertical probe cards typically rely on 'buckling beam' technology, and their travel while largely linear is much more limited by the elastic limits of the contact; vertical technologies therefore provide even less compliance than cantilever systems.

It is well established that a coil spring can provide more deformation with greater force at the same elastic component length. In spring probe applications, the typical maximum compression (or total travel) is 25~35% of spring length. A probe with a ~4mm overall length is well within limits for good electrical performance, and feature at least 1.5 mm of spring free length; as a consequence, the EBSP has 500 microns of compliance available to the test engineer.

A 500% increase in compliance dramatically lessens the burden of tolerance elimination and setup precision and consistency faced by the test engineer. This improvement is directly reflected in increased production yields and shorter setup periods.

The contact force exerted by the interconnect is another key factor which predicts interface reliability. Device leads usually feature a degree of oxidation which must be scrubbed or penetrated to make contact between the contact tip and conductive device lead metal. Spring contact probes typically rely on penetration, since they comply without significant scrub. A conventional spring probe, with its necessarily thick barrel, is generally limited in force to less than 10 grams at usable contact lengths. This is often insufficient, despite the sharpest tip features, to penetrate the oxide layer. Because of its thin barrel wall and correspondingly thicker spring, the EBSP provides a force of ~ 15gf for 0.25mm pitch, which has proven to be much more effective in production test.

Contact resistance (CRES) the sum of the interconnect's bulk resistance and the interface resistance between the tip of the interconnect and the device lead. CRES limits both the current carrying capacity of the interconnect and the range of usable measurements which may be made with the test interface, and is therefore considered as the most significant performance characteristic of a test interconnect for many users.

The EBSP has a CRES approximating 150mOhm, less than half of that typical of cantilever probes. Figure 4 shows the correlations of CRES, force and probe travel. At working travel, the CRES is ~150mOhm and force is 17gf.

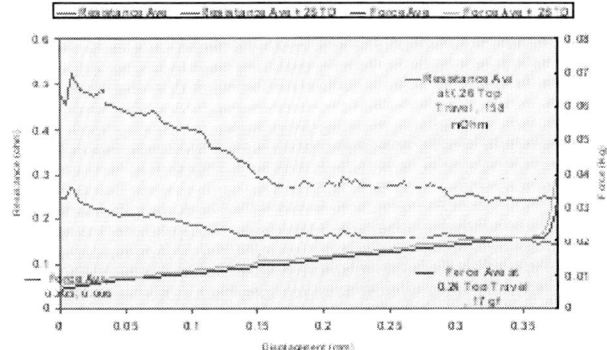

Figure 4. Cres & Force vs, Travel

The mechanical life of the contactor is another major concern in contactor applications. Test tooling is usually valued in terms of cost per cycle – the investment required for the interconnect and successive refurbishments divided by the number of cycles which the contactor may be expected to endure EBSP meets this customer expectation, as shown in Figure 5, and works well up to one million cycles under certain environment.

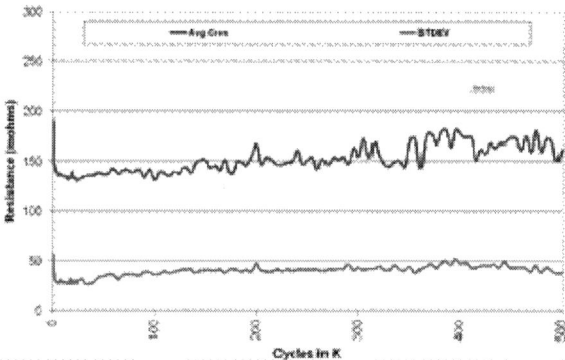

Figure 5. Cres over 500K Cycles

Even more significant is the relatively easy maintenance process associated with the EBSP interposer. The primary failure mode of test interconnects is wear at the contact tip, as successive cleanings of device lead material from the tips of the probes wears away its noble plating layer. When the plunger tip of the EBSP is worn out, this contactor can be disassembled to replace the worn out plunger. This disassembly and assembly is completed by removing screws on the bottom side of the interposer, freeing the retainer plate from the body of the contactor cartridge. It can be done at the production site with minimal training and tooling.

By contrast, both cantilever and vertical probe cards are very complex, requiring highly trained engineers or (particularly in the case of vertical cards) factory refurbishment to effect repair of the card.

Alignment and Pointing Accuracy
A reliable contact between device lead and the contact element requires extremely precise alignment of the contactor to the device under test. As device lead pitch decreases, device lead feature size is reduced accordingly. The misalignment in the x and y axes of the test interface unit therefore become greater in proportion to this feature size; at the moment that the misalignment becomes greater than the radius of the feature, the contact element will completely miss contact with the device lead. In fact, the manufacturing tolerance does not decrease for smaller pitch device and proper alignment must rely on features of contactor.

WLCSP interposers using conventional spring contact probes require that the contact element move freely within cylindrical cavities in the interposer body. A certain minimum gap, or 'working clearance,' must exist between the barrel of the spring probe and the inside diameter of the cavity in the interposer body. This gap allows the spring probe component so guided to tilt, which directly affects alignment of the contact element to the device lead.

Because the EBSP has embedded the barrel tightly inside cavity hole, the gap is eliminated between the barrel and the cavity hole. The EBSP plunger moves inside of the embedded barrel and the good straightness of barrel inside

cavity can maintain good alignment of plunger tips to device pads. Figure 6 is an example of contact marks of EBSP plunger tips on WLCSP balls. All balls have 3~ 4 contact marks located on center area of the device leads, which provides evidence of good alignment of contactor tips to balls.

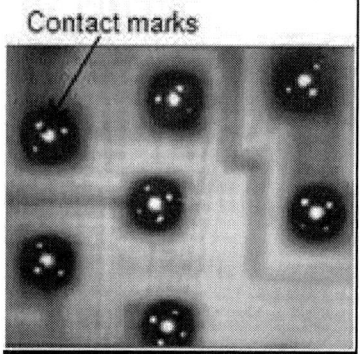

Figure 6. Contact Marks on Device Balls

Alignment of EBSP plunger tips to device pad/balls is related to accumulated tolerances in X-Y direction of all components. Tolerance analysis at socket design stage can predict the possibility of misalignment of the contact elements to the device leads. Although the embedded barrel eliminates the gap of barrel to cavity hole, other factors, especially position accuracy of cavity holes determined by machining process, assist good alignment of the contact element to the device lead. Figures 7 & 8 provide examples of tolerance analysis curves of an exemplary EBSP contactor socket. Considering the gap between the probe plunger inside barrel, the curves show that the EBSP contactor has a pointing accuracy of 0.046mm while a conventional spring probe has a pointing accuracy of ~ 0.06mm. The EBSP contactor therefore permits a ~ 20% improvement in pointing accuracy when compared to a traditional spring probe contactor.

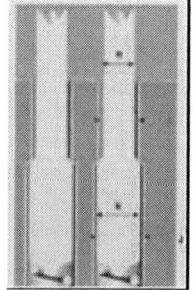

Figure 7. Tolerance Analysis Model

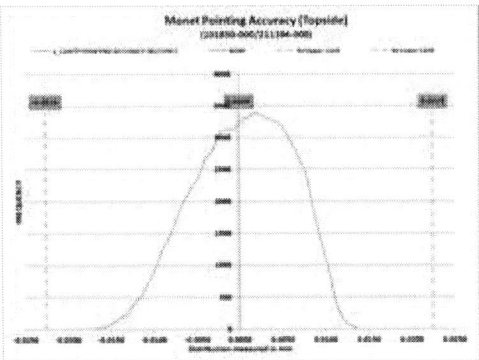

Figure 8. Tolerance Analysis Results

RF Performance

RF performance is another consideration for many contactors for wafer and package testing. As discussed above, the EBSP contactor has a shorter electric path (~ 3.3mm) than other wafer testing probes and traditional probes at the same pitch. Reducing the signal path of contact element is one of the most straightforward approaches to improve the contactor's signal integrity.

The RF performance of an EBSP contactor was simulated using Agilent HFSS software. Figure 9 is the simulation methodology, with one signal probe paired with two grounding probes. As shown in Figure 10, for pitch 0.25mm, the insertion loss of EBSP contactor is 10GHz with -1dB signal loss, which meets requirements of RF performance for most WLCSP and wafer.

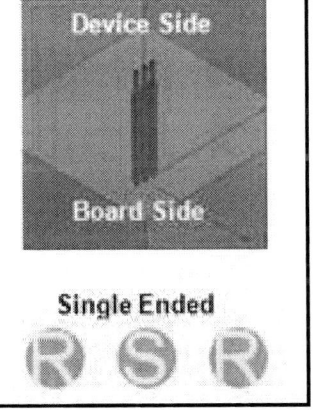

Figure 9. RF Simulation Model

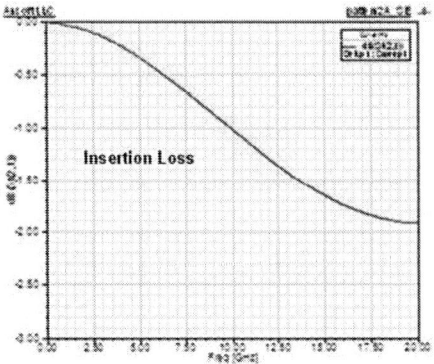

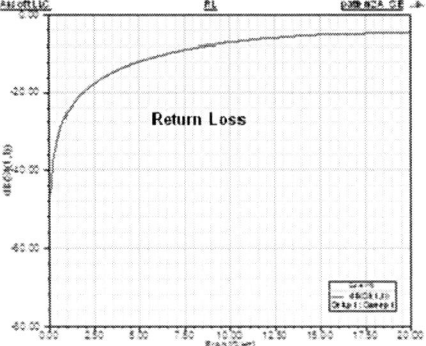

Figure 10. RF Simulation Results

SUMMARY

Spring contact probes have been established as the preferred contact technology for large pitch (0.4mm) WLCSP device testing. However, conventional spring contact probes begin to show serious limitations in applications below 400 μm pitch.

By embedding the probe barrels in the contactor body, significant gains can be made in robustness, alignment, and signal integrity performance. Especially, better pointing accuracy of Embedded Spring Probe Contactor provides more reliable contact between probe tips and WLCSP device pads or balls.

ACKNOWLEDGEMENTS

Special thanks to the following persons and organizations who were instrumental in the creation of this paper: Praba Prabakaran, Khaled Elmadbouly, Brad Henry, Dave Henry, Resty Querubin, Kevin DeFord, Peter Fitzsimmons, Tim Marshall – Interconnect Devices, Inc.

WAFER-LEVEL TESTING CHALLENGE FOR FLIP CHIP AND WAFER-LEVEL PACKAGES

Lim Kok Hwa and Andy Chee
STATS ChipPAC Ltd.
Singapore, Singapore
kokhwa.lim@statschippac.com; kenghwee.chee@statschippac.com

ABSTRACT

Wafer-level packages continue to see strong growth driven by mobile phones, portable players, digital cameras and tablets. All these devices use small form factor and low profile packages such as a wafer-level chip scale package (WLCSP) as it fits the requirements. Conventional flip chip die with solder bump is growing due to the increasing number of new design packages converting from wire bond and new flip chip interconnects such as copper pillar and micro bump are growing as a result of strong demand in 3D stacked ICs. Both WLCSP and flip chip need to be electrically tested in wafer format at some point in the assembly process, either as a Known Good Die (KGD) in 3D ICs or an end product that goes into the PCB of an electronic gadget.

Wafer sort or wafer-level testing was once considered as a method to save packaging cost as this process sorts out bad die before it is assembled into a package. However, today wafer sort or wafer-level testing is an important process for yield enhancement of flip chip packages and a final test requirement for WLCSPs.

The challenge of wafer-level testing has grown significantly due to the increasing complexity of the die or packages. The current technology started to see limitations in hardware and tools. This paper investigates the challenges facing wafer-level testing as well as examining the solutions available to overcome these challenges, identifying the gaps and additional innovation needed to overcome these challenges.

INTRODUCTION

For many years, wafer-level testing or wafer sort was employed for two reasons. Firstly, it is to sort out bad dies to prevent them from being assembled into the final package and this would save the cost of packaging. Secondly, the purpose of wafer-level testing is to provide yield feedback to the wafer fab in time to control the wafer fabrication process [1]. These two purposes are still valid today. However, as packaging technology advances, wafer-level packages such as flip chip and WLCSP changes the wafer test requirements not only for yield enhancement of the final package, but also full coverage of testing as it takes the final form of the end package. These bumped types of wafers need to be contacting at solder bumps instead of pads during test. It seems relatively easy compared to contacting pads as bumps are larger in size and wider in pitch; however, the real situation is more complex as the challenges of testing these bumped chips are different.

Wafer-Level Packaging (WLP) refers to the technology of packaging an integrated circuit at wafer-level and it takes many forms. There are WLCSP, flip chip dies built into flip chip or 3D packages and all these dies or packages need wafer-level testing.

PACKAGES REQUIRE WAFER-LEVEL TESTING

The traditional functions of a semiconductor device package are to provide environmental protection for the die built in it. However, this need of protection becomes less important with the short expected life of the IC and advancement in material used in building the wafer-level package. Wafer-level Chip Scale Packaging (WLCSP), as its name "Chip-scale package" implies, is a wafer-level packaged die with grid array bumps at pitches of 0.4mm and above. This bumped integrated circuits package goes directly onto printed circuit boards of end electronic products such as mobile phones which are one of the fastest growing package types in the semiconductor industry. Fabrication and testing of WLCSPs are done at the wafer level. As the manufacturing cost drops with increased wafer size and reduced die size, cost competitiveness becomes a main motivation to replace traditional QFN packages in mobile devices with WLCSPs. WLCSP is not new; it started with small size and low I/O a decade ago till today with high pin counts and larger package size as shown in below figure 1.

Figure 1: Large WLCSP with I/O above

In the extension of WLCSP, fan-out wafer-level technology overcomes interconnect density limitations on PCBs and reduces the need for bump interconnects [2]. One typical example of fan-out technology is embedded wafer-level BGA (eWLB) show in below figure 2.

Figure 2: Embedded Wafer Level BGA (eWLB)

Flip chip and wire bond are the two standard processes to connect the die to a substrate. The flip chip device continues to see strong growth with an increasing number of new designs converting from wire bond. The reasons for the conversion can be attributed to the advantages of its thermal and electrical properties, lower cost and ability to support high I/O count with a smaller package relative to the die size; as shown in figure 3 below. The connection area required for flip chip is much smaller than what is required for wire bonding because all the I/Os from the chip are connected through the bottom. [3]

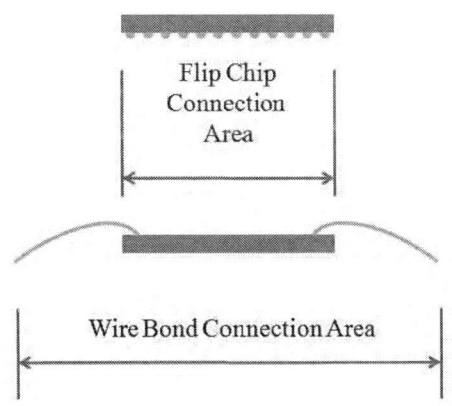

Figure 3: Flip chip connection area compared to wire bond connection area

Flip chip packaging uses area array solder bump configurations and its typical bump pitch is 150μm. Figure 4 shows examples of final flip chip packages in the form of fcBGA and fcFBGA that are in high volume manufacturing now.

Figure 4: Flip Chip packages

As the need for increased I/O signal connections grow; the bump pitch has to decrease. In order to maintain standoff height with decreasing pitch, a solid column type bump has to be employed and this is copper column bump. Flip chip with copper column bump is expected to see greater demand as the need the fine pitch solutions increase. There are two approaches for copper column bumps as shown in figure 5 below: one with a solder cap and the other without. Copper column bump pitch can be as low as 40μm and this makes testing them very difficult.

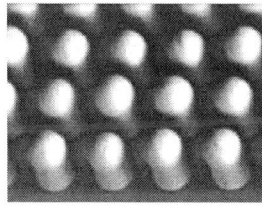

Figure 5: Copper column bump. (Top photo shows solder cap version)

3D packaging is driven by wireless and consumer products that require package level functional integration in a small footprint, low profile and lower in cost. 3D packages come in many different forms. For some, flip chip dies are used in the package and certainly these dies must be known good die (KGD) to ensure the end package quality. Figure 6 shows an example of a 3D package with flip chip die.

Figure 6: Example of 3D package

NEW CHALLENGES IN WAFER-LEVEL TESTING

The paradigm for wafer-level test is changing rapidly. The original intent of final yield enhancement and feedback to the wafer fab for process control for performance wafer sort has expanded. With new package or die types needed to test in wafer level, different challenges have surfaced.

Comparing package handling in a test handler where wafers are effectively all the same, it would appear that the handling of wafers would be an easier task. The actual handling, though, is probably more difficult since wafers are extremely fragile and damaging even one wafer is unacceptable due to the large number of valuable dies lost. The handling equipment or prober needs to be extremely precise and intelligent to handle these fragile wafers.

Wafer-level test handling equipment or probers face significant technical challenges in different markets or wafer types. One of the common issues is the need for high parallelism test. For high parallelism requirements, the prober needs to have accurate alignment capability and very high force test chuck for presenting the wafer to probe card for contacting. There is even a need for full wafer contact of a 300mm wafer with thousands of solder bumps and it becomes a challenge for providing increasing insertion force without damaging the fragile wafer.

In wafer handling, different types of wafers need to be processed in the same prober. We see thin wafers at high warpage that need to be tested. These wafers go through back grinding to the required end thickness before wafer testing on the prober. Typically a standard prober can handle up to 350μm thickness. For wafers thinner than 350μm, there is an increased tendency to warp badly and make handling in a prober very difficult. In this case, a special tool or kit will be needed to handle this type of thin wafer. However, any wafer thickness of less than 150μm is still a challenge for the prober to handle.

With the new type of wafers such as eWLB gaining popularity, the prober now needs to handle wafers with different surfaces or materials. These eWLB wafers are different from normal silicon wafers as the backside of the wafers are a mold compound and they tend to have a higher warpage level with a similar thickness as silicon wafer as shown in figure 7

250μm thin eWLB

Figure 7: eWLB warpage condition

The above mentioned challenges need to be overcome with good handling equipment. Probers now need to have capability and/or options as listed below:

- High force chuck for high parallelism
- Wafer handling arm and chuck to handle thin and warped wafers
- Chuck that can handle different material of wafer such as eWLB wafers with molded material at backside of the wafer
- Intelligence optical alignment for fine pitch bumps
- High speed wafer changing and indexing

As the requirements get more stringent, probers will need to continue improving to meet all the new requirements. The overall end user would need a flexible and cost effective prober to handle different requirements and situations.

PROBE CONTACT CHALLENGES

As the need for increased I/O signal connections grow, the bump pitch has to decrease and maintain a greater standoff height. A solid column type of bump is created either in solid copper column or with an added solder cap.

Fine pitch is always a challenge for probe card geometry and precision. A prober that handles fine pitch wafers needs the precision to achieve an accurate probing. Currently, flip chip die with 150μm pitch is in mass production without issue. Typically this type of bumped wafer uses a vertical probe card with buckling beam type of contact pins. As the pitch of copper column bump scales below 100μm, this type of vertical probe card design cannot scale down to the required pitch. Advanced probe cards with MEMS technology will be needed, however by design, the cost of

MEMS probe cards are very expensive which eventually drives up the overall cost of test. The MEMS design probe pins are also not easily replaceable and this also drives up the cost of maintenance on the probe card and replacement after end of life of these pins. Some of the copper column bumps are even reaching as low as 40µm in pitch which already hits the limit of probe card technology even with MEMS technology.

For copper column bump type of wafers, there are different types of bumps as shown in below figure 8. The materials for these contacts are different. For copper column with solder cap, the contact is on the solder. The requirement is similar to normal solder bump. However, as the thickness of solder cap is low, care must be taken into consideration on the probe depth which must not be too deep. For copper column without solder cap, the challenges are different as the contact is on copper and it is harder to penetrate and thus a sharper probe tip and high force will be needed for good contact [4].

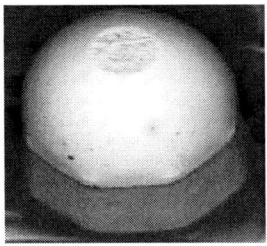

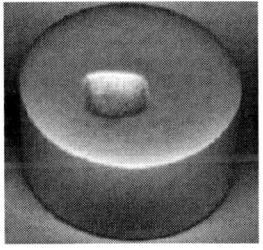

Figure 8: Different copper column bumps and the contacting point after probe

Probe card companies are investigating various approaches to deal with the challenges mentioned above. Probe card companies with traditional buckling beam notice that this technology can never scale to meet the challenges of fine pitch, especially with pitches below 50µm. Several new approaches using MEMS technology can meet the requirement for fine pitch, but until today, MEMS probe cards are still expensive in both initial fabrication cost as well as running cost. Probe cards appear to be lacking in the technology for above challenges, as such probe card companies need to innovate to align with the change in requirements.

PROBE TEST CHALLENGES
In the traditional single die IC manufacturing process, most of the defective dies are eliminated during wafer test, but due to inherent limitations of wafer sort and cost to overcome them, the industry approach is to allow some test escapes in wafer sort but filter these rejects at end during package testing [5]. The emergence of selling Known Good Dies (KGDs) has brought the importance of wafer sort to its highest level. For the reason of ensuring final yield of flip chip or 3D ICs, KGDs are also important as the die itself is a building block of the final package. With the need of KGDs, wafer-level testing is more challenging and it's effectively a "final test" requirement.

To meet the final test requirement, one noticeable trend is the use of final test socket technology into wafer sort with pogo pins type of probe cards. However, the challenge is really the scaling of fine pitch requirements of these types of probe cards as the current spring probe pogo pin is limited to about 0.3mm pitch. New technology is needed here to minimize the spring probe in order to meet fine pitch requirements.

Traditional interfacing hardware uses pogo tower in wafer sort, as illustrated in below figure 9. In the need for eliminating the long signal path between tester and the semiconductor device under test, a direct docking probe method is used which effectively removes the layer in between loadboard and probe pins. Direct docking requires a redesign of probe card, prober and docking mechanism.

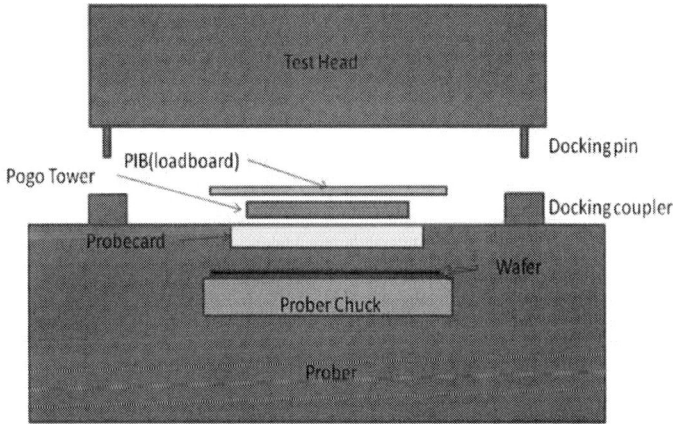

Figure 9: Traditional docking using pogo tower

SUMMARY
Wafer-level testing will need to accommodate the many trends affecting the semiconductor manufacturing industry. The changes in wafer-level testing are driven by wafer-level packages, these include WLCSP and flip chip die. As these packages or die advances; the challenges increase due to their form factor in terms of wafer thickness, size and material, bump material, pitch and shape. Different requirements have complicated the test environment and this is especially true for an outsourced semiconductor company that needs to support many different product types and testing requirements. There is no one solution that can fix all situations. Companies need to manage the complexity by

standardizing process, equipment and tooling so as to minimize overall manufacturing cost.

REFERENCES

[1] Mark Allison, "Wafer Probe Acquires a New Importance in Testing" Chip Scale Review _ May/June 2005
http://www.electroglas.com/PDF/Wafer_Probe_Testing_Im portance.pdf

[2] Aaron Hand Contributing editor -- EDN, August 24, 2010 "Wafer-level packaging pushes past new mobile demands"
http://www.edn.com/article/510248-Wafer_level_packaging_pushes_past_new_mobile_dema nds.php

[3] Chet Palesko and E. Jan Vardaman "Cost Comparison for Flip Chip, Wire Bond, and Wafer-level Packaging," Chip Scale Review, Volume 2, number 1
http://www.chipscalereview.com/archives/0111/index.php

[4] Senthil Theppakuttai, Bahadir Tunaboylu and Bahadir Tunaboylu "Probing Assessment on Fine Pitch Copper Pillar Solder Bumps" IEEE SW Test Workshop
http://www.svprobe.com/SWTW/SV_FPCopperPillarSolder Bumps.pdf

[5] Steve Pateras, "3D-IC Testing with the Mentor Graphics Tessent Platform"
http://www.mentor.com/resources/techpubs/upload/mento rpaper_66946.pdf

PROCESSING, BUMPING AND ASSEMBLY OF SINGLE CHIP PLATED NI/PD OVER ALCAP BOND PADS FOR FLIP CHIP APPLICATIONS AND PROTOTYPING

Brian Lewis and Dan Baldwin
Engent, Inc.
Norcross, GA
dan.baldwin@engentaat.com

Abstract

For Designers and Engineers, it is common during the process development cycle for new products to have limitations on the materials that are available for the prototype work. Most SMT devices are readily available in different formats/solder alloys to satisfy most of the needs for passive components. However, many times IC devices are limited to what is available from the fab or an IC broker. These limitations can mean that die only come in aluminum, wirebond ready I/O metallization or that the silicon wafers are already sawn and exist in single die formats. For applications where advancement in performance or miniaturization is needed, and the benefits of flip chip technology are attractive, then it is not trivial to be able to use these die. In these cases, the process of adding solderable plating technologies to the I/O bond pads is very favorable. The technologies are currently run for wafer lever plating baths, but very little has been done to evaluate single chip plating. Work in plating Ni/Pd onto the ALCAP structure has been performed to evaluate the process and feasibility of processing groups of singulated die with aluminum bond pads.

The work to be detailed in this paper will go through the chemistries used in the plating process onto an aluminum bond pad that makes it suitable for flip chip processes. Several bumping structures, such as solder bumping over this Ni/Pd plating stack up and plating over gold or copper stud bumps, are evaluated. A process for bumping the flip chips is also detailed. The data for shear testing of 10 variations of bumping structures, before and after 500 liquid thermal shock cycles, is detailed. Finally, a comprehensive study for assembly of solder bumped flip chips, with the selective plating process, will be detailed as well as a detailed analysis of the TC reliability of this assembly approach. It will be shown that selective Ni/Pd plating onto singulated, ALCAP bare die can allow, for these typical wirebond die, to be used in a practical approach solder flip chip process and provide reasonable reliability results when compared to a mainstream, wafer processed, solder bumped flip chip die.

Late Submission
Contact the SMTA for this paper.
P: 952-920-7682
E: Patti@smta.org

UNDERSTANDING THE STACKED DIES INTERFACE TEMPERATURE AND ITS INFLUENCE DURING THE 3D IC THERMOCOMPRESSION STACKING PROCESS

R. Daily, G. Capuz, P. Bex, A. Miller
imec
Leuven, Belgium
daily@imec.be

ABSTRACT

Scaling limits in IC manufacturing has made 3D IC integration an important approach. With this comes major fundamental challenges and one of these challenges is successfully stacking N>2 dice with an effective bond both mechanically and electrically.

Thermocompression stacking is a process in the 3D IC integration flow where dies are bonded directly over each other using pressure and heat. Understanding the effects of pressure and heat during the stacking process provides key learning's in achieving a successful bond.

In this paper, we measure the actual temperature in between the interface of two stacked dice during the stacking process. Comparing the materials involved and the changes occurring if set temperatures are modified. We also analyze the influence of these temperature values to the materials involved in the stacking process as a whole.

The experiment set involves die to die (D2D) and die to wafer (D2W) stacking. Experiments are done considering units with and without underfill using no-flow and film types.

We discuss key understanding on the correlation of the interface temperature between the dies during stacking and the programmed set temperature on the tool of the top and bottom chuck. It also touches on the effects on the bond and underfill reactions with these resulting temperatures.

In summary this paper covers tool parameters and material behavior during the thermocompression stacking process. The goal of the paper is to facilitate repeatable successful stacks by understanding one of the major influencing factors in thermocompression stacking, which is temperature.

Key words: 3D, stacking, temperature, chuck, thermocompression, D2D, D2W

INTRODUCTION

Technologies such as 3D integration are concepts being investigated to try to maintain the same level of scaling now that we are starting to approach the limit of Moore's law. There have been both significant achievements and challenges brought about by 3D integration. One of

its key processes, thermocompression bonding (TCB), where individual dies are stacked above each other on a bottom die (D2D) or wafer (D2W) is a challenge in itself. The characterization of the process to enable stacking single/multiple dies above each other efficiently is one of the key factors in its success.

Key to this characterization is understanding each individual factor affecting the process of stacking. Main controllable factors that influence the quality of stacks during TCB are force, temperature, time and environment. Identifying the influence of each individual factor during the stacking process provides insite in making bonds faster or yield more.

The goal of this paper is to understand the influence of temperature as an individual factor on the stacking process. Recipe temperature on a TCB equipment represents the setting that the tool is measuring on a point where the thermocouple will be placed as close to the unit as possible. This does not represent the actual temperature the interface of the stacks are being subjected. Experiments on this paper were performed with the intention of creating a more robust process by assessing the effects of temperature settings against interface temperatures as well as understanding their influence within the stacking process.

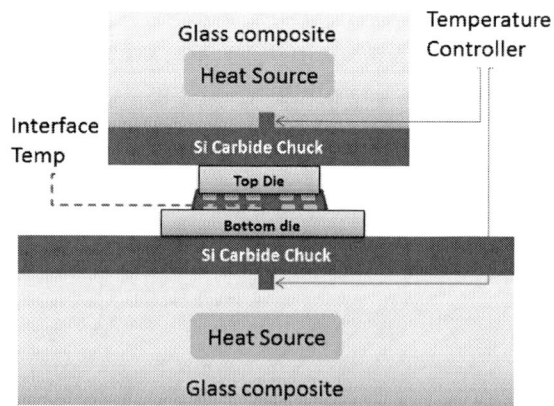

Figure 1. Thermo compression bonding setup

TEST VEHICLE AND EXPERIMENT

When measuring the interface temperature thermocouples of 36 gauge and the ends flattened to 15um fixed in between the top and bottom die/wafer (see figure 2) are used. To analyze the results the temperature readings from the thermocouple are recorded via a data logger.

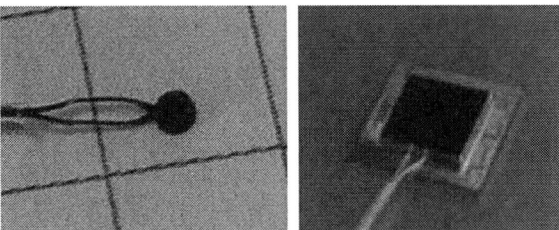

Figure 2. Temperature profiler for a 3D IC stack.

Actual stacks performed during the experiment were run on a test vehicle with 40um pitch and full array bumping (see figure 3).

Figure 3. Maskset A/Maskset B

The µbump diameters of both the top (maskset A) and bottom dies (maskset B) are 25µm. The bump height of the of the top die (maskset A) is 8.5 µm which has a Cu/Sn (5 µm /3.5 µm) metallurgy, while the bottom die (maskset B) only has a 5 µm height Cu (as shown in Figure 4), when stacked together, this results in a gap of around 13.5µm.

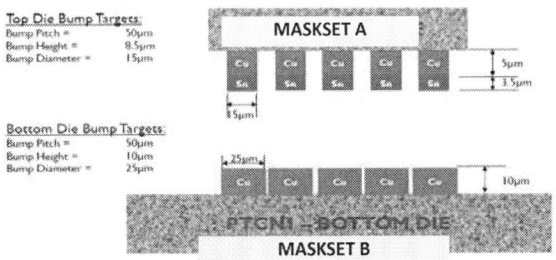

Figure 4. Test vehicle configuration

Pre-dispensed "no-flow" underfill material is used in between the stacks (see figure 5). The underfill has 40% filler content with a curing point of 200°C.

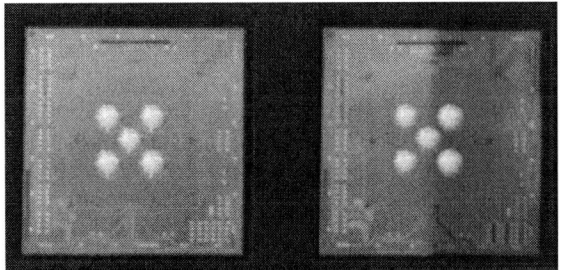

Figure 5. Underfill dispense pattern on device

Experiments were performed on a FC150 high accuracy bonder for D2D and FC300 for D2W, Both are from SET. The FC150 uses lamps to generate the heat, which is directed to a glass composite frame along with a silicon carbide piece as chuck. The thermocouple of the equipment is located underneath the chuck (see figure 1)

The experiments were done in two parts. First is the measurement of the actual interface temperature of set profiles and how it varies with a change in settings of the equipment. Second part is a simulated run with the measured temperature comparing low interface temperature bonding as well as bonding at higher temperatures.

INTERFACE TEMPERATURE EXPERIMENT
Using the thermocouple profiler, measurements was made on the tool according to given set points allowing us to see the actual temperature in between the dies during stacking. (see table 1)

Table 1. Interface temperature experiment

| Run | Targets | | | |
	Top Chuck Temp (°C)	Bottom Chuck Temp (°C)	Dwell time (sec)	Ramp time (sec)
A	290	290	20	50
B	270	70	30	170
C	450	70	5	10
D	350	350	10	10
E	350	70	10	15
F	350	350	20	5
G	350	350	30	5
H	350	200	30	5

The temperature set points were selected in order to find factors with which to enhance the process such as ramp rate, peak temp, and soak time. This is in comparison of the settings of the equipment as well as the actual temperature seen in between stacks. Shown on succeeding graphs.

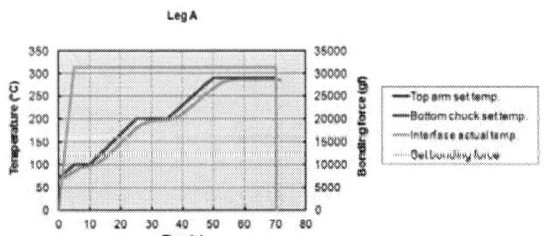

Graph 1.

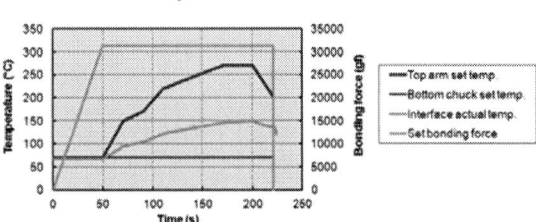

Graph 2.

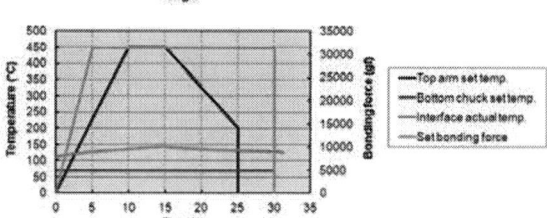

Graph 3.

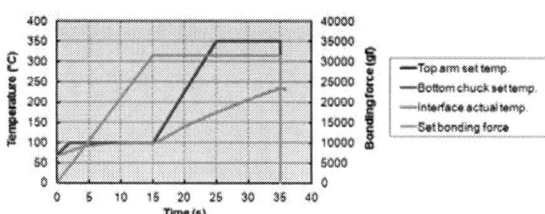

Graph 4.

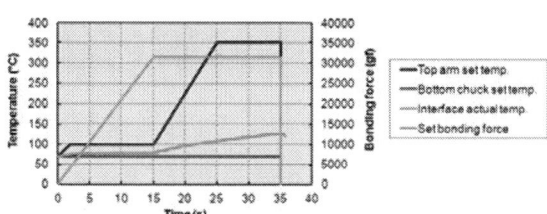

Graph 5.

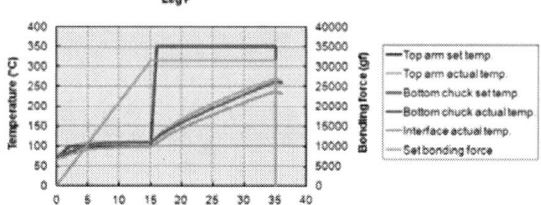

Graph 6.

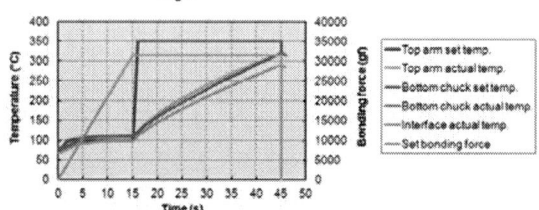

Graph 7.

During the experiment it was noticed that if the set temperature of the top and bottom chuck are similar, the resulting peak temperature will be the same but if they are different the resulting interface temperature is a value in between the temperature setting of the top and bottom chucks. A second observation is that the ramp rates vary as well between the bottom and top chuck. This can be explained by the fact that the top and bottom chuck are of a different thermal mass but this also influences the time for the interface temperature to reach its desired target (see table 2 for results)

Table 2 Results

| Run | Actual | | |
	Interface Temp (°C)	Dwell time (sec)	Ramp time (sec)
A	270	15	55
B	150	20	180
C	140	2	14
D	240	2	20
E	140	2	20
F	270	2	20
G	320	2	30
H	240	5	30

TEST RUN

The previous experiment emphasized the relationship of the top and bottom chuck with the interface temperature as well as the ramp and soak time. The following experiment will show the response of the device given a set of profiles with varying interface temperature. The profiles used are mainly reference profiles for actual device lots. Pressure and time is set constant. Measured interface temperature is shown on Graph 8.

Table 3 Test run settings and results

Run	Samples	Peak temp	Dwell Time	Ramp Time	Electrical Yield passed / samples
1	15	147°C	330 sec	60 sec	09/15
2	15	270°C	330 sec	60 sec	15/15

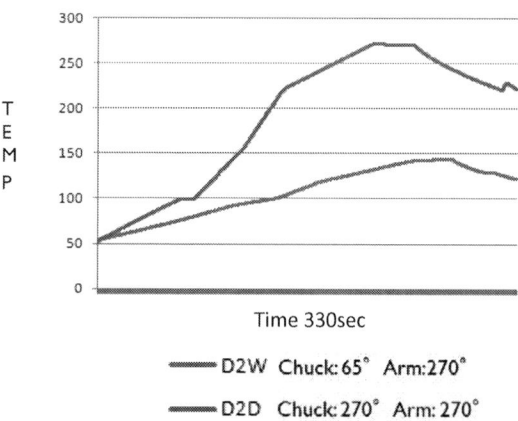

Graph 8. D2D vs. D2W profile

METROLOGY AND VERIFICATION

Units were electrically tested to see effectiveness of the bond. They were also subjected to scanning acoustic microscopy (SAM) to see the effects of the temperature setting on the underfill material used. Infrared inspection was used for detecting bond alignment after stacking. Samples were then cross-sectioned through the middle of the bumps to see the structure of the bond. The cross sections were then analyzed using scanning electron microscopy (SEM).

After electrical testing, comparison of yields showed a difference between the two set points. Both cases yielded electrically but the run with higher temperature showed better results(shown on table 3).

Samples subjected to SAM inspection showed a difference between both set points. The lower temperature setting showed more voids within the underfill compared to the higher set point.

Figure 6. SAM images; left 147°C; right 270°C

Infra red inspection on the alignment marks showed no difference in placemnent accuracy, with the placement being within the tolerance of +/- 3um. Placement accuracy is more influenced by planarity and particles so efforts was made to ensure units were not contaminated with any foreign material and planarity was kept to less than 1um.

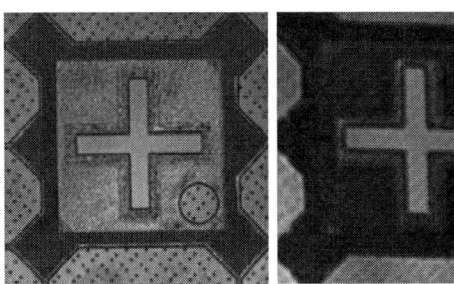

Figure 7. IR images; left 147°C (D2W); right 270°C (D2D)

Units cross-sectioned showed good connection in between bumps. Although on the lower temperature profile the Sn did not melt so it had a "tail" formation on the sides.

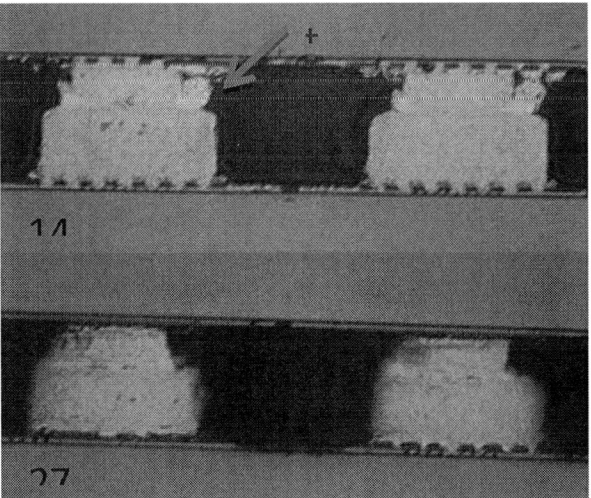

Figure 8. Cross sectional images

Using SEM to verify actual bump interface showed no difference between the low and high temperature profile this is without analysis of intermetallic formation.

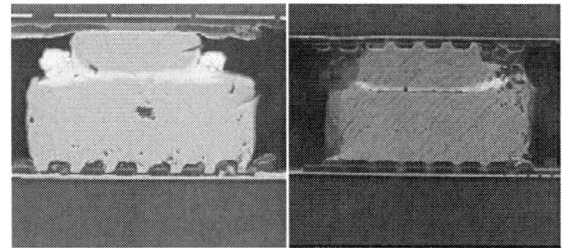

Figure 9. SEM images; left (D2D) 147°C; right (D2W) 270°C

SUMMARY
The interface temperature is dependent on the set point difference between the top and bottom chuck. Consideration must be given also on the location of the thermocouple on the equipment which the tool bases for its set point as well as the material used as chuck in between. Interface temperature has a direct influence on the underfill quality after stack. This will also be dependent on the properties of the underfill to be used. It also has a direct affect on electrical performance. Note that the purpose was not to see reliability of the joint.

CONCLUSION
We have established a relationship between the interface temperature, and set temperature of the tool. This is mainly influenced by the thermocouple location and the type of material used as chucks as well as the size and type of die/ wafer. Also using device dies we have proven the effects on stacking quality by varying the interface temperature. So with this we've concluded that the key factors that influence the interface temperature are the thermal mass of the chucks, ramp rate of the heaters and the thermal mass of the device. Another conclusion is that temperature, given the right bond force, only affects underfill performance as well as joint quality.

FURTHER WORK
The paper focuses on a single part of temperatures influence in the stacking process. There is still further work currently ongoing in understanding its influence such as thermal expansion differences between top and bottom chuck and how to compensate for it with varying ramp temperatures between them (D2D). A second work that is being focused on is creating a formula on the equivalent interface temperature given the material properties involved. This is counterchecked by the method of getting the actual interface temperature used on this paper.

REFERENCES
[1] Alan Huffman et al., Effects of Assembly Process Parameters on the Structure and Thermal Stability of Sn-Capped Cu Bump Bonds, Electronic Components and Technology Conference, pp 1589-1596, 2007

[2] A Munding, et al., Cu/Sn Solid–Liquid Interdiffusion Bonding, Wafer-Level 3D ICs Process Technology, Vol 7, pp 131-169, 2008

[3] Peng Su et al., The Effects of Underfill on the Reliability of Flip Chip Solder Joints, Journal of Electronic Materials, Vol. 28, No. 9, 1999

[4] E.W. Washburn, "The Dynamic of Capillary Flow",
   Physical Review, Vol. 17, pp 273-283, 192

EVALUATING METHODS OF SHIPPING THIN SILICON WAFERS FOR 3D STACKED APPLICATIONS

Richard A. Allen
SEMATECH
Albany, NY, USA
richard.allen@sematech.org
NIST
Gaithersburg, MD, USA
richard.allen@nist.gov

Urmi Ray and Vidhya Ramachandran
Qualcomm
San Diego, CA, USA

Iqbal Ali
SEMATECH
Albany, NY,USA

David Read
Boulder, CO, USA

Andreas Fehkührer and Jürgen Burggraf
EV Group
St. Florian am Inn, Austria

ABSTRACT

An experiment was performed to develop a method for choosing appropriate packaging for shipping 300 mm silicon wafers thinned to 100 μm or less for three-dimensional stacked integrated circuits (3DS-ICs). 3DS-ICs hold the promise of improved performance and/or lower power consumption for a given function by combining multiple chips into a 3D structure. However wafers thinned to 100 μm or less, which may be sourced from fabrication facilities anywhere in the world, must be collected in a single location for integration into 3D stacks. The methods evaluated were based on the procedure specified in ISO 2248:1985, entitled "Packaging – Complete, filled transport packages – Vertical impact test by dropping." Four types of wafer packaging systems were tested. Wafers 50 μm and 100 μm thick and drop heights of 800 mm and 1200 mm were selected. A few wafers fractured during some of the tests, mainly those wafers with significant edge defects.

Key words: Drop tests, finite element modeling, temporary wafer bonding, three-dimensional stacked integrated circuits (3DS-IC), wafer bonding, wafer shipping.

INTRODUCTION

Three-dimensional stacked integrated circuits (3DS-ICs) hold the promise of improved performance and/or lower power consumption for a given function. Since multiple chips are combined into a 3D structure, different functions can be integrated that cannot typically be fabricated using a single process; for example, a combination of memory, logic, RF, optoelectronics, and/or MEMS could be merged into a single device stack. These wafers are typically thinned to less than 100 μm. The thinning process involves the following steps (certain processes reverse steps 1 and 2):

1. Temporarily bonding the device wafer face-down to a carrier wafer
2. Removing the contoured edge using an edge-trim process
3. Mechanical grinding, followed by chemical mechanical polishing, of the back surface to the desired thickness
4. (Optional) TSV reveal and patterning of redistribution layer (RDL)
5. Applying the back surface of the device wafer to dicing tape on a standard metal or plastic dicing frame
6. Debonding the carrier wafer from the device wafer

Steps 1 and 5 are typically done on exactly complementary tools, at a single location, necessitating that the thinned

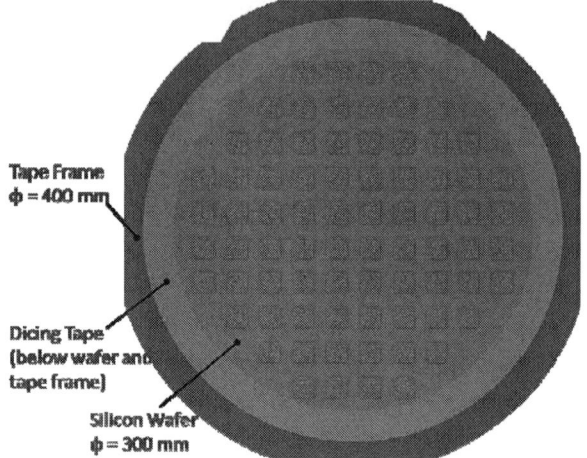

Figure 1. 300 mm wafer mounted on tape frame using dicing tape. The tape frame is a 400 mm diameter ring with flats and notches. The locations and dimensions of these feature are defined exactly in SEMI G74 and G87 and enable automated handling. The dicing tape is first attached to the back surface of the tape frame with the adhesive side of the tape facing the front. The wafer is subsequently attached to the adhesive side of the tape.

wafers be shipped on tape rather than shipped as bonded wafers before debonding. These thin and fragile wafers, originating from different processes in factories anywhere in the world, must be delivered to a single site for integration into 3D stacks. The means of shipping these thinned wafers becomes an enabling technology for high-volume manufacturing of 3DS-ICs.

The dimensions of silicon wafers first became standardized in the 1970s when the first edition of SEMI M1, entitled "Specifications for Polished Single Crystal Silicon Wafers" [1], was published. Through numerous revisions, the diameters have expanded from 2" (50.8 mm)[1] to 450 mm. Since single crystal silicon is a brittle material [2], to minimize breakage during processing and transport, the standard for wafer thickness has increased in lockstep with increases in diameter from 0.011" (279 µm) for 1" wafers to 925 µm for 450 mm wafers. The development of 3DS-IC technologies reverses this process, necessitating extremely thin, large diameter wafers.

SEMI's Thin Wafer Handling Task Force of the 3DS-IC Committee is developing a document, "Guide for Multi-Wafer Transport and Storage Containers for Thin Wafers," to help address issues with shipping thin wafers. To support the development of this guide, a series of experiments using thinned wafers and different classes of shipping containers is being performed. The wafers were thinned to 50 µm or

_____

[1] Standard dimensions for 2" and 3" diameter wafers follow U.S. customary units; standard dimensions for 100 mm and larger diameter wafers follow SI units

100 µm and the shipping configurations include those that hold multiple wafers horizontally or vertically; also tested were tray or clamshell configurations that can be densely stacked. In this paper, we present results from these experiments following the procedure specified in ISO 2248:1985 "Packaging – Complete, filled transport packages – Vertical impact test by dropping" [3]. In addition, results from finite-element modeling of important elements of existing wafer shipping systems are also included.

SHIPPING SYSTEM DESCRIPTION
Thin Wafers on Tape Frames
According to the process described earlier, wafers are mounted on one of two types of tape frames: metal or plastic. Metal frames must conform to SEMI G74 [4]; plastic frames must conform to SEMI G87 [5]. Figure 1 shows a wafer on a tape frame. Two of the shipping boxes support only the tape frame; the remaining two types support the wafer from the tape surface as well as the tape frame.

Wafer Shipper Types
Several types of commercially available wafer shipping boxes were suggested by their manufacturers to the Thin Wafer Handling Task Force for use with thin wafers. This experiment did not consider any prototype shipping boxes designed specifically for thin 300 mm silicon wafers. Four types of shippers were tested.

Multi-Wafer Horizontal Shipping Boxes
One type of horizontal multi-wafer shipping box was used in this experiment: the coin-stack shipping box. This box, shown in Figure 2, can hold up to 13 wafers on tape frames.

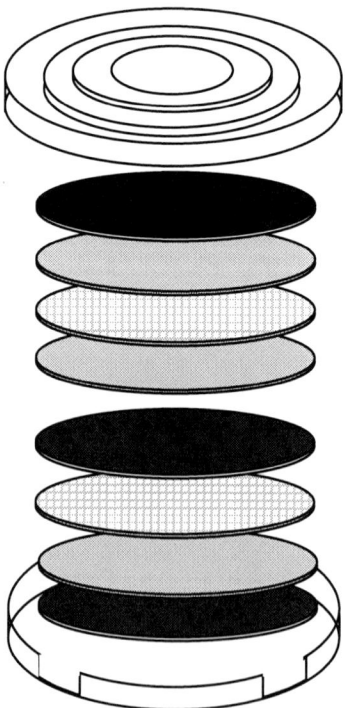

Figure 2. Horizontal coin-stack multiple wafer shipping box

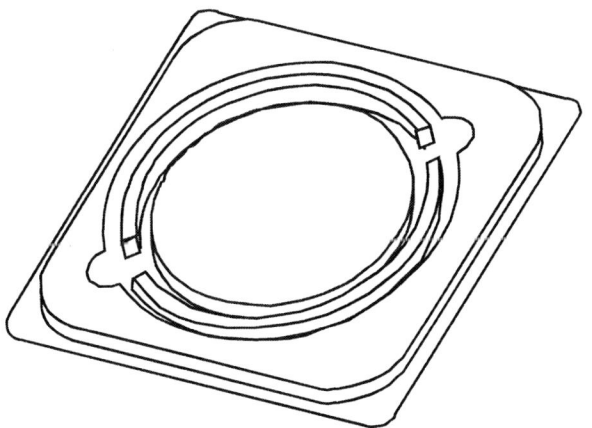

Figure 3. Horizontal stackable tray shipping box

It is the subject of a prospective SEMI standard, which was recently balloted as SEMI document 5295.

Horizontal Stackable Shipping Box
Two types of horizontal, stackable shipping boxes were included: trays and clamshells. Each is designed so that the wafer, which is on tape on the frame, is placed on a support surface and retained by a second component of the shipping box. Each of these shipping boxes is designed for a conformal fit into a 432 mm (17") x 432 mm (17") cardboard box, which represents the entirety of the secondary packaging materials.

Stackable trays are designed for wafers on a tape frame to be placed tape side down in the tray. The tape frame is held in place by putting a second tray on top of the first. Additional wafers and trays can be added to the stack. A total of n + 1 trays can be used to ship n wafers; a drawing of one of these trays is shown in Figure 3.

The clamshell box holds an individual wafer on its frame and supports the frame rigidly; it also supports the wafer indirectly through the tape. A schematic of a clamshell box such as was used in these experiments is shown in Figure 4.

Vertical Shipping Box
One type of vertical shipping box was used. It has slots for 13 tape frames; these slots are specifically fabricated to hold either metal tape frame (1.5 mm thick) or plastic tape frames (2.5 mm thick). This system is shown in Figure 5.

Secondary Packaging
Secondary packaging refers to the materials that protect the shipping box. They typically include an outer cardboard box and inner liner materials that hold the shipping box in place and dampen part of the force of any impact. When the shipping box manufacturer provided secondary packaging materials or recommended packaging materials, these were used. Otherwise, the drop test was performed using secondary packaging materials chosen according to common shipping practices.

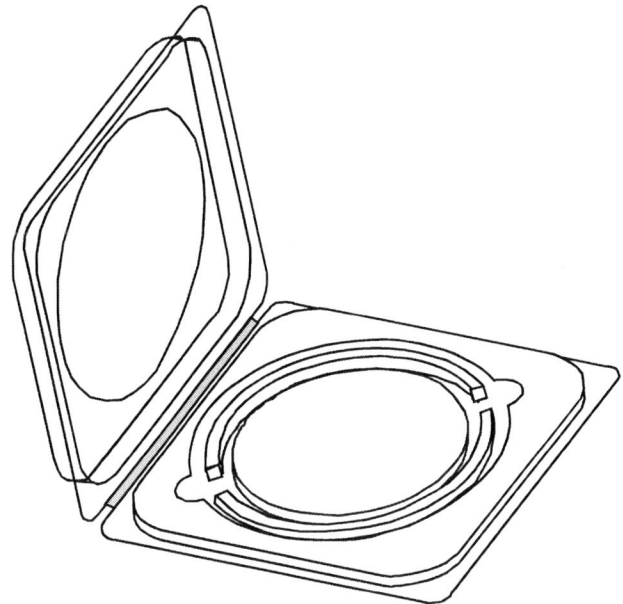

Figure 4. Horizontal clamshell stackable shipping box

MEASURMENT PROCEDURE
Wafer Processing
Two hundred and fifty unpatterned wafers were processed in multiple lots. Several temporary bonding process flows were used, the details of which are beyond the scope of this paper. After processing, 83 wafers mounted on dicing tape on tape frames underwent the drop tests. Of these, 47 were 100 µm thick and the remaining 37 were 50 µm thick. The 100 µm thick wafers could be further split into two groups based on the quality of the edge trim. Thirty-eight wafers had visibly poor edge quality, as shown in Figure 6a. The remaining 9, plus all of the 50 µm thick wafers, had much smoother edge quality, as shown in Figure 6b.

Packaging procedure
The wafers were placed in the primary package(s). For shipping boxes that could accommodate multiple wafers, all slots were filled with either wafers on tape frames or tape frames with tape but no wafer. Table 1 shows the different packaging configurations used in the experiment.

Sensors
Commercial-grade, single-use shock threshold sensors measured how much impact was delivered to the primary packaging. Sensors, with trigger values from 10 g to 75 g, were attached to the exterior of the primary package. These shock sensors, shown schematically in Figure 7, irreversibly change color, typically to red, when exposed to acceleration exceeding a target value in the axis, or axes, of sensitivity. The tube-style sensors used are sensitive to accelerations perpendicular to their long axis. One or more sensors were attached to the primary package during each drop test, oriented such that the sensitive axes of the sensors were aligned to the drop axis. The g-values were chosen to narrow on the expected impact value, e.g., so that at least one sensor would trigger and at least one would not.

DROP SET	Shipping Box	Wafer Thickness	Tape Frame
1	Vertical, multi	50	Metal
2	Vertical, multi	100	Metal
3	Coin Stack	50	Metal
4	Coin Stack	100	Metal
5	Coin Stack	50	Plastic
6	Coin Stack	100	Plastic
7	Tray	50	Metal
8	Tray	100	Metal
9	Tray	50	Plastic
10	Tray	100	Plastic
11	Clamshell	50	Metal
12	Clamshell	100	Metal
13	Clamshell	50	Plastic
14	Clamshell	100	Plastic
		*thickness in μm	

Table 1. Design of Experiment Matrix

Drop test

The attitude of the drop is defined in IS) 2248:1985 [3] as the orientation of the package on impact. A rectangular box has a total of six unique faces, twelve edges and eight corners. However, since both the silicon wafers and the packaging systems exhibit certain degrees of rotational symmetry, several orientations were not tested. ISO 2248:1985 does not define a preferred drop height, but leaves this up to the user. For this experiment, two drop heights were chosen: 800 mm and 1200 mm. These roughly correspond to an unprotected impact force of 100 g (100 times earth gravity) and 150 g, respectively.

A video recording was made of each drop to verify that the correct height and attitude were used. After each drop, the package was carefully opened and the contents inspected for damage to the inner packaging, shipping box, or wafers. In addition, if any or all of the shock sensors were triggered, this information was recorded and the sensors were replaced.

Data recording

A data sheet was developed to record the data required by ISO 2248:1985 plus any additional data relevant to the particular experiment, such as wafer descriptions and locations of shock sensors.

MEASURMENT RESULTS

The drop tests were performed at two separate times based on wafer availability; the first tests used only the poorer quality 100 μm thick wafers; the second tests included all types of both 100 μm and 50 μm wafers. During the second tests, both types of 100 μm wafers were included side-by-side to provide evidence of whether edge quality affects the fragility of similar wafers.

Overall, few wafers were broken during the drop tests. Further, none of the shipping boxes failed completely in any drop, *i.e.*, no single drop test cracked all the wafers. In the tests in which multiple wafers failed, those wafers were generally not adjacent to or even near one another, as might be expected if the forces impacting the wafers were higher in one region of the shipping box than another, e.g., if the delivered force were higher nearer the point of impact between the secondary packaging and the drop surface.

Key observations include the following:
1) Wafers with poor edge conditions were more likely to break than those with smooth edges.

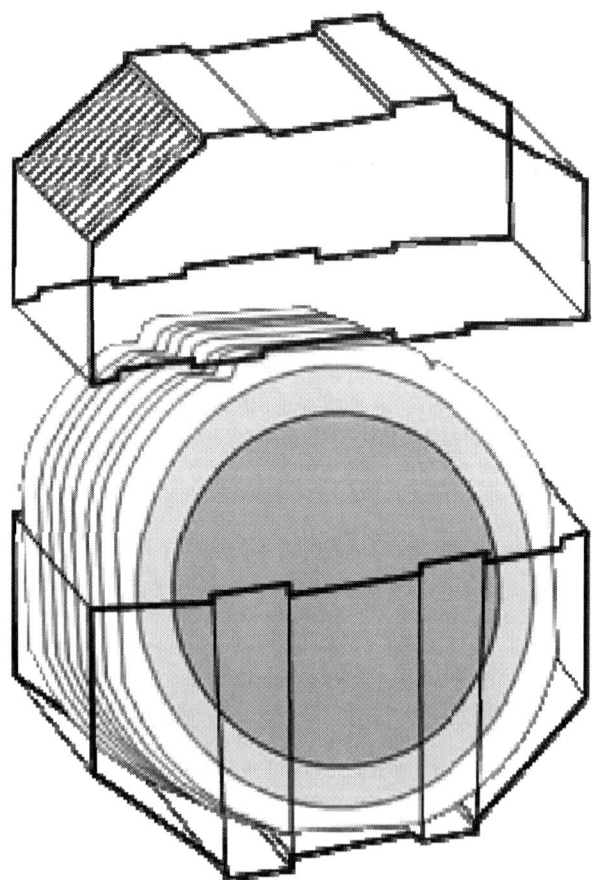

Figure 5. Vertical multiple wafer shipping box

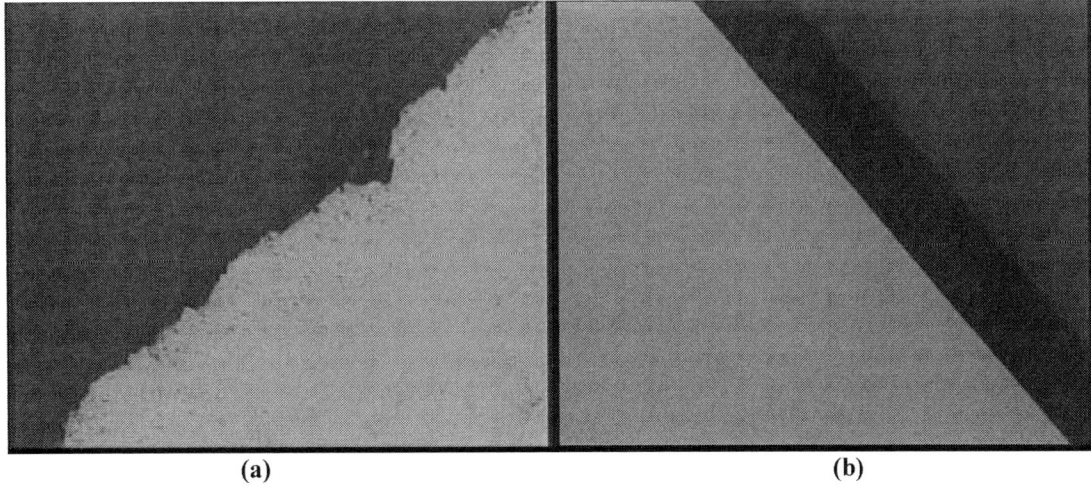

Figure 6. Edge quality of typical wafer from Lot 1 (a) and Lot 2b (b) after debonding.

2) Breaks appeared to originate from defects at the edges; see Figure 8a for a crack initiating from a large edge defect on a 100 μm wafer.

3) The drop from 1200 mm represents a severe case for shipped wafers. Visible damage to the shipping systems was observed as follows for 1200 mm drops:

    a. In one drop in which a 50 μm wafer broke the force was such that a triangular piece of the wafer delaminated from the tape in a central region of the wafer (Figure 8b).

    b. In another drop, one of the shipping boxes cracked.

    c. In multiple cases, internal secondary packaging exhibited significant deformation on impact.

Secondary packaging appeared to play a significant role in the results. That is, the shipping boxes are stiff and the tape rings with the attached wafers are held firmly in place by all configurations; in some configurations the wafer is also held firmly in place against a surface. This means that the primary packaging provided only minimal cushioning, while the secondary packaging provided the majority.

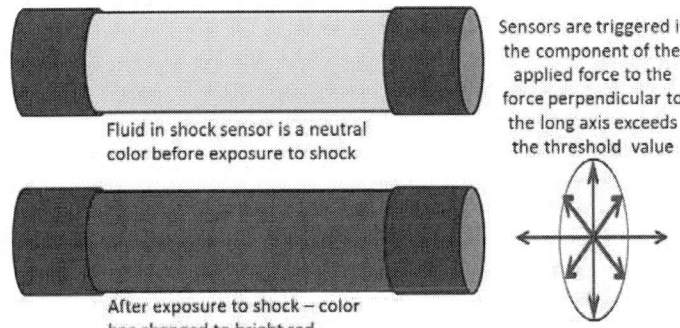

Fluid in shock sensor is a neutral color before exposure to shock

After exposure to shock – color has changed to bright red

Sensors are triggered if the component of the applied force to the force perpendicular to the long axis exceeds the threshold value

Figure 7. Diagram of function of shock sensor

FINITE ELEMENT ANALYSIS OF THIN WAFERS ON TAPE FRAME

Cook [2] gave a "design stress" of 130 MPa for well-polished, full thickness wafers. However, no technique for measuring stress during a drop test of thin silicon wafers has been reported. Therefore, finite element analysis (FEA) was used to estimate the stress. The wafer-tape-metal frame system, with its standardized dimensions and known material properties, was modeled exactly. A typical maximum acceleration of 75 g was estimated from the shock threshold sensors, considering both those that did trigger and those that did not. This value should be taken as only a rough estimate of the shock applied to the wafer-tape-frame systems. Because no time history of the deceleration of the frames was available, an "ideal cushioning" assumption was made: a constant deceleration of 75 g was applied for a time period sufficient to bring the frame to a stop. With these assumptions, the calculated maximum principal stress at the wafer center was less than half of Cook's design stress of 130 MPa. The assumptions used in the calculation prevent a more precise result. However, because none of the wafers failed in the center, a stress value well below 130 MPa is consistent with expectations. The FEA produced stresses near the wafer edge that varied considerably, but with a maximum value about half the value in the center of the wafer. This low stress value can be reconciled with the experimental results by assuming that the edge defects, as shown in Figure 6a, lower the critical stress near the wafer edge to a level well below Cook's value. Some support for this interpretation can be found in the experimental results on the second lot of wafers, in which few wafer failures occurred. The failures that did occur were associated with damage to the packaging systems.

CONCLUSIONS

We have shown a process for evaluating methods of shipping thinned 300 mm wafers. Several

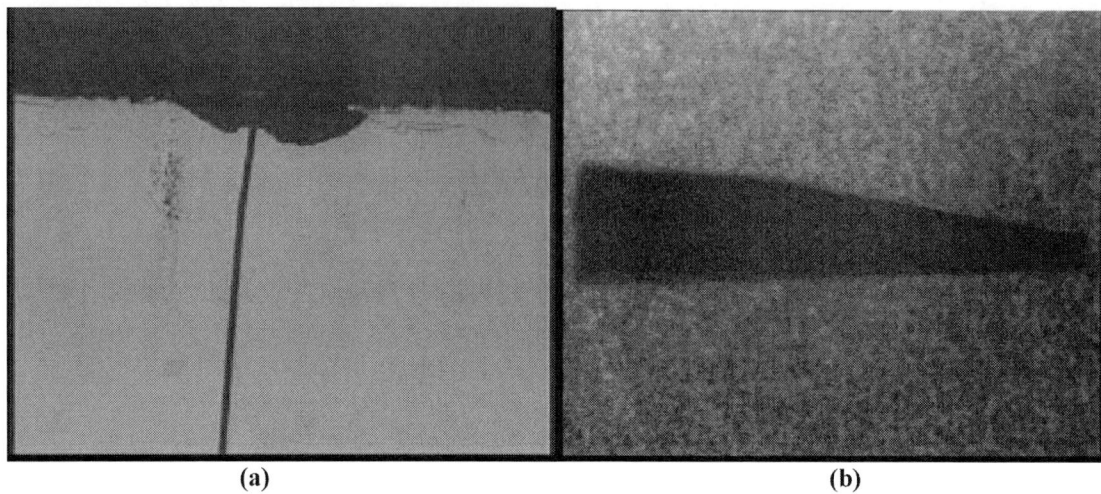

(a)   (b)

Figure 8. Break example from Lot 1, extending from a region of poor edge quality (left) and a break on a Lot 2b wafer, where the force delivered to the wafer was enough to detach a triangular piece of the wafer from the dicing tape (right).

different classes of tape frame wafer shippers were suggested for use in this application; all gave acceptable results.

The fragility of the wafers should be considered when choosing the shipping configuration. It was clear from the results that the likelihood of breakage for any specific drop was increased with poorer quality edges.

The results suggest that perhaps the key differentiating factor among the shipping systems tested is the performance of the secondary packaging.

Estimated stress values at the wafer centers and edges in these drop tests, calculated by FEA of the wafer-tape-frame system with the use of boundary conditions roughly estimated to correspond to the present set of tests, may be interpreted as showing that the "design stress" level for these thinned wafers is significantly reduced by defects at the wafer edges. For wafers with severe edge defects, the level is well below the literature value [2] of 130 MPa for well-polished, full thickness wafers.

Based on the results, all of the options evaluated handled both wafer thicknesses adequately. Ultimately the shipping method needs to be qualified by each user. The standard that will result from D5175 is intended to provide a guideline for an experimental procedure that can be used to make this decision.

ACKNOWLEGEMENTS
The authors would like to thank Raghu Chaware of Xilinx and Janet Cassard of NIST for discussions and Dr. Stephanie Hooker, Chief of the NIST Materials Reliability Division, for technical support.

REFERENCES
[1] Specification for Polished Single Crystal Silicon Wafers, SEMI M1, SEMI, San Jose, California, U.S.A., www.semi.org.

[2] R.F. Cook, Strength and sharp contact fracture of silicon, J. Mater. Sci. 41 (2006) 841-872.

[3] Packaging – Complete, filled transport packages – Vertical impact test by dropping, ISO 2248:1985, International Organization for Standardization, Geneva, Switzerland.

[4] Specification for Tape Frame for 300 mm Wafers, SEMI G74, SEMI, San Jose, California, U.S.A., www.semi.org.

[5] Specification for Plastic Tape Frame for 300 mm Wafer, SEMI G87, SEMI, San Jose, California, U.S.A., www.semi.org.

3D PACKAGING- SYNTHETIC QUARTZ SUBSTRATE AND INTERPOSERS FOR HIGH FREQUENCY APPLICATIONS

Vern Stygar[#1], Tim Mobley[*2]

[#]Asahi Glass Corporation, [*]nMode Solutions, Inc.

Hillsboro OR, Tucson AZ

[1]vstygar@agem.com

[2]tmobley@nmodesolutions.com

ABSTRACT

Mobile communications and demands for instant data drive the need for smaller form factors with high performance and long battery life. Superior RF (Radio Frequency) substrates and 3D packages allowing the integration of higher bandwidth data into the mobile device have long been desired for accomplishing this goal. This paper introduces a new low-loss substrate that can be used as a standalone platform for RF design as well as a system for next generation 3D packaging for die-to-die interconnect. The next generation of mobile communication devices will operate at much higher frequencies to allow video on demand and video vis a vis RF communication. However, as frequencies rise, the need to reduce interconnects and integrate IC's continue to be in demand. Advanced packaging materials and processes will need to be developed, validated and implemented.

INTRODUCTION

Higher frequency designs now being implemented by electrical engineers are approaching millimeter wave lengths which impose interesting challenges for high volume standard substrates. Engineers have been able to work around certain problems but they are demanding new technology solutions as the next generation of devices are being introduced to the marketplace. These devices operate at >3GHz, transferring data >40 gigabytes/sec and server farms approaching Tera bits/second. This paper introduces a new set of substrates and technologies that are here to exceed the requirements of electrical and mechanical performance as well as being low cost.

Developing a substrate and packaging technology that is low cost, highly reliable, and available in high volume manufacturing and has design rules that enable the electrical engineer to design and simulate is not a trivial task. The complexity stems from the number of design elements that need to be satisfied. 1) Stable and predictable Dielectric Constant (Er) and Loss Tangent from 2-40 GHz, 2) Low cost metallization with high conductivity, 3) design rules and multiple manufacturing vendors that can fabricate designs into high volume products, 4) Substrates not effected by moisture and Er (Dielectric Constant) that are independent of temperature and frequency up to 50GHz [4],

and, finally, 5) the substrate of choice must be available in a wide variety of thicknesses and sizes. In addition, the substrate must be compatible with current manufacturing equipment to allow easy adoption by the manufacturer.

Standard PWB laminates have enjoyed an almost overwhelming spectrum of new designs largely because engineers have been able to overcome high board losses. One of the reasons is that they are using advanced techniques for patterning and plating over thick Cu on the PWB. The challenge that is disruptive to this low cost board solution is being driven by high I/O count digital IC's with very high clock speeds. The PWB fabrication cost and complexity, however, is being driven higher by this dynamic and, as a result, manufacturers resort to combining board materials that are suitable for microvias, which also demonstrate good losses for RF design (i.e. Panasonic Megtron 6). What was once a $300 board is now greater than $15,000 per board. This results in combining packaging design and PWB board design which has many challenges. This paper offers a new solution to these challenges. We will describe the available substrates for high frequency applications, metallization, and the testing of those substrates at high frequency. The results of this testing then will be compared and contrasted for suitability at high frequency.

TECHNICAL APPROACH

Board/Packaging Material

As board design and packaging design techniques and methods converge, designers are looking for a middle ground where technology can offer benefits of both. Ceramic technology has been a solid performer and a good compromise in this area. However, recent demands of large WLP (wafer level packaging), where flatness across wafers the size of 200-300mm in diameter is essential, has become too stringent for ceramic and PWB technologies. This is where glass/quartz is far superior and, combined with low loss characteristics, offers the designer capabilities never before been realized until now. PCB materials and techniques are another solution. But as mentioned previously, these traditional solutions are cost prohibitive when the feature sizes of vias and traces are <4mils. Figure 1 lists a table of traditional materials used for packaging, board design, and WLP.

Substrate	Dielectric Constant @18 GHz	Loss Tangent (10⁻⁴) @18 GHz	Surface Roughness (Angstroms)
Alumina	9.8	7	500
Synthetic Quartz	3.8	<1	10
Glass Sodium Free	5.9	30	300-1000
Soda Lime Glass	6.72	170	300-1000
Silicon	11.7 -12.7	150	2500
PCB(ceramic/Teflon)	2.94	14	3000
Sapphire	11.5	<1	10-100

Figure 1. RF Characteristics of Typical Materials

Typical substrate materials used for circuit design have printed wiring board (PWB) with copper foil, Alumina fired with tungsten and plated up with silver or gold, and a hybrid combination of hydrocarbon polymer and ceramic.

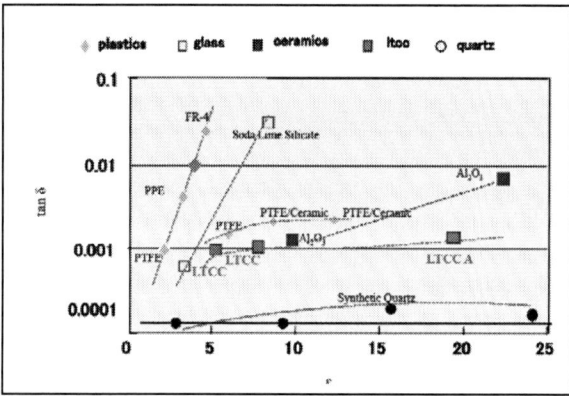

Figure 2. Loss tangent data for board/package materials.

The dielectric losses of these materials can be seen in the graph in Figure 2. These materials have been the choice of substrate materials for the past 50 years. Advances in electronics, design, form, and function have pushed the standard materials to the limit. Figure 2 shows the typical materials used in typical designs. However, as frequency increases, new materials need to be tested and qualified for high production use. Ceramic and glass materials have the lowest insertion loss. They have an added advantage of being hermetically and thermally stable throughout a wide variation of temperatures. Several types of alumina exist from standard Alumina to a form of Alumina called "Superstrate." Synthetic quartz from AGC called AQ has low dielectric loss and has no piezo electric response usually associated with natural quartz (i.e. has the same losses in all directions)

Several types of glass exist on the market today but the glass that is sodium-free has the best frequency response. The reason lies with the electro negativity of the electron cloud loosely held by the sodium ion. During signal propagation, the electron cloud is distorted and results in the absorption of the energy, thus attenuating the signal. In sodium-free glass there is very little perturbation.

Surface smoothness is typically overlooked in high frequency applications. Surface roughness impacts signal integrity as a phenomenon known as skin effect [1]. As the frequency rises the signal predominantly propagates near the surface of the metallization resulting in an intrinsic rise in resistance. At frequencies greater than 30 GHz, the skin effect can be as high as 30% as shown in studies by Chai [2], et al.

Design Parameters
Figures 3 and 4 call out the design features that are currently achievable on glass and quartz. New developments are ongoing to achieve smaller diameter vias with tighter pitches in thinner and thicker glass and quartz for unique applications.

Parameter	0.3mm thick Glass	0.25mm thick Quartz
Via Diameter (d) (um)	50	50
Via Pitch (p) (um)	130	90
Via Conductors	Ag or Cu (not plated)	Ag or Cu (not plated)
Via Capture Pad (cp) (um)	d + 20	d + 20
Trace Width (w) (um)	10	10
Trace Gap (g) (um)	20	20
Trace Conductors	Cu or Ni/Cr/Au	Cu or Ni/Cr/Au

Figure 3. Design Parameters

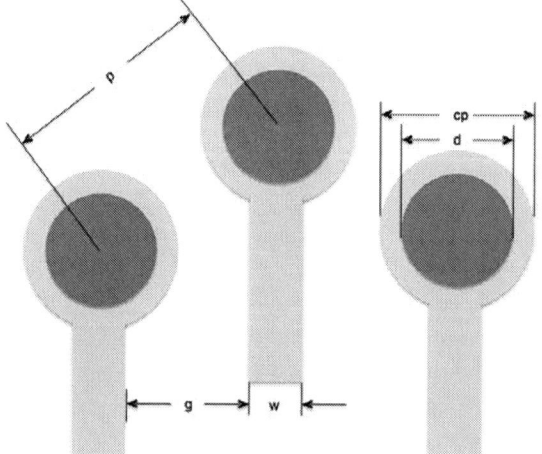

Figure 4. Design parameters for traces and vias

APPLICATIONS

3D Packaging

Adoption of high performance glass interposers are being driven by the need of electronics manufacturers where interconnects continue to shrink and battery life and heat are major concerns. Typical applications, such as computers, servers, graphic chips, and optical electronic moems are the main driving force. Because of the 3D nature of silicon and the need for precise placement and alignment, Glass/Quartz provides the most thermally stable substrate. In addition, advancements made in metallization (such as copper), fine-line width/spacing have now made these applications achievable at low cost while not compromising reliability.

THE GRAPHIC ON THIS PAGE WAS NOT AVAILABLE ON THE E-MEDIA VERSION OF THIS PAPER.

Figure 5. TGV (Thru-Glass Via) Interposer

There are many ongoing industry efforts to support TSV (Thru-Silicon Vias) for interposers. This seems to be a logical fit since TSV's are being used on the wafer for routing the interconnect to the bottom of the die instead of traditional wire bond pads. The issues that we take with this approach are several, including, 1.) reliability is compromised at the PWB attachment point, 2.) the cost is higher due to the fact that they are processed in a Silicon Foundry clean room when fabricating the vias for the Interposers, and 3.) the Cu plating to create the Silicon vias has a significant mismatch to the PWB when soldered or epoxied. This causes cracking along the side walls which, in turn, entraps liquids from subsequent processes and later burns out during the thermal cycle of component attachment. These issues are typically not detected at the device level but further down stream at the component assembler or OEM (Original Equipment Manufacturer). Yield losses become progressively more expensive as the component moves downstream. Glass is a natural fit to resolve all of these issues when combined with a hermetically sealed via and used as a separator between the PWB and the Silicon IC.

RF Components

Synthetic Quartz, with its extremely low loss and low coefficient of thermal expansion, provides the ideal 3-D interposer for the RF designer who requires both high frequency performance and a stable substrate. Synthetic quartz with a surface roughness approaching single digit angstrom levels essentially negates skin effects in the millimeter wave range of use[3]. Figures 6 and 7 show a component designed and fabricated on low loss synthetic quartz. The backside has solid metalization while the top side shows the design patterns with ground vias. There is a shadow that can be observed under the circuit traces which is due to the angle of the camera, lighting, and the high transparency quality of the synthetic quartz.

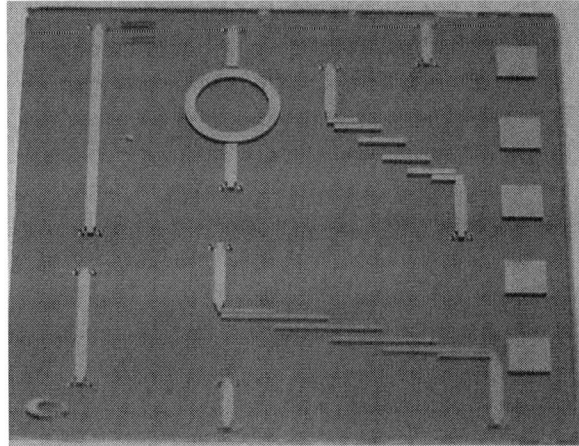

Figure 6 AQ1 Fabrication Coupon of RF Components

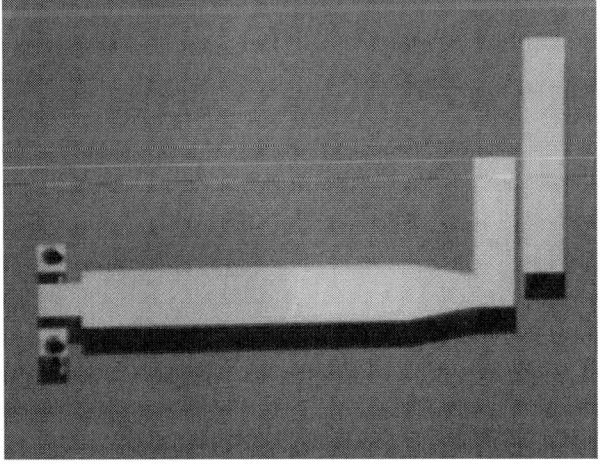

Figure 7. RF Feed line showing the ground vias

Figures 8 and 9 show the simulation results from Sonnet for both an 18GHz and 36GHz edge coupled filter. The combination of the low Er (i.e. wide traces), low surface roughness, low dielectric loss, and fine line widths/gaps, provides the designer an opportunity to design RF components with superior performance at low cost.

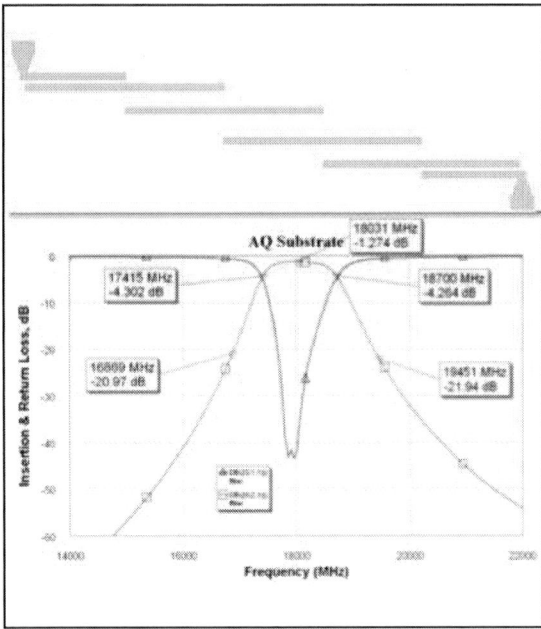

Figure 8. 18GHz Filter design on AQ1

The designs were implemented on 10mil thick AQ1 (low loss synthetic quartz) and 10 mil thick Alumina Superstrate from Coorstek. The measurement results in Figure 9 show that the AQ1 is quite superior to the Alumina Superstrate.

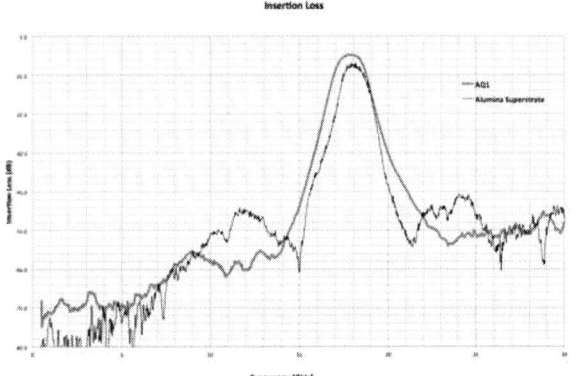

Figure 9. Measurement results of the 18GHz filter for AQ1 and Alumina Superstrate

The combination of lower dielectric constant (i.e. wider lines) and superior loss tangent gives the AQ1 a significant advantage over Alumina as well as the PTFE based materials (i.e. Rogers 6002). Figure 9 shows a 3-4dB improvement over Alumina, which is significant at 18GHz, when combining with amplifiers and other active signal processing IC's. The differences between the model and the measured are primarily attributed to the via connections to ground, as they were not included in the Sonnet model.

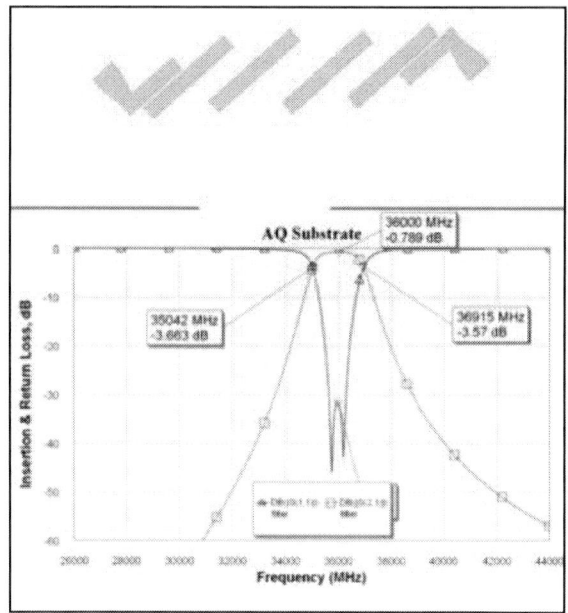

Figure 10. 36GHz Filter design on AQ1

RF Board Design
Glass/Quartz Interposers allow the ability to batch process utilizing LCD glass technology such as G4.5 (730 mm by 920 mm). Because of the stability of glass/quartz, scaling up of circuit design is feasible allowing designer to employ and leverage extremely large scale with flatness and roughness levels, unheard of for PWB and Ceramic board materials. Layout of circuit designs and rendering into precise substrates utilizing techniques developed in LCD technology drives cost down, and enables the designer to utilize the transparency of the glass in methods involved integration of electrical circuits, thereby eliminating interfaces and cost.

CONCLUSIONS
A new platform of packaging and board materials is upon us. This paper demonstrated designs that were achieved as 3D packages that we are calling TGV's. They are superior to Silicon TSV's and, as RF components, superior to the traditional board materials such as Alumina Superstrate. These new glass and quartz material platforms provide the designer with a low cost solution while providing hermetic, highly reliable packages and boards. Why sacrifice the reliability, as TSV's do, when glass and quartz materials are hermetic and can be used at lower cost?

Further work planned requires characterizing TGV's, and multi-layer RDL (Redistribution Layers) and surface roughness effects at 40-100 GHz. Thermal cycling and hermeticity testing of vias to demonstrate the robustness of the TGV system are currently underway.

REFERENCES

[1] D. M. Pozar, *Microwave Engineering*, New York, John Wiley & Sons, 1998.

[2] Liang Chai, Aziz Shaikh, and Vern Stygar "LTCC Systems for Wireless and Photonic Packaging applications", EMAPS 2002 The 4th International Symposium on Electronics Materials and Packaging 2002

[3] J.W. Reynolds, P.A. LaFrance, J.C. Rautio, and A.F. Horn, III, "Effect of conductor profile on the insertion loss, propagation constant, and dispersion in thin high frequency transmission lines," DesignCon 2010.

[4] W.W. Snell, Jr "Low-Loss Microstrip Filters Developed by Frequency Scaling" Lucent-Acatel Manuscript January 21, 1971

CO-DESIGN STRATEGIES FOR MEMS PACKAGING

Mary Ann Maher
SoftMEMS LLC
Santa Clara, CA, USA
maryann@softmems.com

Sebastien Cases
SoftMEMS EURL
Grenoble, France

ABSTRACT
MEMS based products use advanced packaging techniques and combine heterogeneous fabrication processes to create systems including multiple MEMS sensors/actuators, analog and digital circuitry, micro-controllers, power sources and sensor fusion algorithms. MEMS packages act to encapsulate MEMS devices, but in addition, the package must allow exposure to external stimuli such as pressure, light or fluids. Special purpose structures are often added to MEMS packages to provide mechanical stability and isolation or to create a high quality vacuum environment. A major challenge in the design of MEMS systems is the performance analysis of the packages and the simulation of the 3D mechanical, thermal and electrical interactions between the packaging and the individual device components.

New co-design tools enable MEMS and IC designers to incorporate the analysis of packaging effects into IC and MEMS component designs and allow them to collaborate with package designers to optimize system performance and trade-off requirements. As a result, co-design reduces time to market for new products and increases system performance. This paper discusses the modeling of important 3D packaging effects and wafer level packaging for MEMS and will describe computer aided design (CAD) solutions for MEMS packaging. Many of the same MEMS package design challenges are also present in the design of new IC packaging and the software solutions described in this paper may be applied to their design.

I. INTRODUCTION
The design of integrated circuits (ICs) is simplified by the use of many standard materials, standard connections and standard packaging. Creating packaging for MEMS based systems poses many challenges as standards are few and the distinction between device and package may be blurred as MEMS packaging can play an important part of a MEMS product's function. Due to the diversity of MEMS, design and analysis in multiple physical domains such as thermal, mechanical, electronic, magnetic, and fluidics is required. Bottlenecks in bringing MEMS-based products to market stem from errors made in the design of packaging caused by

problems in package-device interactions in these multiple domains. Co-design and co-simulation of the MEMS and its packaging from the design start is a key enabler to avoid these problems.

II. MEMS-SPECIFIC PACKAGE DESIGN ISSUES
In an IC, the packaging works to protect and isolate devices from the environment. In packaging MEMS, in addition to protecting the device, the package must allow external stimuli to interact with it. For example, pressure sensors and microphones need a hole to input pressure. Blood processing devices need inlet ports for the blood to enter. Optical devices must have a transparent opening. These constraints add to the complexity of design, incur additional cost, and contribute to the uniqueness of the packaging solution as different types of inputs require different packaging. There is little standardization even within a class of devices.

On the other hand, some MEMS devices need to have additional environmental protection compared to IC devices, for their best performance. This constraint is especially important in high frequency, high performance devices. Devices such as accelerometers, gyroscopes, timing devices such as resonators, micro-mirrors, and infrared sensors require special care with respect to environmental effects. Devices requiring a high quality factor require a high quality hermetic environment. These types of devices often utilize a wafer level cap and the packaging is done in a special vacuum controlled environment.

For MEMS structures, the package thus is an important part of overall device structure and often device performance cannot be calculated without taking into account effects determined by the package such as vacuum level. Because of this relationship, package-device co-design is important.

Another very important constraint on MEMS packages is managing temperature effects. Automotive and industrial MEMS sensors must operate over a large temperature range. IC device performance is a function of temperature through the sensitivity of electronic device outputs to temperature changes. MEMS devices also have this sensitivity and additional constraints.

The TCE (temperature coefficient of expansion) mismatch in packaging materials can cause temperature changes to become mechanical problems in MEMS packages. Because of the mismatch, package materials expand at different rates causing stress to develop within the package. This stress can be transmitted through the package to the MEMS device and cause the device to malfunction or be out of specification. For example, in piezo-resistive acceleration sensors and pressure sensors, piezo-resistors sense the stress developed by motion or due to pressure changes. The stress from packaging acts as a mechanical parasitic that adds to the stress output. Stress can also cause miss-alignments or bending in MEMS devices interfering with proper operation such as misalignment in optical fiber grooves.

High performance inertial devices are sensitive to shock and vibration since they include devices that are moving. Special isolation structures may need to be added increasing package cost and complexity. Figure 1 shows an example of two types of MEMS specific packaging enhancements designed to address two important MEMS packaging issues. The left hand side shows an example of a typical package. In the middle example, an oven control structure has been added to control the temperature of the die and enable the MEMS structure to operate at a fixed temperature, reducing sensitivity to the outside environment. On the right side, an example of a vibration isolation structure is shown that has been placed on top of the oven control structure. Figure 2 shows a device cross-section showing how the MEMS device is placed on top of the isolation platform.

Introducing these additional structures into the package design can greatly improve device performance, however, the package and device must be co-designed carefully. Co-simulation of the device and package is important so that not only the performance enhancements can be predicted but that any unwanted effects due to the introduction of the new structures can also be predicted. Calculating the resonant frequencies of each added component of the packaging is important, for moving MEMS structures for example, so that exciting unwanted resonant modes is avoided.

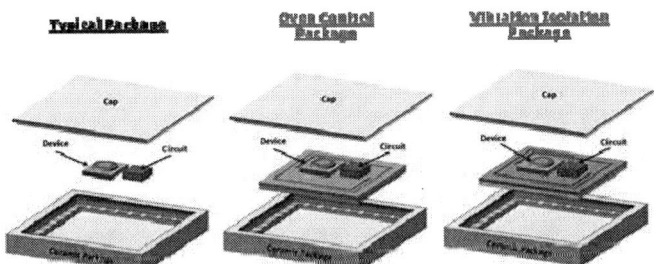

Figure 1: MEMS Specific packaging structures courtesy of ePack.

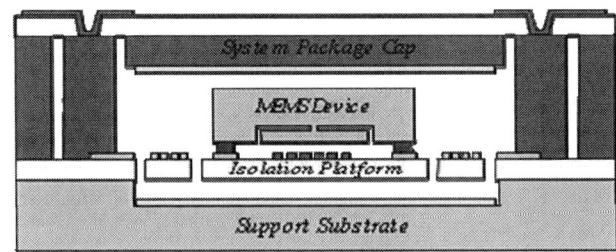

Figure 2: Cross-section of MEMS packaging structure showing the isolation platform courtesy of ePack.

III. MEMS-CO-DESIGN EXAMPLES

In addition to the functions described above, MEMS packaging may also be used to transmit signals as in the MEMSIC thermal accelerometer. The device shown in Figure 3 below consists of a MEMS accelerometer and CMOS electronics on the same die created using a standard CMOS process. The working principle is shown in Figure 4. The device consists of several polysilicon heaters which create heated gas bubbles. When acceleration is applied to the device, a heated gas bubble will move and be detected by thermopile temperature sensors also on the chip. A cavity is machined out underneath the MEMS device to facilitate airflow and the device is packaged using wafer level packaging to create an air cavity. As part of the design process, designers must optimize the cavity depth and simulate the temperature and airflow in the package using 3D computational fluid dynamics as shown in Figure 5. In this way, the MEMS device design, electronics and package are interdependent and co-design produces an optimal result.

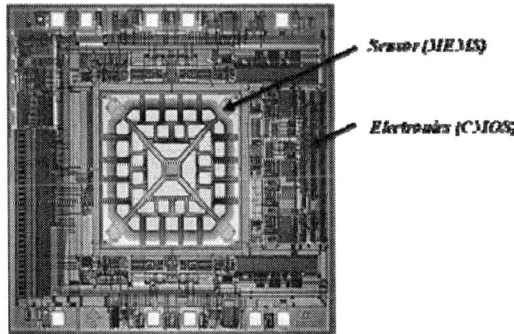

Figure 3: MEMSIC Thermal acceleration sensor courtesy of MEMSIC.

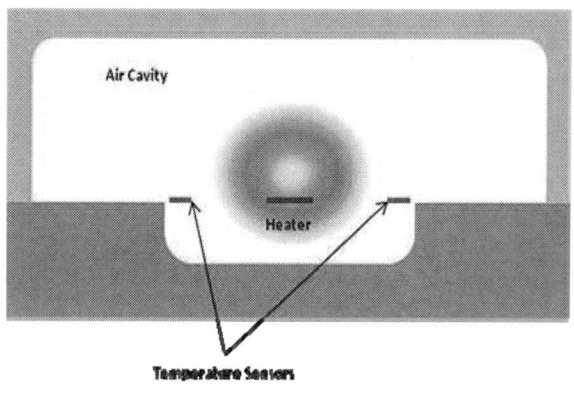

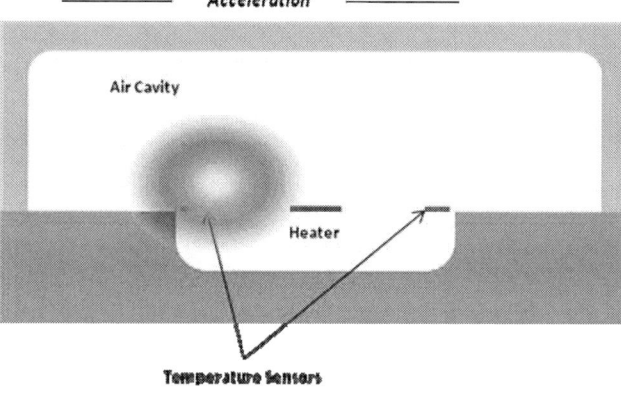

Figure 4: Principle of operation of the MEMSIC thermal acceleration sensor. Courtesy of MEMSIC.

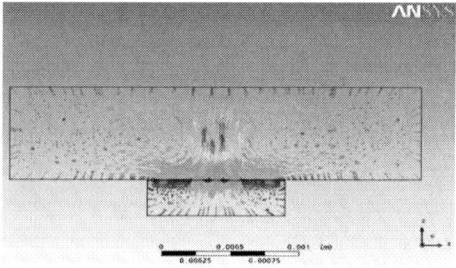

Figure 5. Simulation of air flow in the package of the MEMSIC Acceleration sensor. Courtesy of MEMSIC.

IV. MEMS CO-SIMULATION EXAMPLE

Another example of how the packaging is integral to a MEMS product design can be seen in devices with moving MEMS structures. An important MEMS device is the energy harvester. An example of a state of the art energy harvester from MicroGen Systems with its concept packaging is shown in Figure 6. The energy harvester uses a MEMS cantilever with piezo-electric material deposited on it to convert or "harvest" vibrational energy and converts it into an electrical signal. The stress on the cantilever due to its motion is converted by the piezo-electric material into electrical charges that appear across the device's electrodes. The quality factor of the MEMs motion is determined by the air damping and hence vacuum level in the package. The

design of the package is also impacted by the extent of travel of the moving structure. So again, co-design and co-simulation is important in the design of this product. Coupled fluid-structure interaction simulations can be used to evaluate the damping factor. Coupled piezo-electric-mechanical simulations incorporating the damping must be done to determine the MEMS motion and hence its output voltage and generated power.

Figure 6: MEMS Piezo-electric energy harvester in a wafer level package courtesy of MicroGen Systems.

V. MEMS-PACKAGING ECOSYSTEM

As described before, MEMS packaging is application and MEMS device specific, so custom development is often required. Customized tooling may be needed and the development time can be long. The lines have blurred between device fabrication and packaging and where various packaging steps are done and by whom. For example, many MEMS packaging steps are now done in the MEMS foundry under clean room conditions and other traditional steps are done in a less "clean" typical packaging and assembly environment. Materials that were traditionally not found in an IC fab are being used as part of MEMS device and packaging steps in the MEMS fab.

Companies specializing in MEMS packaging are joining the MEMS ecosystem to offer packaging services. The SMART center at Lorain Country Community College for example, is a MEMS packaging development foundry. These specialty packaging facilities are intermediate between a university environment, and a MEMS foundry, or traditional IC packaging house such as Amkor or ASE. They offer package prototyping and pre-production runs. Prototyping is important as new products usually require new package concepts. There is a mantra in MEMS – "one product, one process, one package".

Much of the infrastructure built for packaging and assembly of integrated circuits cannot be used for MEMS. A lot of the automation used in the IC industry for handling device such as pick and place is rarely used in MEMS due to their fragility. Because of their diversity in application and working principle, testing is also nonstandard and requires special equipment. Equipment manufacturers are starting to offer MEMS-specific handling, assembly and test equipment and the specialty packaging facilities like the SMART Center cut across the boundaries of various activities to encompass back-end wafer processing, package, assembly, testing, and reliability. Co-design software is also used by these facilities to find design errors and shorten development time.

VI. CONCLUSIONS

Due to the expansion of the MEMS market and its robust growth rate compared to the semiconductor market there is much interest in MEMS packaging. Yole Development estimates a $10B market for MEMS with $1.4B related to packaging in 2011. Packaging and test strategies are important intellectual property and a competitive advantage in the MEMS world. But the desire to have the lowest cost and highest yield are leading some MEMS makers to use IC-type packaging when they can to take advantage of the years of manufacturing engineering that have gone on in the IC industry. However, the "one product, one process, one package" truism will be with the industry for a while.

Many of the wafer level packaging techniques that have long been used to package MEMS devices have recently started to be used in the IC industry and the techniques used to create working designs and lessons learned can be transferred to the wafer level packaging of integrated circuit devices. There are of course, some important differences in the packaging specifications. In the IC world, the pitch of TSVs and their depth for example is much tighter than in the MEMS world. A variety of materials and combinations of materials are used to fabricate TSVs by different vendors in the MEMS industry. In contrast, the IC industry seems to use only a few combinations.

Many of the simulation tools, however, can be utilized to address similar design and simulation issues in IC packaging. Thermal issues are important for chip-stacking as there is less room to dissipate heat. Mechanical rigidity issues are important as dies/wafers are thinned and stacked. Thermo-mechanical modeling for stress engineering can be used as it is in the MEMS industry. Fabrication process simulation is used by MEMS designers to determine the performance characteristics and create virtual models of their devices and these tools can be used in the same way for the design of a new via structure and to find fabrication errors and manufacturability issues. The simulator emulates the building up of materials used to construct the via in the foundry.

The future of MEMS packaging holds new promises and new challenges. New packaging techniques such as the use of heterogeneous 3D integration and flexible substrates are enabling MEMS products to have enhanced performance, reduced size and enter new markets. For example, new

biological sensors such as wearable body sensors require packaging using flexible substrates. Implanted devices have additional requirements on form factor and accessibility to biological "signals". Materials selection for bio-compatibility as well as for their mechanical properties is also key issues. A typical design team consists of materials scientists, fabrication experts, mechanical designer, fluid flow modelers, packaging experts, surgeon, and after-care physician. So, the package design considerations must be part of a series of complex design decisions made by a diverse design team. These packages bring new design challenges such as characterization of new packaging materials as their material properties may not be well known. Since a lot of the real cost of a MEMS component can be in the testing, packaging, and assembly, co-design is very important to the overall success of MEMS products.

YIELD AND STRENGTH OF METAL WAFER-LEVEL MEMS DEVICE SEALING USING AL, AU, OR TI

K. Schjølberg-Henriksen, PhD; E. Poppe
SINTEF ICT, MiNaLab
Oslo, Norway
karisc@sintef.no

A.S. Moen, E. Fasting
Oslo and Akershus University College of Applied Sciences
Oslo, Norway

ABSTRACT

Metal-based wafer-level bonding seems well suited for hermetic wafer-level packaging at low cost. We have compared the performance of Au-Au thermocompression bonding, Al-Al thermocompression bonding, and Ti-Si bonding, applying the same test wafer design for all three processes. Device yield, bond strength, and electric resistance of the bond were investigated. Both Au-Au and Al-Al thermocompression bonded wafers had a yield of 100%. The yield from Ti-Si bonded wafers was significantly lower. Au-Au and Al-Al bonds had bond strengths above 30 MPa and ranging as high as 159 MPa. Ti-Si bonds had strengths between 13 and 23 MPa. The electrical resistance of the bond seal was similar for all investigated bonds. The current study indicates that Au-Au hermocompression bonding and Al-Al thermocompression bonding are well suited for wafer-level packaging of MEMS devices in industrial production, while Ti-Si bonding seems less well suited.

Key words: MEMS packaging, wafer-level sealing, metal bonding

INTRODUCTION

Micro electro-mechanical systems (MEMS) enable sensitive and reliable devices at low cost due to the advantages of batch processing. However, sealing or packaging of the individual devices may induce significant costs. Wafer-level packaging lowers these costs substantially. Glass-frit bonding [1], anodic bonding [2] and fusion bonding [3] are well known and widely used techniques for wafer-level packaging and sealing of MEMS devices. Recently, metal-based wafer bonding has received attention as technology for hermetic wafer-level sealing that also facilitates vertical integration [4]. Both diffusion-based and eutectic metal bonding techniques have been investigated.

Several considerations must be made in order to choose a bonding method that is suitable for a specific device. These considerations include material compatibility, process temperature, process time, and limitations on applied bond pressure. Thermal and electrical properties of the resulting bonded seal may also be important. Metal sealing processes that have been investigated recently include bonding using

copper [4 – 6], aluminium [6 – 8], titanium [9, 10], and gold [11 – 15]. Copper, aluminium, and gold are bonded by metal interdiffusion, while the bonding mechanism of Titanium to Silicon seems to be a formation of a thin layer of an amorphous structure with titanium and silicon [10].

Aluminium and titanium both have the advantage over copper and gold of being CMOS-compatible materials. Reported bonding temperatures are lower for gold, titanium, and copper than the 400 – 550°C reported for aluminium [4, 6, 7]. The reported bonding time tends to be lower for copper and gold than the 45 – 60 minutes reported for Al-Al and Ti-Si bonding. The requirement for applied pressure varies substantially between reports. The aluminium oxide formed on aluminium surfaces is reported to be a potential challenge for successful Al-Al thermocompression bonding [6].

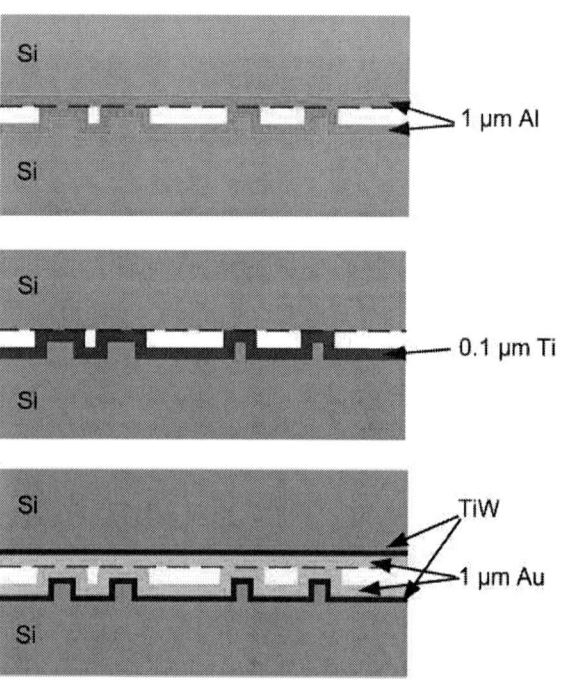

Figure 1: Cross-section of test chips showing the bond materials of the seals compared in this study. The bonded interface is indicated by a dashed line.

This paper reports on investigation of device yield, seal strengths, and electric resistance in bonding seals realized by metal-based wafer-level bonding. Three metal systems were selected: Au-Au thermocompression bonding, Al-Al thermocompression bonding, and Ti-Si bonding.

Figure 1 illustrates the selected metal systems. The same test wafer design and the same bonding equipment were used for all laminated wafer pairs, enabling a comparison of the three selected processes.

EXPERIMENTAL
Chip Design
Four chips with square frames suitable for device encapsulation were designed. The outer dimensions of the frames were 3×3 mm^2. Frame widths of 100, 200, and 400 μm with straight corners were designed and named F100, F200, and F400. A frame of width 200 μm with rounded corners (F200R) was also designed, as shown in Figure 2. In addition, two chips were designed with a mesa of side edge 50 or 100 μm in the center of the chip. These chips were named M50 and M100. The six chip types and their nominal bond surface are listed in Table 1Table 1. Figure 3 shows the distribution of chips across the wafer. On the whole wafer, there were 70 pcs of each mesa type chip and 54 pcs of each frame type chip. Each chip was 6×6 mm^2.

Table 1: Overview of the six different chip designs.

Chip ID	Description	Bond area [mm^2]
M50	Mesa, side edge 50 μm	0,0025
M100	Mesa, side edge 100 μm	0,01
F100	Frame, width 100 μm	1,16
F200	Frame, width 200 μm	2,24
F200R	Frame, width 200 μm, rounded corners	2,14
F400	Frame, width 400 μm	4,16

Test Wafers
Double-side polished, p-type wafers with resistivity 0.2 – 0.8 Ωcm and diameter 150 mm were used. The protruding frames and mesas were realized in the silicon by 6 μm deep reactive ion etching applying an AMS 200 I-Prod etcher (Alcatel). Wafers to be used for electrical measurements were implanted on both sides with 2×10^{15} cm^{-2} boron at 50 keV and activated for 60 minutes at 900°C in N$_2$.

Bond metal was sputter deposited on wafers with frame pattern and smooth silicon wafers. For Al-Al thermo-compression bonding, 1.0 μm of Al was deposited on both wafer types. For Au-Au thermo-compression bonding, a TiW adhesion layer and 1.0 μm Au was deposited on both wafer types. For Ti-Si bonding, 0.1 μm Ti was deposited on wafers with protruding silicon structures. Prior to bonding, the Ti and Si surfaces were pre-treated as described in [9]. The Si surface was cleaned in a Piranha solution, rinsed in DI water, and treated in 10% HF for 30 seconds. The Ti surface was cleaned in 1:1 DIW : 37% HCl for 30 seconds, then rinsed in DIW and spin-rinse-dried.

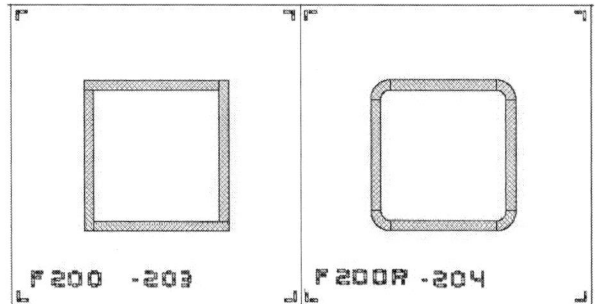

Figure 2: Chips with frame width 200 μm and straight corners (left) and rounded corners (right).

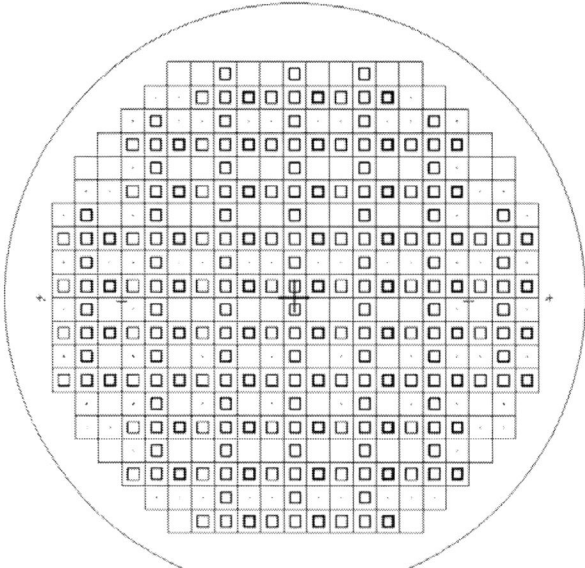

Figure 3: Wafer design with chip distribution.

Table 2: Overview of bond parameters for the wafer pairs in this study. The bond pressure is the applied bond force divided by the bond area.

Pair ID	Metal layers	Bond parameters
Au350	TiW + 1 μm Au	350°C, 30 min, 21.6 MPa
Au400	TiW + 1 μm Au	400°C, 15 min, 21.6 MPa
Au450	TiW + 1 μm Au	450°C, 30 min, 21.6 MPa
Ti350a	0.1 μm Ti	350°C, 60 min, 21.6 MPa
Ti350b	0.1 μm Ti	350°C, 60 min, 21.6 MPa
Al550a	1 μm Al	550°C, 60 min, 33.9 MPa
Al550b	1 μm Al	550°C, 60 min, 33.9 MPa

The wafers with frame patterns were bonded to smooth wafers in an SB6e wafer bonder (Suss). The applied bond times, temperatures, and pressures for the bonded wafer pairs in this study are listed in
Table 2. The bonding was performed in a vacuum ambient with a chamber pressure below 10^{-3} mbar. The chamber was flushed with N$_2$ prior to bonding Au-Au and Al-Al wafers. Before bonding Ti-Si, the chamber was flushed with Ar to minimize the presence of gases that react with Ti. Laminates to be used for electrical measurements were metallised with 1.0 μm Al for electrical contacts on both sides of the chip and sintered at 350°C for 30 minutes. Two different metal contacts were used, as illustrated in Figure 4. All wafer

laminates were diced into individual chips of size 6 × 6 mm², each chip containing one frame.

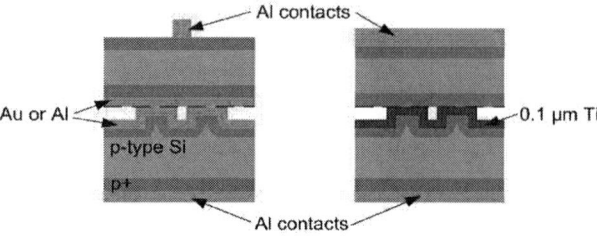

Figure 4: Cross-section of chips used for electrical measurements showing the electrical contact of the Al-Al and Au-Au laminated chips (left) and the Ti-Si laminated chips (right).

Measurement Methods

The yield of the bonding process was defined as the percentage of non-delaminated chips after dicing. The percentage was calculated for each of the six chip types. The yield was calculated on full wafers with 356 chips on wafers Au400, Au450, Ti350a and Al 550a. A piece containing three rows of chips (See Figure 1) was used for electrical testing of wafer Al 550b, so that only 90% of this wafer was used for yield measurements. The wafer Ti350b was partly broken during measurement, and the yield was calculated based on the surviving 30% of the wafer.

The bond strength was measured by pull testing of bonded chips. Diced chips were glued to screws, which were mounted in a MiniMat 2000 (Rheometric Inc) pull tester. The elongation versus applied force was recorded, and the force for which fracture occurred was noted. A minimum of 3 chips and a maximum of 13 chips were tested of each type. The bond strength was calculated as the fracture force divided by the nominal bond area. After fracture, the fractured surfaces were inspected visually and in optical microscope. The fractures were characterized as cohesive, adhesive, or mixed. The material of cohesive fractures and the interface that delaminated during adhesive fractures were noted.

Chips from the implanted wafers were used for measurement of electrical resistance across the bonded interfaces. The current through the chip was measured when ramping the voltage from 0 – 1 V applying a Power device Analyzer/Curve tracer B1505A (Agilent). The I-V curves were recorded and the resistance was calculated. Four chips of each type were tested.

RESULTS

The yields for the different frame designs and wafer laminates are shown in Figure 5 – Figure 7. The Ti-Si bonds had the lowest yield, ranging from 0 – 43% on wafer Ti350a. The yield on wafer Ti350b was higher, but those yield calculations were based on 110 chips only. The Al-Al bonds had the highest yield, ranging from 92 to 100% on the two investigated wafers with 356 and 317 chips. The yield for the Au-Au bonds was 100% on wafer Au450, and ranged from 53 – 76% on wafer Au400.

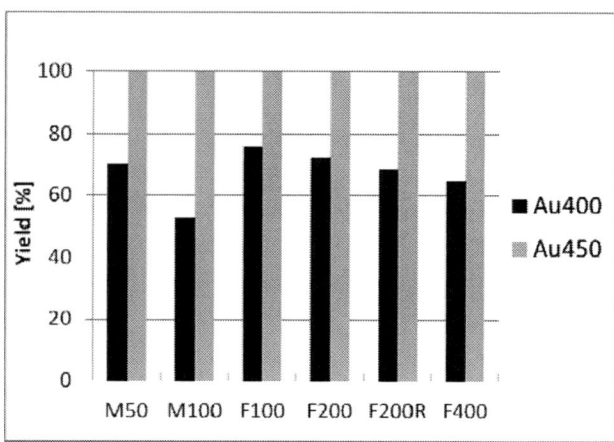

Figure 5: Yield for the four frame designs on wafers Au400 and Au450, laminated with Au-Au thermo-compression bonding. The yield calculations were based on 356 chips for both wafers.

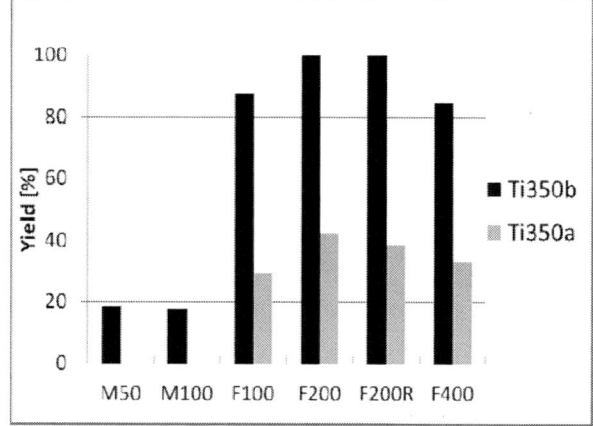

Figure 6: Yield for the four frame chip designs on wafers Ti350a and Ti350b. The yield calculations were based on 356 and 110 chips, respectively.

The fracture forces and the average bond strengths of the tested chips are plotted in Figure 8 and Figure 9. Wafer Au450 had the highest level of fracture force, with all four chip types having an average fracture force above 145 N. The fracture forces on wafers Au400 and Al550 were around 60N, except for chips F200R on wafer Al550, which had a fracture force of 233 N, which was the highest value of all chip types and wafers. The lowst fracture forces were around 40 N, obtained on wafer Ti350. The fracture forces and calculated bond strengths show that wider bond frames did not result in stronger chip seals. Table 2 lists the average values and the standard deviation of the bond strength for each wafer and chip type.

34

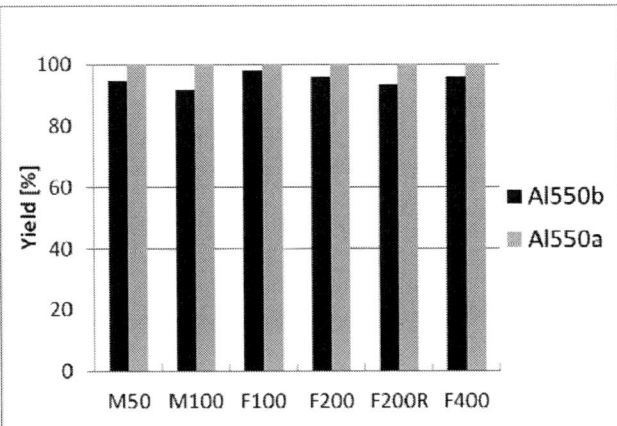

Figure 7: Yield for the four frame designs on wafers Al550a and Al550b. The yield calculations were based on 356 and 317 chips, respectively.

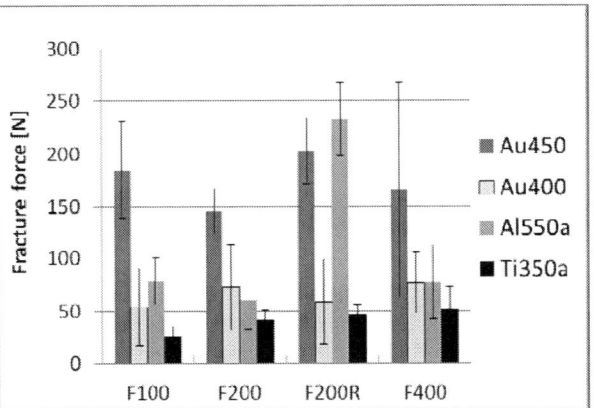

Figure 8: Average and standard deviaton of the value for the force at which fracture occurred. All four frame designs from wafers Au400, Au450, Ti350a, and Al550a are plotted.

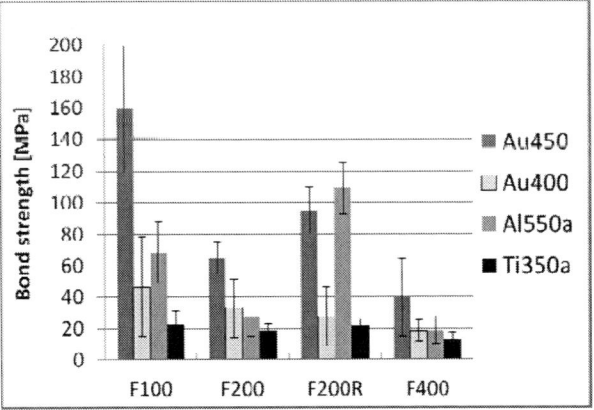

Figure 9: Average and standard deviaton of bond strength of the four frame designs from wafers Au400, Au450, Ti350a, and Al550a.

Table 3: Average bond strength values and standard deviation for the four frame designs from wafers Au400, Au450, Ti350a, and Al550a.

Wafer ID	Bond strength [MPa]			
	F100	F200	F200R	F400
Au400	47 ± 32	33 ± 18	28 ± 19	19 ± 7
Au450	159 ± 40	65 ± 10	95 ± 15	40 ± 25
Ti350a	23 ± 8	19 ± 4	22 ± 4	13 ± 5
Al550a	68 ± 19	27 ± 12	109 ± 16	19 ± 8

An example of a recorded I-V curve for the resistance assessment is seen in Figure 10, and the resulting average of the measured resistance values are plotted in Figure 10.

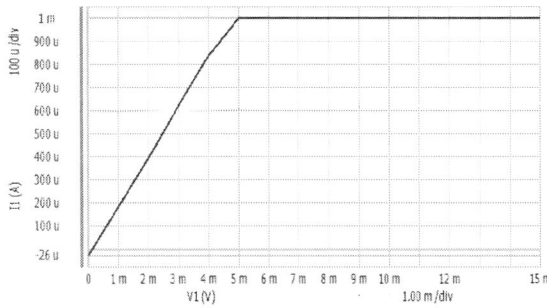

Figure 10: Example of I-V curve recorded in chip with cross-section as shown in Figure 4 (left).

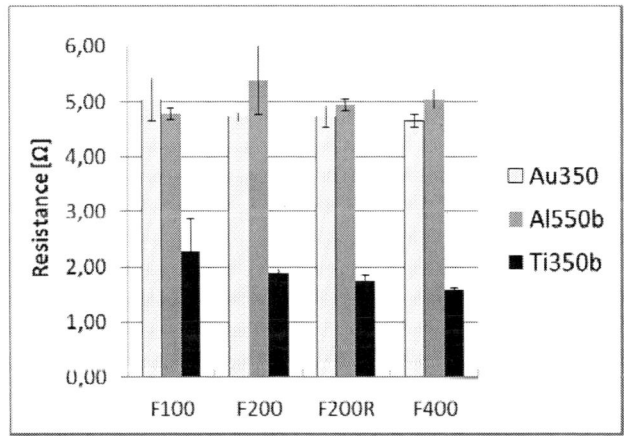

Figure 11: Average resistance values and standard deviation plotted by chip type and lamination method for chips from wafers Au350, Al 550b, and Ti 350b. Four chips of each type were measured.

Pictures of typical fractured surfaces after pull testing are shown in Figure 12 - Figure 15. On wafer Ti350a, adhesive fracture at the interface between the Ti film and the bulk Si occurred on nearly all the chips. Usually, the adhesive fracture occurred both at the interface between the originally sputtered Ti interface and the bulk Si and at the interface between the bonded Ti and the bulk Si. An example with the adhesive fracture occurring approximately 50% at each interface is seen in Figure 12. Cohesive fractures occurring in the bulk silicon were seen in about half of the F100 chips and all the F200R chips on wafer Ti350a. The cohesive bulk silicon fractures occurred in 15 – 50% of the bond frame.

Figure 12: Fracture surfaces of F200 chip from wafer Ti350a. Adhesive fracture at the Ti-Si interface occurred at both Ti-Si interfaces: at the originally Ti-sputtered side (right) and at the Ti-Si bonded side (left). A cohesive fracture in the bulk Si is also seen in frame, occurring in approximately 5% of the bonded frame.

Figure 13: Fracture surface of F 200R chip from wafer Au400a. The adhesive fracture at the TiW-Si interface exposed the bulk silicon in the bonded area. A cohesive fracture in the bulk silicon is also seen, covering about 5% of the bonded frame area.

The fracture surfaces of chips from wafer Au400 were a mix of adhesive fracture at the TiW-Si interface, cohesive Si fracture, and cohesive Au fracture. Some adhesive fracture at the Si-TiW interface was observed in more than half of the chips on wafer Au400. Cohesive fractures in the bulk silicon were seen in nearly all the chips, but the amount varied from 5% to 100% of the bonded area. On wafer Au450, adhesive fracture at the TiW-Si interface was seen on one chip only. Nearly all the chips had cohesive fracture in the bulk silicon. The amount ranged from 20 to 100% of the bonded area. Cohesive Au fracture was seen in approximately half of the chips from wafer Au450. Figure 13 and Figure 14 show two examples of fracture surfaces from wafer Au400 with 5% and 60% cohesive fracture in the bulk silicon, respectively. In the chips that had cohesive bulk silicon fracture in a large percentage of the bonded area, the visual appearance of the unbonded gold had changed. In these chips, the unbounded gold seemed less brilliant and smooth, as seen in Figure 14. In most of the chips from wafers Au400 and Au450, the gold film delaminated at a small distance from the chip edges during dicing. This is seen in Figures 12 and 13. The film delaminated at the interface between the TiW film and the bulk Si.

Figure 14: Fracture surface of F 200R chip from wafer Au400a. A cohesive fracture in the bulk silicon is seen in approximately 60% of the bonded area. The remaining 40% appears to have adhesive fracture in the TiW – Si interface. The visual appearance of the gold is significantly changed in the un-bonded area near the bonded frame area.

Figure 15 shows a typical fracture surface from wafer Al550a. All the pull tested chips from this wafer had cohesive fracture in the bulk silicon in 100% of the bonded area.

Figure 15: Fracture surface of F100 chip from wafer Al550a. Cohesive fracture in the bulk silicon occurred in 100% of the bonded area.

DISCUSSION

The three examined bonding metals had different yield. Of the three investigated metal systems, the Al-Al thermo-compression bonds had highest yield, and the Ti-Si bonds had lowest yield. The yield on wafer Ti350a would be unacceptable for devices in production. Wafers Ti350a and Ti350b were bonded with the same parameters, but Ti350b had significantly higher yield. The difference in yield indicates that the Ti-Si bonding process was less repeatable than the Al-Al thermo-compression bonding process. The higher yield on wafer Ti350b could also be related to the lower number of tested chips on this wafer. The yield results from wafers Al550a and Al550b are in good agreement with the work of Yun et al [8], who reported die yields of 96.8 – 100%. This result compares well with the 93 – 100% yield found for the frames in the present study.

The yield was 100% on wafer Au450, and varied between 53% and 76% on wafer Au400. This could indicate that the yield increased with increasing the bonding temperature and time from 15 minutes at 400°C to 30 minutes at 450°C. However, Au-Au thermo-compression bonding with high yield has been reported to occur at 300 – 350°C [12, 14, 15]. We therefore think that the lower yield on wafer Au400 was related to an erroneous dicing parameter, which was applied for wafer Au400 only, and could have induced excess stress and chip delamination during dicing of wafer Au400.

The bond strength was found to vary with frame design and applied bond metal. The highest fracture force, 233 N, was obtained for chips of type F200R, bonded by Al-Al thermocompression bonding. The highest bond strength was obtained for Au-Au frames of 100 µm width bonded at 450°C, with an average strength of 159 MPa. The Ti-Si bonds were the weakest bonds in the study, with average strengths ranging from 13 to 23 MPa. The fracture force was similar for the different chip types on each wafer. This indicates that the bonds could sustain a certain strain, and that the yield strain did not increase with increasing bond frame width. For Au-Au and Al-Al bonds, frames of 100 µm width and 200 µm wide frames with rounded corners seemed to have the highest bond strength. This fact may be of interest in future seal design. It is assumed that bond strengths above 10 MPa are usually desired for device seals in industrial products. All three investigated bonds, Au-Au, Al-Al, and Ti-Si, can be designed to meet that criterion.

The Au-Au bond strengths from both wafers Au400 and Au450 are significantly higher than the ~10 MPa reported by Taklo *et al.* [12] and the 20-25 MPa reported by Kurotaki *et al.* [14]. The difference in bond strength could be due to differences in the design of the applied test chips, which is likely to have given different stress concentrations, and therefore different fracture forces.

The fracture surfaces give indications of the state and properties of the bonding seals. The occurrence of cohesive fracture in the bulk silicon indicates that the bonded seal is of a strength that is comparable to the bulk materials. Strong bonds therefore seem to have been formed on wafers Al550a and Au450. The bonds on wafer Ti350a had adhesive fracture at the interfaces between the Ti and the bulk silicon. Our observed delamination at the sputtered Ti/Si interface indicates that improved adhesion of the sputtered Ti film could improve the bond strength of the Ti-Si bonds. Yu *et al.* [9] found the bonded interface between Ti and Si to be stronger than the interface between sputtered Ti and thermally grown SiO_2.

The resistance measurements in Figure 11 showed that the resistance across the chips was below 5.5 Ω for all measured chips. The fact that no significant difference was observed with respect to bonded area, shows that the resistance in the metal bond was insignificant compared to the resistance in the bulk silicon of the upper and lower wafers. Based on the chip geometry and silicon resistance, the theoretical resistance of the bulk silicon is calculated to be between 0.1 and 28 Ω. The measured values are between these two limits, as expected. The lower resistance of the Ti-Si bonded chips (see Figure 11) is due to the different design in the top metal contact, and does not indicate a difference in the resistance of the three investigated bonding metals.

CONCLUSION

Au-Au thermo-compression bonding, Al-Al thermo-compression bonding, and Ti-Si bonding have been compared with respect to bond yield, bond strength, and electrical resistance of the bonded interface. Both Au-Au and Al-Al bonded wafers had a yield of 100%. The yield of Ti-Si bonded wafers was significantly lower, and ranged from 0 to 43%, depending on the test chip design. The highest fracture force was 233 N, obtained on Al-Al bonded frames of width 200 µm and with rounded corners. The highest bond strength was obtained on Au-Au frames of 100 µm width bonded at 450°C, with an average strength of 159 MPa. The bond strength of Ti-Si bonded chips ranged from 13 to 23 MPa, depending on the test chip design. The electrical resistance of the bonded interface was found to be below 5.5 Ω for all investigated bonds. No significant difference in resistance was found between the bonding metals. The current study indicates that Au-Au thermo-compression bonding and Al-Al thermo-compression bonding are well suited for wafer-level packaging of MEMS devices in industrial production, while Ti-Si bonding seems less well suited.

ACKNOWLEDGEMENTS

The authors wish to thank colleagues at SINTEF MiNaLab for fruitful suggestions and discussions. This work was supported by the Norwegian Research Council through the MSENS project, contract No 210601/O30.

REFERENCES

[1] R. Knechtel, M. Wiemer, J. Frömel, "Wafer level encapsulation of microsystems using glass frit bonding", *Microsyst Technol* **12** (5), pp. 468–472, 2006.

[2] D. Lapadatu, B. Blixhavn, R. Holm, T. Kvisterøy, "SAR500 - A high-precision high-stability butterfly gyroscope with north seeking capability", *Proc IEEE PLANS Position Locat Navig Symp*, Indian Wells, CA, 4-6 May 2010, pp. 6-13.

[3] J. Kyynäräinen, J. Saarilahti, H. Kattelus, A. Kärkkäinen, T. Meinander, A.Oja, P. Pekko, H.Seppä , M. Suhonen, H. Kuisma, S. Ruotsalainen, M. Tilli, "A 3D micromechanical compass", *Sensors and Actuators A* 142 pp. 561–568, 2008.

[4] S. Farrens, "Metal Based wafer level packaging" Suss white paper, 2005, http://www.suss.com/en/media/technical-publications/wafer-level-packaging.html

[5] L. DiCioccio, P. Gueguen, R. Taibi, D. Landru, G. Gaudin, C. Chappaz, F. Rieutord, F. de Crecy, I. Radu, L.L. Chapelon, L. Clavelier, "An overview of patterned metal/dielectric surface bonding: mechanism, alignment and characterization", *J. Electrochem. Soc*, **158** (6), pp. P81-P86, 2011.

[6] V. Dragoi, G. Mittendorfer, J. Burggraf, M. Wimplinger, "Metal thermocompression wafer bonding for 3D integration and MEMS applications", *ECS Transactions* **33** (4), pp. 27-35, 2010.

[7] J. Martin, "Wafer capping of MEMS with fab-friendly metals", *Proc SPIE* **6463,** pp. 64630M1 – 6, 2007.

[8] C.H. Yun, J.R. Martin, L. Chen, T.J. Frey, "Clean and conductive wafer bonding for MEMS", *ECS Transactions* **16** (8), pp. 117-124, 2008.

[9] J. Yu, Y.M. Wang, J-Q Lu, R.J. Gutmann, "Low-temperature silicon wafer bonding with titanium", *Electrochem Soc Proceedings* **2005-06**, pp. 311 – 318, 2005.

[10] J. Yu, Y. Wang, A.W. Haberl, H. Bakhru, J-Q- Lu, R.J. Gutmann, "Mechanisms of low-temperature Ti/Si-based wafer bonding", *Mater Res Soc Proc* **863** pp. 387 – 392, 2005.

[11] C.H. Tsau, S.M. Spearing, M.A. Schmidt, "Characterization of wafer-level thermocompression bonds" *J. Microelectromech Systems* **13** (6), pp. 963 – 971, 2004.

[12] M.M.V. Taklo, P. Storås, K. Schjølberg-Henriksen, H.K. Hasting, H. Jakobsen, "Strong, high-yield and low-temperature thermocompression silicon wafer-level bonding with gold", *J. Micromech Microeng* **14**, pp. 884 – 890, 2004.

[13] G-S Park, Y-K Kim, K-K Paek, J-S Kim, J-H Lee, B-K Ju, "Low-temperature silicon wafer-scale thermocompression bonding using electroplated gold layers in hermetic packaging" *Electrochem Solid-State Letters*, **8** (12) pp. G330 –G332, 2005.

[14] H. Kurotaki, H. Shinohara, H. Kobayashi, J. Mizuno, S. Shoji, "Study of low-temperature wafer bonding with Au-Au bonding technique", *Proc MicroNano*, June 3-5, Kowloon, Hong Kong, 2008, pp. 747 – 748.

[15] R. Fraux, J. Baron, "STMicroelectronics' innovation in wafer-to-wafer bonding techniques shrinks MEMS die size and cost", *3D packaging* No 21, pp. 24 – 27, 2011.

SEALING DISPENSING FOR MEMS WAFER CAPPING

Heakyoung Park
Nordson ASYMTEK
Carlsbad, CA, USA
heakyoung.park@nordson.com

ABSTRACT

The MEMS industry has gained momentum recently with significant unit growth, especially in the consumer market. While traditional MEMS devices like automotive accelerators have established high-volume manufacturing processes and packages with high reliability, most MEMS devices have been fragmented in packaging because of their unique requirements and small volumes, resulting in high packaging costs. In the cost-sensitive consumer market, emerging devices such as MEMS microphones, accelerometers, and gyroscopes in mobile devices have rapidly increased their production volume and chased lower packaging cost while accepting relatively less reliability than traditional devices. This paper will present technical requirements for dispensing sealant, volumetric accuracy, and motion systems for MEMS wafer capping to meet packaging trends and will also address manufacturing cost reductions.

Key words: MEMS, dispensing sealant, wafer capping, volumetric accuracy, motion systems

INTRODUCTION

Wafer capping technology in MEMS production has gradually become popular in many MEMS devices such as inertial sensors, oscillators, and microfluidic packages. During production, MEMS release timing is very critical. Wafer dicing and die handing are difficult after release because the MEMS devices become fragile. Wafer capping provides MEMS structure protection after release during the dicing process. A cavity wafer covers the MEMS structure by bonding the cavity wafer to the MEMS die wafer, and the two bonded wafers are diced together. In addition wafer capping can provide a vacuum or low pressure with the right material, and a hermetic or almost hermetic seal can be realized.

However, there are manufacturing challenges. First, wafer capping requires precise alignment. Second, a variety of bonding methods have technical drawbacks. Stencil or screen printing is one way to deposit sealing lines to cap wafers but consumes a relatively wide area as sealing width and may not work if the sealing line must go onto the MEMS die wafer which is not flat surface. Dispensing, an alternative to printing, can apply a conductive or non-conductive sealing line on the MEMS die wafer even if it has fragile MEMS structure. Precise alignment is still a challenge regardless of various capping methods.

Cap wafer screen printing method

Cap wafer fabrication is done to make a cavity using different etching techniques such as deep-silicon reactive ion etch (DRIE). Bonding material, like seal glass, is applied to a cap wafer as shown in Figure 1[1]. The material is applied to only the bonding area. Then the MEMS wafer and cap wafer are aligned together and bonding is completed by heating them to the fusion point of 430 °C. The bonded wafers are diced together. In some cases cap structures are partially etched and thus wire bonding pads are exposed as shown in Figure 2.

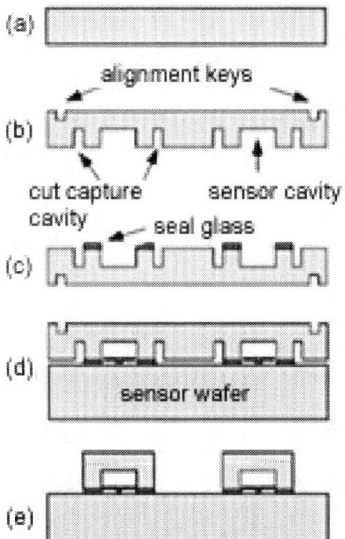

Figure 1. ADI Process [1]

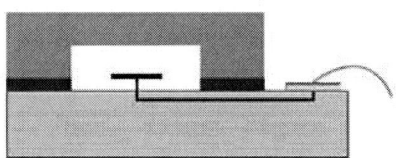

Figure 2. Bond pad exposed after capping

Wafer capping dispensing method

Some structures require sealant in MEMS Si wafer and it is difficult to use a printing method on a device wafer that contains movable parts. Dispensing is attractive due to its flexibility to apply sealant in various shapes especially for areas with geometric constraint or three dimensional

39

structure surfaces. For example, when the getter material is applied to the cap wafer, sealant material needs to be applied on the MEMS wafer. Sealant material like silver paste, bonding adhesive, UV curable material, or low-temperature solder, is applied to the MEMS wafer side. Rectangular shape sealant is common. Then the cap wafer such as a glass or silicon wafer is aligned with the MEMS wafer. When glass cap wafer is used with UV curable material, UV light penetrates the glass wafer to cure the adhesive [2]. Once the base and top wafers are bonded together, dicing is performed at wafer level while the released MEMS structure is protected. Figure 3 illustrates the sealing dispensing method.

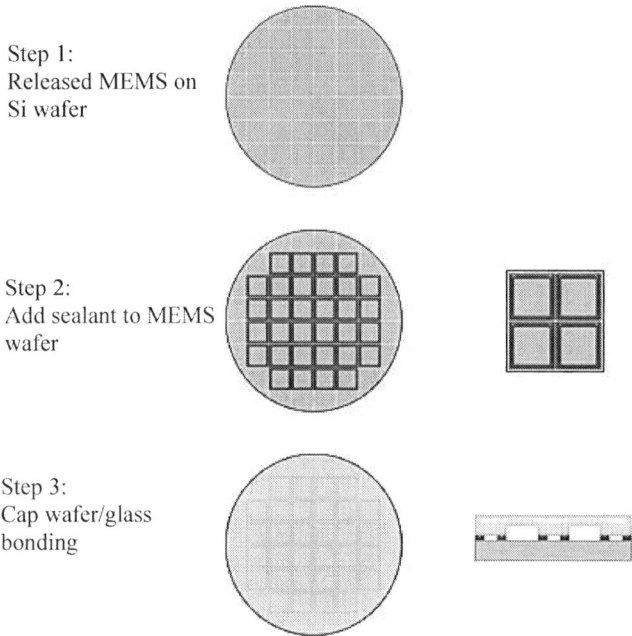

Figure 3. Wafer capping dispensing method

DISPENSING REQUIREMENTS
There are many challenges to meet packaging requirements for MEMS devices. High yield and throughput are critical to realize low-cost solutions in volume manufacturing environments. The following section details three main topics. First, the choice for dispense sealant is important for its end use. The right sealant material for the end application is selected based on defined life time, reliability concerns, and functionality of the device. Second, consistent volumetric accuracy is a must to achieve a process yield required for volume production. Lastly, the throughput impact of sealing dispense will be studied using a specific case example. Throughput study will guide manufacturer to choose right system and applicators based on their specific application requirements.

1) Sealant material
Many varieties of capping technologies are used today based on reliability requirements ranging from solder bonding, anodic bonding, eutectic bonding and adhesive bonding. Each method has different hermetic levels and

different tools for bonding. The MEMS designer will choose the right material for their packaging requirements based on lifetime of the device. Dispensing can be one option if liquid-type sealant is used, like epoxy adhesive, UV cure material, silver paste and solder paste. Adhesive may not meet the hermetic requirements for medical or military applications. However, consumer devices can accept less reliable but cost-effective approaches, however, time-to-market could be much more critical. Wafer capping is attractive in high-volume production because yield can be improved by MEMS immediate protection, and handling is easier during dicing, pick and place, and wire bonding.

Figure 4 shows an example of conductive adhesive jetting on wafer to create uniform thin lines of 0.3 mm width.

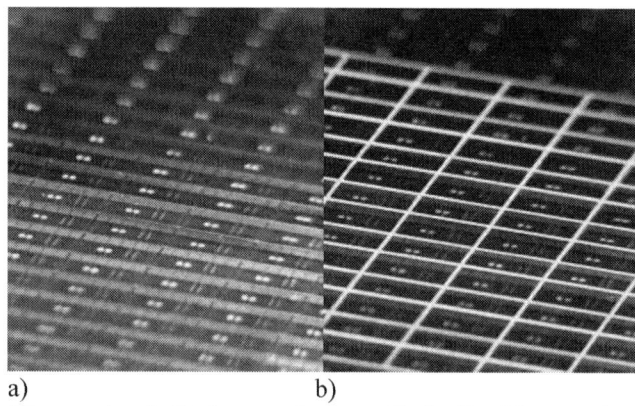

a)   b)

Figure 4. a) Before sealant applied; b) After sealant dispensing

2) Volumetric accuracy
There are many variables to control the flow of the material to dispense tight sealant lines. Fluid pressure, needle diameter, length of needle, and fluid viscosity determine volumetric flowrate shown in Figure 5.

$$Q = \frac{\pi d^4 \Delta P}{128 \mu L}$$

d = inside diameter of needle or nozzle
μ = fluid viscosity
L = length of needle or nozzle
ΔP = pressure drop
Q = volumetric flow rate

Figure 5. Volumetric flowrate - Poiseuille's law

Many varieties of materials have a pot life and viscosity will change over time. This changes the flowrate for a given setup. Along with flowrate change, dispensing volume changes if dispensing parameters are fixed. Thus dispensing parameters need to be adjusted to keep dispensing volume consistent, which calls for the use of a flowrate compensation tool. Dispensing uniform lines can be a challenge due to material characteristics if a flowrate compensation tool is not applied. Figure 6 shows an example of two-part silicone flowrate changes over a 4-hour

40

period (after fluid mixed) while dispensing parameters were fixed.

Once flowrate over time is characterized, the weight calibration tool can be applied at a given frequency for routine maintenance. Examples of variables for calibration are dispensing on-time, fluid pressure, or the number of shots (when jetting), depending on available tools and chosen applicators.

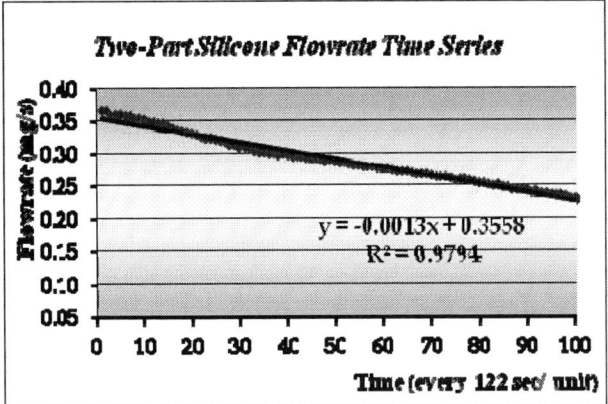

Figure 6. Flowrate change over time

Dispensing by auger or time-pressure valves
Certain materials are suitable for needle-type dispensing, like solder paste or dam epoxy material. There have been many efforts to create uniform thin lines and to dispense consistent volume at the knitting point for sealing. Sealant tends to come out slowly from the needle at the beginning and more excessively at the end of the sealing line. These inherent effects make it challenging to dispense a uniform knitting point. When the needle dispenses the sealant on the surface, controls at the starting and ending points are critical to make volumetrically consistent dispensing. Figure 7 shows the typical dispensing point when dispensing parameters were not optimized on the left (a) and improved case on the right (b)

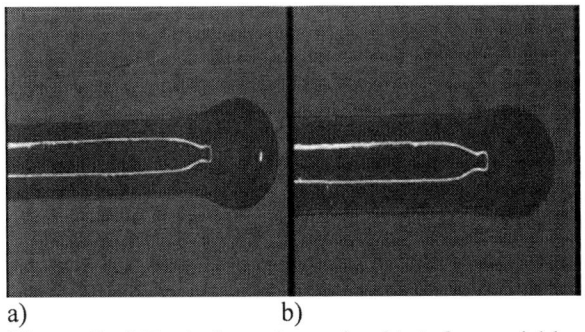

a)   b)

Figure 7. a) Typical starting point; b) A few variables were adjusted to minimize "dog-bone" effect

Impacts on the sealing line volume control
• Response time
Fluid will start to flow once fluid pressure is on. The duration between the valve turning on and sealant coming down to the surface depends on viscosity of fluid, fluid

pressure, valve motor speed and signal delay from the system. Somewhat uncontrollable flow time through the needle tip can change the volumetric consistency. Initial flow of sealant material takes time to reach the surface.

• Coordination with motion control
The dispense head starts to accelerate at the first dispensing location to reach dispensing velocity. It requires a certain distance in order to reach pre-defined dispensing velocity. At the first starting point, the system will put down more material at given flowrate because applicator velocity is initially slow. Plus, the applicator decelerates to stop at the end position and thus more fluid will be dispensed at a given flowrate by ramping down applicator velocity. For instance, if a 5mm line is dispensed from start to finish at given motion profile (Figure 8) to reach 35mm/s dispensing velocity, it takes 0.173s to dispense one straight line. During this time 0.53mm distance is used for acceleration (31ms) and 3.95mm distance is used for constant velocity.

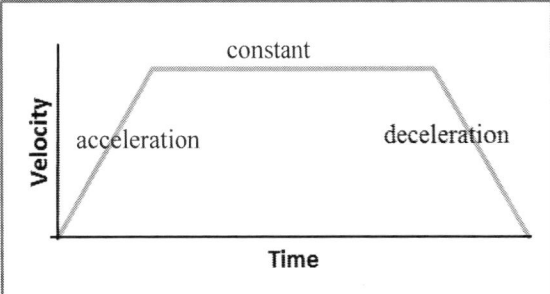

Figure 8. Motion profile

• Fluid break-off
When the needle is close enough to the substrate surface with certain dispense gap (the distance between needle tip and substrate) during dispensing, fluid is still in contact with the surface. Fluid can break-off when the needle moves up. The amount of fluid in the needle tip and the amount of fluid on the surface are difficult to control when the fluid break-off point is inconsistent.

Solutions:
There are two issues to address: knitting point and inconsistent volume. Precise time control of turning fluid pressure or valve motor before the dispense head reaches the first dispensing location or turning on delay time before dispensing begins would minimize the "dog-bone" shape line. This timing is critical to control the amount of material volume. Timing control of pressure off before the end of the sealing line could eliminate the excessive amount of sealant at the knitting point due to slow response time of the flow similar to start location. In addition, motion parameters should provide needle move-up distance, acceleration and velocity control for consistent fluid break-off. Those three parameters can be optimized based on sealant type. The shortest needle move-up distance with optimized maximum acceleration and velocity for a given application setup would result in high throughput if needle move-up and -down time is a significant portion of total dispensing time.

Jetting small uniform droplets

Another dispensing applicator can eliminate the "dog-bone" issue fundamentally: the jet applicator. Jet dispensing delivers small droplets, without contacting the substrate. High energy is delivered by a piston to the tip of the nozzle and fluid is ejected through the small nozzle orifice as Figure 9 shows. An array of small dots can be connected to form a uniform line as shown in Figure 10. Precise valve control of a millisecond achieves consistent volumetric accuracy, and eliminates "dog-bone"-shaped lines. Jetting can be optimized by moving the dispense head to compensate for acceleration distance. Thus, at the starting point, dispensing already reaches constant velocity when it begins to dispense dots continuously.

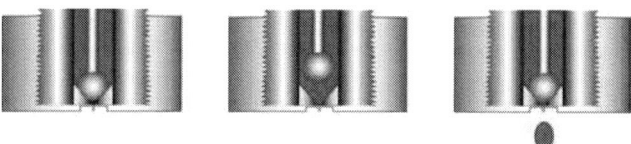

Figure 9. Mechanical jet applicator ejecting small droplets. The jet uses a pneumatic piston with a ball tip to raise the piston by air pressure and to push fluid through a narrow orifice at the jet nozzle. As the ball tip on the end of the piston engages in a seat at the nozzle, the fluid is energized and shoots a droplet from the end of the jet [3].

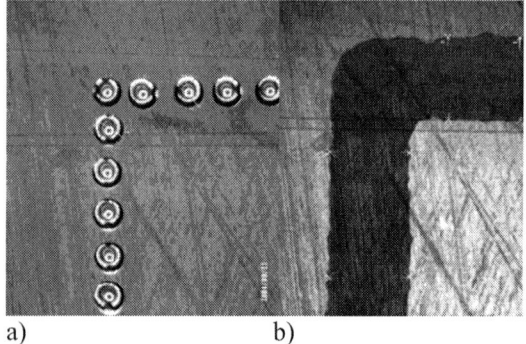

a)   b)

Figure 10. a) Arrays of dots right after landing; b) Dots connect to form a continuous line after laminating with top glass

If a sealant fluid contains fillers, then the choice of nozzle size will be limited by particle size distribution. Larger nozzle choices due to filler size can make it challenging to put down a fine line width. Figure 11 shows the droplet of non-filled material forming 100um diameter dot in the air using non-filled material. Thus a nozzle with small inner diameter of 50um was used.

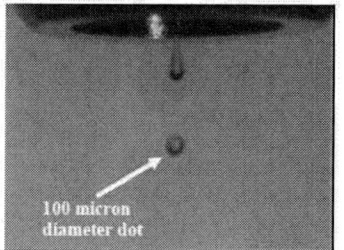

Figure 11. Jetting non-filled material 100 um diameter dot in the air

Potential issues

If sealant volume control is inconsistent and more sealant is applied in one line, then excessive material can contaminate wire bonding pads or other components near the sealant material after capping. This is a typical failure and re-work can be difficult. When sealant is disconnected due to material void or line width is inconsistent due to un-optimized parameters, this will be a critical failure when it requires certain hermeticity or low vacuum after capping (Figure 12).

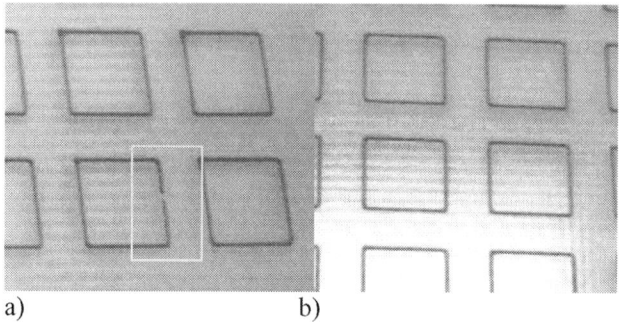

a)   b)

Figure 12. a) Disconnected sealant line (yellow box); b) Optimized sealant line

3) Motion systems – throughput requirements

This section evaluates the impact on throughput by changing motion parameters such as maximum velocity. In addition two different applicators were used for the experiments. The sealing line width requirement is driven by the geometric constraint of a package structure. Thus a certain needle/nozzle inner diameter, fluid pressure, line speed, and others were selected for this study.

Experiment setup:

Two different motion parameters were set up. Motion 1 used normal mode and motion 2 applied high-performance parameters. Motion 1 with a single applicator was used as a baseline. Table 1 shows the setup. "Dual" in this case means moving two applicators simultaneously that dispense different MEMS sealing lines in parallel.

Table 1

	Motion Type	Applicator
Scenario1	Motion 1	Single
Scenario2	Motion 2	Single
Scenario3	Motion 1	Dual
Scenario4	Motion 2	Dual

Height sensing and vision:
It was assumed that the wafer had good alignment and flatness. Thus three points of height sensing (HS) per substrate and two fiducials (alignment mark) were applied.

Dispensing dimensions:
5mm X 5mm square sealing line was used per unit. One panel included 100 units. Each unit spacing was 5mm pitch.

Non-contact jetting case:
Putting down sealant in the desired location can be a challenge when there are tight tolerance and dimension requirements. It is important that the dispensing system provides constant velocity during dispense in order to put down the droplet at consistent distance intervals. Cycle time per unit is shown in Figure 13. Total time of handling and dispensing includes handling, vision, height sensing and dispensing time.

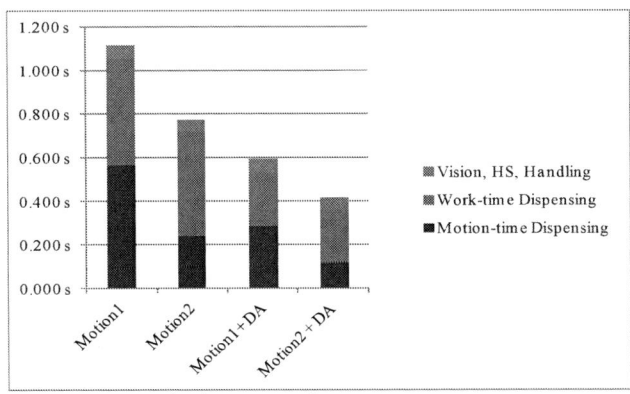

Figure 14. Cycle time per unit

$$\text{Cycle time per unit} = \frac{\text{Total time of handling and dispensing}}{\text{Number of units per tray}}$$

Figure 13. Cycle time

Applicator velocity during dispensing was determined by flowrate and other requirements such as line width. In this case it required a 0.3mm line width and thus a small orifice nozzle was needed to make the small volume droplet. Work-time dispensing means the duration of the applicator putting down sealant material. Applicator velocity was locked to 42mm/s during dispensing patterns of square sealing units as work-time dispensing. Motion-time dispensing is defined by non-dispense move time between square sealing units. Non-dispense moves of 5mm were used in this experiment as each unit pitch and two different motion parameters showed significant difference in cycle time by reducing motion-time dispensing. Figure 14 shows that high throughput can be achieved by high-performance motion parameters or normal motion parameters with dual applicators. A large portion of motion-time dispensing (51%) is allocated in Motion 1 and thus a dispensing system with high-performance motion parameters can increase unit per hour (UPH) in this case. Certain dispensing system accuracy is required for the dual applicators to meet tight geometry constraints.

Surface contact dispensing case:
The same dimension parts were used for the throughput comparison. This applicator needed to use a small dispense gap in order to put down the sealant in a consistent manner. The goal was to achieve 0.3mm line width using an auger valve with needle. Typical line dispensing velocity is 25mm/s to ensure uniform line width. Figure 15 shows that the majority of time allocation per part was driven by work-time dispensing due to slow applicator velocity during dispensing required for this sealing dispense. Plus, the needle required move-up time to provide clean fluid break-off from the needle. In this case high-performance motion parameters did not improve throughput significantly.

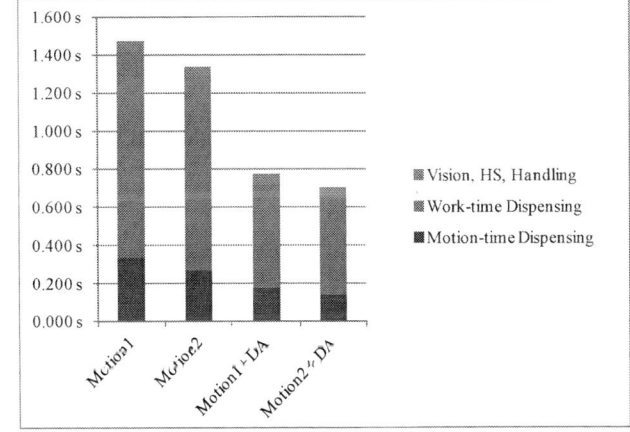

Figure 15. Cycle time per unit

Each applicator with different part dimensions and spacing has a different cycle time. In addition cycle time by using vision, height sense, dispensing work and motion time will vary greatly. For instance if singulated parts are placed in arrays and this requires height sensing and finding fiducials for each part to compensate the alignment, then vision and height sensing time allocation will be much greater than the previous two cases. The material choice will determine the right applicator. For instance, solder paste may require a different auger valve for different solder mesh types, and this will determine the smallest needle inner diameter to eliminate plugging.

Choosing the right sealant material and achieving dispensing volumetric accuracy will result in high yield. Cycle time analysis by breaking down the time usage provides accurate understanding of how a dispensing system is used. This will help further optimize parameters such as applicator velocity impacting total cycle time. Certain applications could realize improved UPH by utilizing a dispensing system with high-performance motion parameters. Thus high yield and high throughput efforts will drive down the cost per unit.

CONCLUSIONS

MEMS wafer capping has been seen in many devices such as inertial sensors, oscillators, and microfluidic packages. This paper presented technical requirements for dispensing sealant, volumetric accuracy, and motion systems. Sealant material choice depends on the lifetime of the end application, reliability, and functionality. This paper also addressed challenges of volumetric accuracy when dispensing sealant and then presented solutions to improve dispensing consistency. Throughput optimization was studied by using specific MEMS applications by varying motion control parameters and dispensing parameters.

FUTURE WORK

Future work will include investigating the impact on the throughput model due to size reduction efforts in packaging. More dies can be produced per given wafer size, which means there is more dispensing area per wafer. In addition, sealing line requirements will be more challenging, and so further work will address these demands.

ACKNOWLEDGMENTS

The author would like to thank Akira Morita for providing insights of the contents, Roberta Forster-Smith for editing, and Jay Sibley for graphics of the paper.

REFERENCES

1. L.E. Felton, N.Hablutzel, W.A Webster, K.P. Harney, "Chip Scale Packaging of a MEMS Accelerometer," ECTC, May 2009.
2. A. Morita, "Dispensing Advantages for MEMS Wafer Capping," Chip Scale Review, Nov 2011.
3. S. J. Adamson , M. Peterson, "Enabling high density System in Package (SiP) manufacturing and consumer electronics devices through the use of jetting technology to minimize substrate area for underfill," SMTAI, Oct 2006.

LOW STRESS THICK FILM PHOTOPATTERNABLE THICK FILM SILICONES FOR LARGE DIE WAFER LEVEL APPLICATIONS

Herman Meynen
Dow Corning Europe S.A.
Parc Industriel - Zone C- Rue Jules Bordet
7180 Seneffe, Belgium
herman.meynen@dowcorning.com

Ranjith John, Ken Weidner, Craig Yeakle, Mike Bourbina, Arianne Tan, Brian Russell, James Rosson
Dow Corning Corporation
2200 W. Salzburg Road
Midland, MI, 48686-0994, USA

ABSTRACT

Materials of Dow Corning® brand WL-55XX Photopatternable Silicones can easily be coated onto different substrates at thicknesses ranging from 6 to 45 μm for a single coating, and patterned using standard photolithographic processes. Unique to these materials is that the required exposure dose is film thickness independent, allowing that a thinner film of 10 μm requires the same dose as a thick film of 40 μm, maintaining the throughput of the mask aligner. The patterned features provide sidewalls to facilitate the step coverage of the metallization processes.

The cure process window is broad as it allows curing of the material at 180°C with further potential of lower temperatures during longer cure times. Cured at low temperatures (<250°C), they provide colorless, transparent, low modulus films that are moisture resistant and acid free. They show very little shrinkage during cure (~2%), do not require high temperature processing, and provide a very low residual film stress. The heat dissipation in devices causes an increase in stress over time, while this unique class of materials maintain almost the same stress level.

Key words: photopatternable silicones, low stress, stress buffer .

INTRODUCTION

Silicones are well known for their excellent thermal resistance, chemical resistance and optical properties in the electronics market. They have been commonly used as a protective coating, encapsulant, adhesive and as Thermal Interface Material (TIM) for electronic modules. These materials are for example used in the automotive market and more recently as encapsulants for Light Emitting Diodes

(LED). In older technologies, silicones were used on the die level as glob top material to reduce the stress induced by the Epoxy Molding Compound (EMC). Other micro-electronics applications where silicones are applied are die attach, lidseal and Thermal Interface Material (TIM1).

The focus of the current paper is to discuss the photopatternable silicones which allow film coatings of 6 μm to 45 μm with a very low stress. Photopatternable silicones were introduced a decade ago as a Silicone Under the Bump (SUB) solution for enhancing the reliability [1-5]. The solution did demonstrate improved reliability as these materials have a very low stress.

The introduction of low-k materials (k=2.5-2.9) which have a weak mechanical strength, was the driver [7-8]. These materials increase the tensile stress on the wafer compared to the compressive stresses created by the traditional CVD oxides used for older technology nodes [6]. Many photopatternable dielectric materials traditionally used as stress buffers, such as benzocyclobutene (BCB) or polyimide (PI) [10-12], may possess unacceptably high residual stress values for large die size devices while silicone materials have considerably, lower residual tensile stress [4]. These low modulus, highly flexible photopatternable silicone materials are able to absorb deformations and reduce stresses created in the device by the Coefficient of Thermal Expansion (CTE) mismatches of different materials [4-5].

Key material properties for the next generation of devices, therefore, will include flexibility, low stress, high thermal stability and low temperature cure. Materials with these properties will also create new opportunities in emerging areas such as for wafer level packaging, MEMS and optical wafer level applications.

Thermo-mechanical stress failure [13] is a critical issue faced by large die sizes such as those found in power electronics devices, particularly Insulated Gate Bipolar Transistors (IGBT's). These large device dies are faced with high stress environments due to the organic materials used as passivation.

The existing silicone technology used in these applications suffered from a number of shortcomings such as room temperature shelf-life and reproducibility. Dow Corning is advancing technology on a new line of materials to address these issues.

RESULTS
Experiments
Experiments were conducted by spin coating the photopatternable silicones on 150 and 200 mm wafers and characterizing the coating performance as well as the mechanical and electrical properties. In case of electrical measurements, low resistance (<0.05 ohm/cm) silicon wafers were used whereas high resistance double side polished silicon wafers were used for mechanical property measurements.

A DNS 80A spin coater was used with manual static dispense of the silicone materials. Bakes on a hotplate were executed in an air environment.

Wafers were UV exposed (broadband) with an EVG mask aligner.

The development of the patterns was done on a standalone Karl Suss RC8 spin coater.

All thickness measurements were done with a semi-automated optical spectrometer F50 from Filmetrics.
The stress curves were generated using a Tencor KLA Flexus tool.

For Thermo Gravimetric Analysis (TGA), a TA instruments Q 500 TGA was used in air. The sample was heated at 10 degrees per minute to the final temperature and held isothermal for 60 minutes.

The electrical measurements have been obtained using a CVMF- HP4194, CVMF-Signatone Quiet Chuck and Dark Box, CVMF-SMU K1237, CV6-Capacitance Meter K1590, and CV6-HP4140.

Process
The processing sequence consists of a spin coating step, a soft bake step to evaporate the solvent, a UV exposure using a broadband mask aligner, a Post Exposure Bake (PEB), a development step using an organic solvent and finally a hard bake step to cure the silicone material (fig. 1).

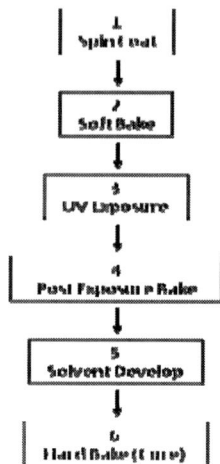

Figure 1, process sequence for negative tone photopatternable silicone

Table 1, overview of processing conditions for Dow Corning® brand WL-55XX Photopatternable Silicones.

Process Step	Recommended Conditions
1. Spin Coat *a) Spread* *b) Spin* *c) Edge Bead Removal*	200-500 rpm/10-25 s. 1000-3000 rpm/30 s. 700 -900 rpm/20-30 s.
2. Soft-bake	Hot Plate 110°C/120 s
3. UV Exposure	i-line or Broadband 1000-1400 mJ/cm^2
4. Post-Exposure Bake	Hot Plate 130-150°C/120 s
5. Development	Butyl Acetate Solvent Double Puddle Development Process
6. Hard Bake - Cure	Oven (Air, N$_2$) 180°C/60 min. or 250°C/60 min.

Spin coating process: There is no pre-treatment applied to the silicon wafer prior dispensing the material dynamic or static. The spin coating recipe consists of a spread step first to distribute the material properly across the wafer prior the higher spin speed step is applied to set the thickness. Due to hardware limitations only manual static dispense direct from the bottles was applied. Spin coating was done in all experiments with an open spin cup. Spin curves for 10 and 40 μm thick film solutions can be observed in below figures 2 and 3.

46

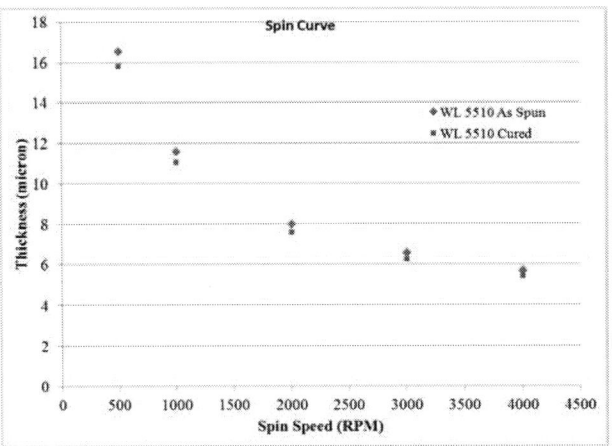

Figure 2, spin curve for solution targeting 10 μm thick films – Dow Corning® WL-5510 Photopatternable Silicone for as spun and after a hard bake of 180°C, 1hour in air.

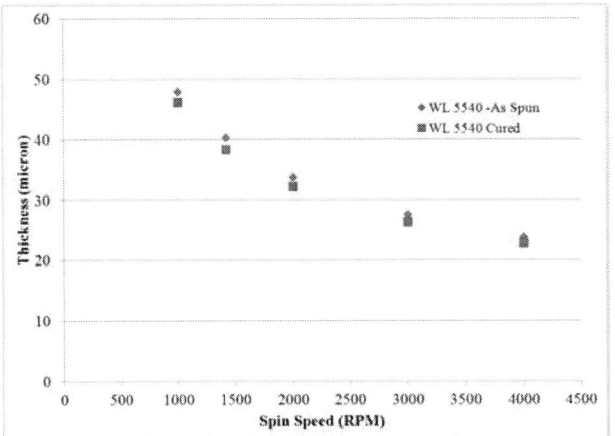

Figure 3, spin curve for solution targeting 40 μm thick films – Dow Corning®WL-5540 Photopatternable Silicone for as spun and after a hard bake of 180°C, 1 hour in air.

The uniformity is 0.7% for 10 μm thick films both as spun and 0.7% after cure at 180°C, 1 hour.

For thicker films, uniformity is slightly worse. The uniformity is 1.5% for 40 μm thick films both as spun and 1.3% after cure at 180°C, 1 hour.

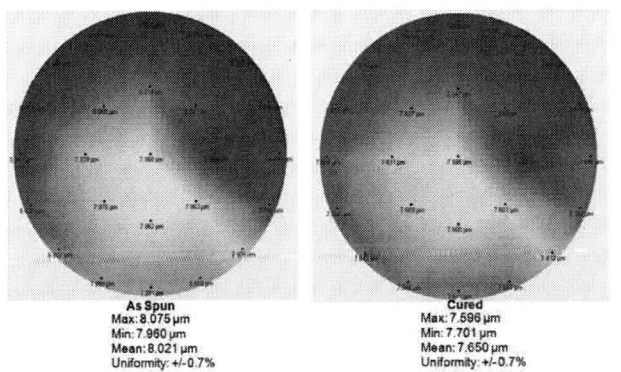

Figure 4, uniformity across the wafer of 8 μm thick films spin coated at 2000 rpm – Dow Corning® WL-5510 – 200 mm wafer

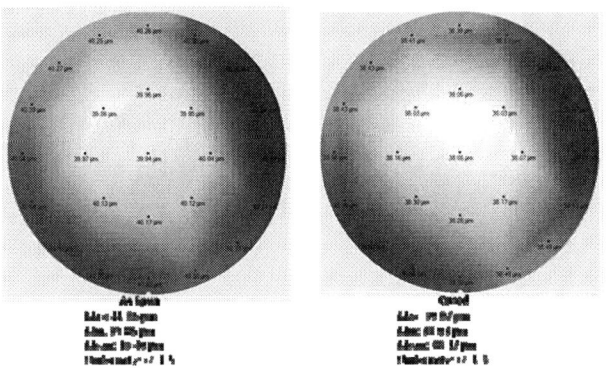

Figure 5, uniformity across the wafer of 40 μm thick films spin coated at 1420 rpm – Dow Corning® WL-5540 – 200 mm wafer

The soft bake is applied to remove the solvent of the solution and was fixed in all experiments at 110°C, 120s on a hotplate in air.

Exposure dose was varied within the following range: 1000 – 1400 mJ/cm^2. The proximity gap is dependent on the film thickness. The dose is independent of the thickness, so thick and thin films are exposed with the same dose which is quite different compared with other materials. This allows maintaining good throughput of mask aligner. The explanation for this phenomenon is due to the fact that the films are not absorbing UV radiations in the top layer but allow penetrating across the whole film thickness. The UV exposed areas are more selective during development. The required minimum dose is 1000 mJ/cm^2 to create sufficient selectivity between exposed and unexposed areas and to maintain 80-90% of the spin coated thickness in combination with a proper Post Exposure Bake (PEB).

The PEB is required to complete the cure reaction that was catalyzed by the UV exposure. This is performed on a hotplate in air at a temperature between 130-145°C for 120s.

Double puddle development was applied for all experiments where patterns were created using butyl acetate. This solvent has a higher flash point compared to with hexamethyldisiloxane while having good drying properties compared to higher boiling point solvents like mesitylene and Negative Resist Developer (NRD).

A puddle of solvent is created on the wafer. The puddle stays for 30s and is removed by a 10s spin while the solvent is sprayed on the wafer. This is followed by another solvent dispense of butyl acetate which remains 30s on the wafer, after removal, a final rinse is applied after which a spin dry is completed.

Next a final bake is applied in a conventional oven in an air environment. Different curing temperatures have been used from 180°C to 250°C, from 60 to 30 minutes.

Via resolution as low as 15 μm in diameter have been demonstrated. Figure 6 is the 15 μm via on 10 μm film and figure 7 is a 30 μm line-space on a 40 μm film. There is some loss of material as shown in the figure 7, which can be modified by optimizing the PEB and UV

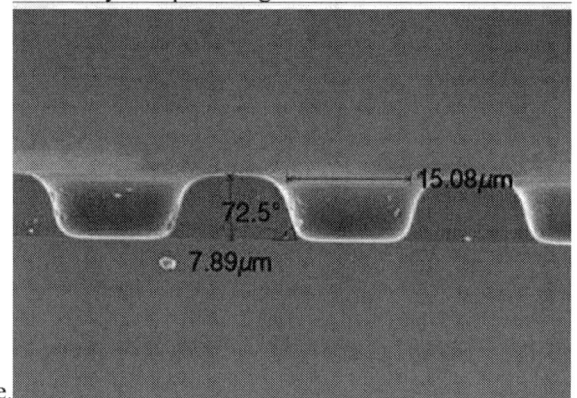

Figure 6, SEM cross section of a 15 μm via in a 10 μm thick film.

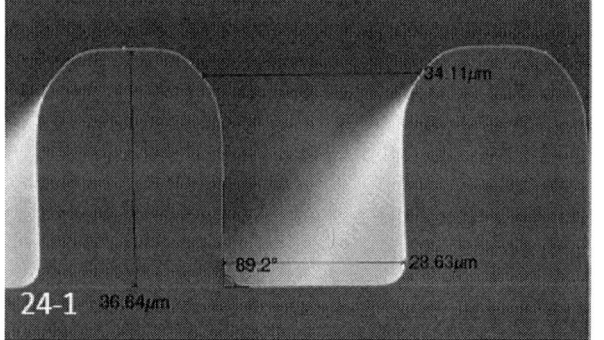

Figure 7, SEM cross section of a 30 μm lines/spaces in a 40 μm thick film.

The stability of the solution at room temperature exposure is an important parameter for acceptance of the technology in the fabs. Comparison was made of material which was stored at room temperature for 5 months and material stored in a refrigerator. The results demonstrate that there is no impact on the resolution while the older technology resulted in no patternability. Optimization of the formulation did provide this benefit which creates much better flexibility in usage of the material both for R&D, prototyping and volume manufacturing.

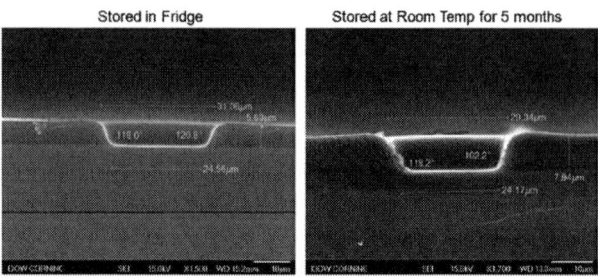

Figure 8, shelf life study comparing patterned features of Dow Corning® WL-5510 stored in refrigerated and room temp conditions.

Creating patterns is one thing but there is also a need for rework if something goes wrong with the spin coating process or when there is a misalignment during exposure or when used as a protection coating to remove after wet or dry etching steps.

Experiments have been executed with 16 μm thick cured photopatternable silicone films. The curing temperature of these films was 180°C for 1 hour in air. Test sample with cured PhotoPatternable silicone was placed in a beaker containing Dynasolve 711. The beaker was then placed in a warm water bath where the temperature of the bath was maintained at 40°C. The films were easily removed after 30 minutes.

None cured films prior PEB can be removed in solvents like butyl acetate or hexamethyldisiloxane.

Material properties
The materials – Dow Corning® brand WL-55XX Photopatternable Silicones - discussed in this paper provide a low stress and low temperature cure. Experiments were executed where material was exposed at room temperature and ramp to 300°C in N_2 with a temperature ramp rate of 2.3°C/min and ramped down to room temperature. The stress has been analyzed for material cured at 180°C for 60 minutes in air and the stress curve is shown in Figure 9.

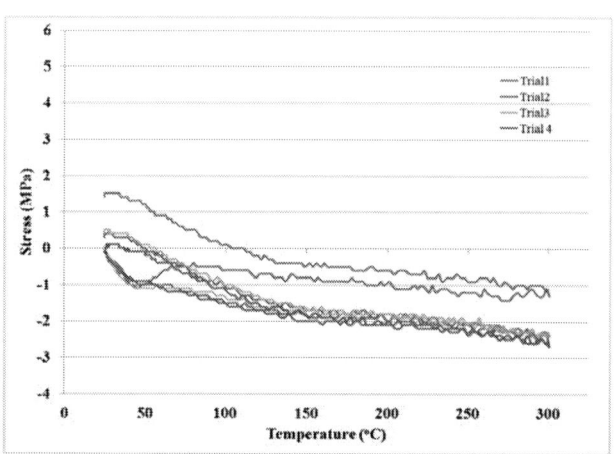

Figure 9, stress as a function of temperature up to 300°C and cool down to room temperature, for 40 μm thick films – Dow Corning® WL-5540 – cured at 180°C for 60 minutes

The results show extreme low stress behavior, at room temperature the stress is nearly 0MPa. The stress is slightly increasing after ramping to 300°C and cooling down to room temperature. This is unique compared to other materials. It demonstrates that this family of photo-patternable materials has a low stress behavior even after multiple ramps to temperatures above cure temperature. It provides potential to be integrated in a passivation module as a stress buffer material where it may have a dual function, first a mask function to dry etch the bondpads and after removing etch polymer residues, it still has potential to act as a stress buffer layer.

The Modulus and Coefficient of Thermal Expansion (CTE) were measured by Dynamic-mechanical analysis and Thermo-mechanical analysis. Bulk sample about 2 mm thick was prepared by curing the material from room temperature to 150°C over a period of days to ensure uniformity of sample. The Modulus at room temperature is 29.3MPa creating a soft material, providing flexibility with the integration of the material in the application.
The Coefficient of Thermal Expansion is 245μm/m/°C. This value is higher compared with organic materials like polyimide or benzocyclobutene. The integration capability has been proven for the older silicone technology with similar mechanical properties [1-5, 7]. It is not a single mechanical property that defines if a material can be integrated or not but the combination of the mechanical properties that makes that a dielectric material can be integrated successful. It has been demonstrated that for a Silicone Under the Bump (SUB) configuration that reliability of the build device was clearly improved [1-3].

Thermal stability was also measured by Thermal Gravimetric Analysis (TGA) up to 300°C in air. It is obvious that not just thermal effects play a role but also that at higher temperatures the oxidation of the surface is playing a role during TGA. It is difficult to subtract these effects as also in a N_2 environment as there are potential some ppm of O_2 present.

Table 2, TGA in air for cured material at 180C for 1hr holding time of 1 hour at each temperature

Material	180°C (wt%)	200°C (wt%)	225°C (wt%)	250°C (wt%)	300°C (wt%)
WL-5540	0.90	0.63	1.15	2.42	9.35

Iso-thermal degradation at 300°C is initiated after 10 minutes at 300°C. The weight loss is 9.35% when hold for 60 minutes in air.

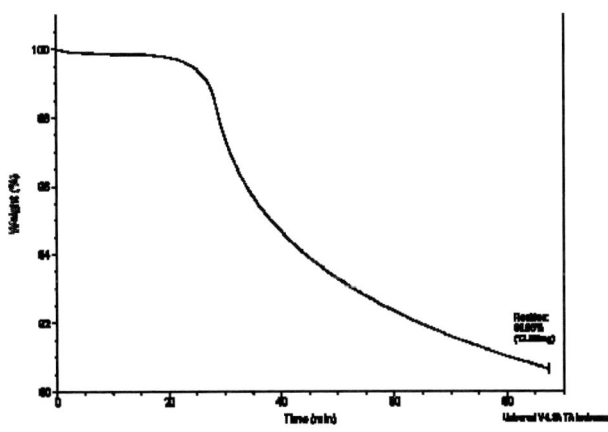

Figure 10, TGA of Dow Corning® WL-5500 cured at 180°C, heated to 300°C and held isothermal for 60 minutes

Electrical properties:
The dielectric constant was measured for different film thickness. These films are cured at 180°C, 60 minutes in air. The results for film thickness of 8 to 38 μm, show very reproducible values, 3.02 ± 0.04 for 5 different locations across the wafer.

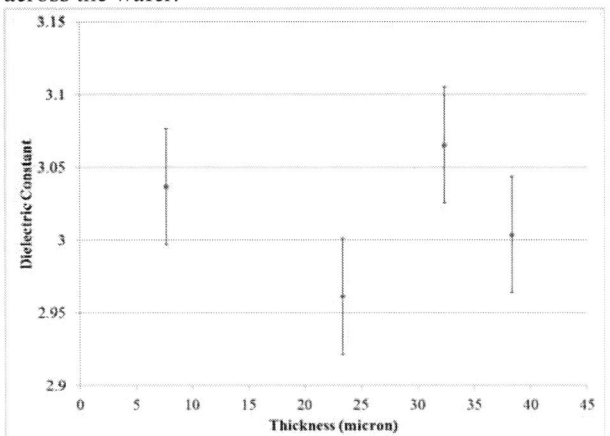

Figure 11, dielectric constant for different film thickness on 5 wafer locations for films cured at 180°C

Breakdown voltage was measured by fabricating capacitors using aluminum. The results obtained, varied from 39-53 kV/mm, for material cured at 180°C, 60 minutes these are comparable with the performances of silicones used in electronic modules and assembly silicone materials like die encapsulant and die attach materials.

CONCLUSIONS
Silicone technology for negative tone photopatternable thick dielectric materials (Dow Corning® WL-5510 & Dow Corning® WL-5540) was successful developed with long-term room temperature stability targeting stress buffer applications for large die sizes like Insulated Gate Bipolar Transistors (IGBT).

The fabrication process of these patterns has been explained. The process consist of the following steps: spin coating, soft bake, UV exposure, PEB, development and cure to create low stress films and patterns.

Very thick films are possible as the UV exposure (broadband) dose for thin and thick films is identical.

Excellent room temperature stability was demonstrated of more than 5 months, which allows flexibility both during qualification, prototyping and volume manufacturing.
A rework method was explained.

Technology demonstrates very low stress even multiple temperature ramps are executed up to 300°C. A rather flat hysteresis is obtained so stress in not increased a lot after the multiple temperature ramps.

Electrical results show good dielectric behavior of the material.

ACKNOWLEDGEMENTS

The authors would like to thank the analytical group of Dow Corning Corporation for the support provided for the various material property analyses.

REFERENCES

[1] B.R. Harkness, G.B. Gardner, J.S. Alger, M.R. Cummings, J.L. Princing, Y. Lee, H. Meynen, M. Gonzalez, B. Vandevelde, C. Witters, E. Beyne, Photopatternable silicone compositions for electronics packaging applications, SPIE2004 Microlithography, to be published.

[2] M. Gonzalez, B. Vandevelde, C. Winters, E. Beyne, Y.J. Lee, L. Larson, B.R. Harkness, M. Mohamed, H. Meynen, E. Vanlathem, An analysis of the reliability of a wafer level package (WLP) using a silicone under the bump (SUB) configuration, Proceedings of the 53rd Electronic Components and Technology Conference, 2003, pp. 857-863.

[3] H. Hedler, T. Meyer, W. Leiberg, R. Irsigler, Elastic Bump Wafer Level Packaging Platform (not only) for memory products, IMAPS 2003

[4] G.B. Gardner, B.R. Harkness, E. Ohare, H. Meynen, M. Vanden Bulcke, M. Gonzalez, E. Beyne, Integration of a low stress patternable silicone into a wafer level package, Proceedings of the 54rd Electronic Components and Technology Conference, 2004

[5] H. Meynen, M. Vanden Bulcke, M. Gonzalez, B. Harkness, G. Gardner, J. Sudbury-Holtschlag, B. Vandevelde, C. Winters and E. Beyne, Ultra Low Stress and Low Temperature Patternable Silicone Materials for Applications within Microelectronics, Proceedings of the European Workshop on Materials for Advanced Metallisation 2004, p.212-218

[6] T. Matsuda, M.J. Shapiro, S.V. Nguygen, Dual Frequency plasma CVD fluorsilicate glass deposition for 0.25 μm interlevel dielectric, Proceedings of first international Dielectrics for VLSI/ULSI Multilevel Interconnection Conference (DUMIC) 1995, p.22-28.

[7] B. Zhong, H. Meynen, F. Iacopi, K. Weidner, S. Mailhoutre, E. Moyer, C. Bargeron, P. Schalk, A. Peck, M. Van Hove, K. Maex, "A New Ultra-Low K ILD material Based On Organic-Inorganic Hybrid Resins" – Materials Research Symposium. Vol 716 – 2002

[8] H. Meynen, J.N. Bremmer, F. Iacopi, R. Donaton, H. Struyf, M. Lepage, J. Van Aelst, W. Boullart, M. Van Hove, S. Vanhaelemeersch, K. Maex , Integration of Porous Inorganic XLK™ 20-340 in Damascene Structures", Semicon West 2001

[9] Mathieu Vanden Bulcke, M. Gonzalez, B. Vandevelde, C. Winters, E. Beyne, L. Larson, B.R. Harkness, G. Gardner, M. Mohamed, J. Sudbury-Holtschlag and H. Meynen, Introducing a silicone under the bump configuration for stress relief in a wafer level package, Proceedings of the Electronics Packaging Technology Conference (EPTC), Singapore, December 11-12, p. 380-384, 2003

[10] Janggil Kim, N. Rolland, P.A. Rolland, S. Bouwstra, Thermo-Mechanical simulation of BCB membrane thin-film package, Thermal, Mechanical & Multi-Physics Simulation, and Experiments in Microelectronics and Microsystems (EuroSimE) , 2010 11th International Conference, 26-28 April 2010

[11] R. Ekwal Sah, Silicon Nitride, Silicon Dioxide and Emerging Dielectrics 9, Issue 3, p. 714 fig. 9

[12] Takashi Nishino, M. Kotera and K. Nakamae, Residual stress evaluation in aromatic polyimides by X-ray diffraction, Polyimides and Other High Temperature Polymers, Vol 1, P. 65-77, 2001

[13] M. Bouarrondj, Z. Khatir, J.P. Ousten, L. Dupont, S. Lefebvre and F. Badel, Comparison of stress distributions and failure modes during thermal cycling and power cycling on high power IGBT modules, Power Electronics and Applications 2007, European Conference

A NEW SINGLE WAFER CLEANING TECHNOLOGY FOR ADVANCED PACKAGING APPLICATIONS

Richard Peters, Travis Acra, Spencer Hochstetler, Kimberly Pollard, Keith Cox, Don Pfettscher
Dynaloy, LLC,
Indianapolis, IN
richardpeters@dynaloy.com

Thorsten Matthias, Thomas Glinsner, Martin Schmidbauer
EV Group
St. Florian am Inn, Austria

ABSTRACT

We have developed a novel single wafer cleaning technology for photoresist and photoresist residue removal. Development has specifically targeted the emerging needs of wafer level packaging: removing thick, highly cross-linked films such as photoresists and fluxes while maintaining the pre-clean integrity of the solder bump, exposed metals, under-bump metallization, and dielectric layers. The platform, called CoatsClean™, is a combination of both process and chemical technology featuring significantly reduced chemical usage, point-of-use heating, short process times, wafer-to-wafer consistency, and process flexibility in a single bowl tool. In this paper, the technology is described and results are shown that demonstrate the capability to remove both thick liquid and dry film photoresist on wafers with different solder bump types including eutectic, lead-free, and copper pillar.

Keywords: Resist strip, single wafer, packaging, dry film

INTRODUCTION

We have developed a novel single wafer cleaning technology that targets the needs of wafer level packaging: removing thick, highly cross-linked films such as photoresists and fluxes while maintaining the pre-clean integrity of the solder bump, exposed metals, under-bump metallization, and dielectric layers. The platform, called CoatsClean™, is a combination of both process and chemical technology. The innovation started from the insight that wafer cleaning is a chemical process, and the technology is the result of the design and engineering of the optimal chemistry and process for wafer cleaning. This technology features significantly reduced chemical usage, point-of-use heating, and short process times in a single bowl tool. In addition to environmental sustainability, the reduced chemical usage allows the use of fresh, unused solution on every wafer leading to wafer-to-wafer consistency and a cost of ownership that can be lower than other immersion or single wafer tool processes. This technology provides flexibility in photoresist cleaning processes including the ability to balance resist removal with materials compatibility, increased stability of chemical formulations, and the ability to run multiple wafer types and chemistries on the same tool. In this presentation, the CoatsClean technology will be described and results will be shown that demonstrate the capability to remove both thick liquid and dry film photoresist for the production of advanced packaging applications.

PROCESS DESCRIPTION

The CoatsClean process is performed using a newly developed EVG300RS single-wafer photoresist stripping system designed expressly for its implementation. Photoresist is removed using organic solvent based formulated strippers that are designed for the unique characteristics of low solution usage and short cleaning times, and the chemistry is optimized for each application to maximize cleaning performance. The process is a multi-step process performed in a single bowl, which enables a small tool footprint. The process flow is shown in Figure 1. Wafers are coated in a single shot dispense with the formulated stripper using only the minimum volume needed to allow complete resist removal, resulting in significantly reduced chemical usage per wafer compared to immersion or single wafer spray tools. Next, using point-of-use heating, the formulation is heated on the wafer. Figure 2 shows a typical temperature profile of the liquid formulation on the wafer during heating. Point-of-use (POU) heating offers flexibility to process different wafer types at different temperatures on the same tool and in the same bowl with no transition set up or cost. After heating, the wafer is rinsed to remove formulation and resist residue and prepare the surface for the next step in the fabrication process. Finally, the wafer is dried by spin drying. In addition to reduced chemical usage, the use of fresh, unused solution on every wafer leads to wafer-to-wafer consistency and increased stability of chemical formulations because the chemicals stored in the tool are held at room temperature rather than at elevated cleaning temperatures. Overall, the technology provides a new approach to photoresist removal and wafer cleaning using custom designed chemistry combined with an optimal process that provides environmental sustainability and a lower cost of ownership compared to traditional resist strip processes.

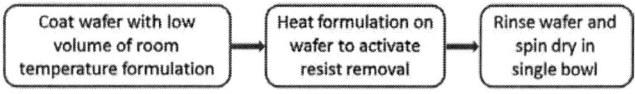

Figure 1. Single wafer clean process flow diagram.

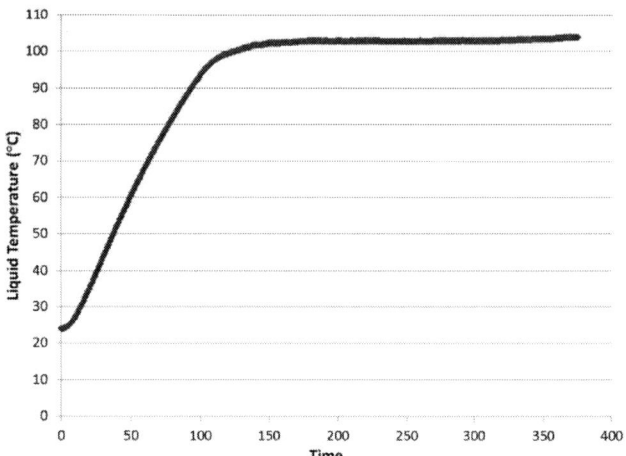

Figure 2. Liquid temperature on wafer vs. time (sec). Liquid temperature was measured with a thermocouple probe attached to wafer surface and covered with liquid formulation. Liquid was heated to a target temperature then maintained at the target temperature.

SYSTEM AND PROCESS ADVANTAGES

The CoatsClean™ resist strip platform has several advantages compared to immersion or single wafer spray systems. The use of a single bowl for performing all process steps enables a small tool footprint, which increases utilization of cleanroom space leading to reduced fixed costs. By comparison, immersion systems may require tools with large footprints containing multiple tanks for different process liquids. Immersion systems also have increased safety concerns due to the potential for chemical exposure created by large open tanks of heated solvent and pumping of heated solvent through the tool. These concerns are eliminated with the small volumes used and POU heating. Single wafer spray systems generally have multiple process modules for different steps of the process. Both immersion and single wafer spray systems typically have recycle loops that require filters, adding additional cost. Filters are not required using the CoatsClean process. Additionally, the small footprint for each process bowl allows multiple bowls to be configured together in one tool (see Figure 3). The number of bowls in a tool can be optimized to balance wafer throughput, cleanroom space, cost, and process flexibility, such as the ability to process different wafer types in separate bowls on the same tool with no downtime or set up cost.

The use of fresh chemical on every wafer improves wafer-to-wafer consistency. With immersion systems, the chemical composition of the formulation varies with time. Components of the formulation degrade over time, especially while held at elevated temperatures in an immersion tank. Additionally, the level of photoresist

residue in the tank increases with time, which has an adverse effect on cleaning performance. Figure 4 shows a plot with a representative concentration of the active component in a resist stripping formulation over time for both an immersion process and a CoatsClean process. The active component concentration is monotonically decreasing for the immersion process, but is constant for the CoatsClean process. For the immersion case, the varying chemical composition leads to variable cleaning performance and can result in poor wafer cleaning, especially as the bath life increases. Figure 5 shows a comparison of the cleaning performance after an extended period of time, where the cleaning performance deteriorates for the immersion case.

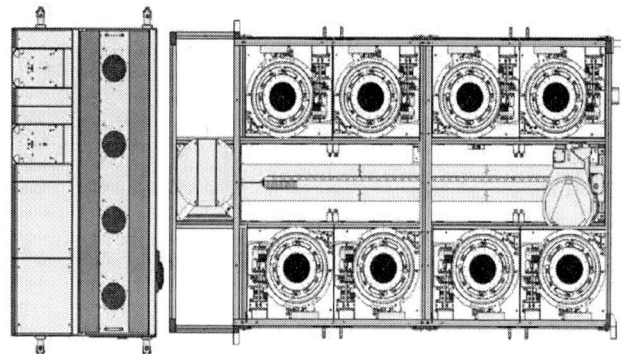

Figure 3. EVG resist strip platform showing configuration with 8 process bowls.

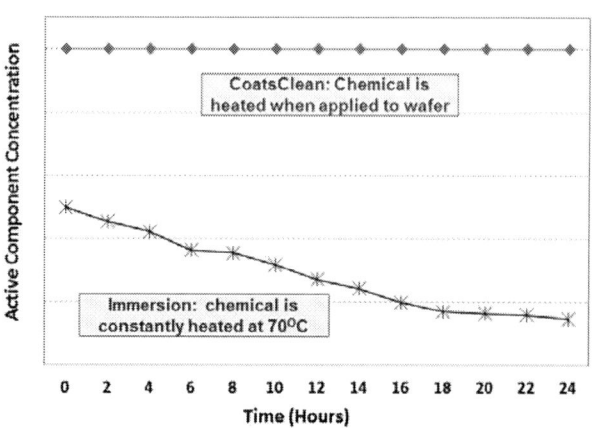

Figure 4. Active component concentration in chemical formulation vs. time for an immersion process and a CoastClean process. The concentration monotonically decreases in the immersion process but remains constant for the CoatsClean process due to the use of fresh, unused chemical on every wafer that is heated only after it is applied to the wafer.

PHOTORESIST STRIPPING FOR PACKAGING APPLICATIONS

We have successfully removed thick dry film and spin-on photoresists from bumped wafers with different solder types. For a typical process, a single bumped wafer is first coated with a solvent-based formulation. The formulation is next heated on the wafer to the desired temperature, and then held at that temperature for a specified time. Once the photoresist is dissolved, the wafer is rinsed to remove formulation, dissolved photoresist, and residue. Finally, the wafer is dried by spin drying. The process has been used to clean several different wafer types for both 200mm and 300mm wafers. Figure 6 shows representative optical microscopy and SEM images from wafers cleaned using CoatsClean technology, where all wafers had complete resist strip. After the resist was completely stripped from the wafers, the wafers underwent a wet acid-based etch step to remove the copper field metal. The copper field metal etch is a good indicator of the resist removal performance as well as the copper surface finish. Any remaining resist residue will block the copper etch step. Additionally, a poor surface finish of the copper field metal after resist strip can adversely impact the copper etch step. Figure 7 shows optical microscopy images after copper field metal etch of the same wafers shown in Figure 6. In all cases, the copper field metal was completely etched, indicating complete resist removal and a compatible copper surface finish.

CONCLUSIONS

We have developed a novel single wafer cleaning technology that meets the needs for wafer level packaging. The CoatsClean technology was used successfully to remove photoresist from bumped wafers with both thick dry film and spin-on photoresists with bumps with different solder types. Complete copper field metal etch was achieved indicating complete resist removal as well as a compatible copper surface finish. Overall, the CoatsClean technology provides a new approach to photoresist removal and wafer cleaning using custom designed chemistry combined with an optimal process on a flexible tool with a small footprint, environmental sustainability, and reduced EHS concerns to provide a lower cost of ownership compared to traditional resist strip processes.

ACKNOWLEDGEMENTS

The authors would like to acknowledge the following people for their contribution to this paper: Ronald Holzleitner and Mike Cain from EVG, Mike Quillen, San Sutanto, Max Ye, and Mike Phenis from Dynaloy, Allan Hilton and Alan Huffman from RTI for test wafers.

Immersion

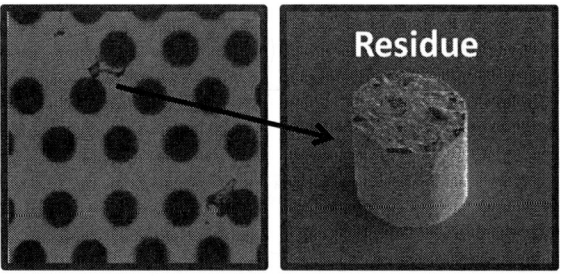

CoatsClean

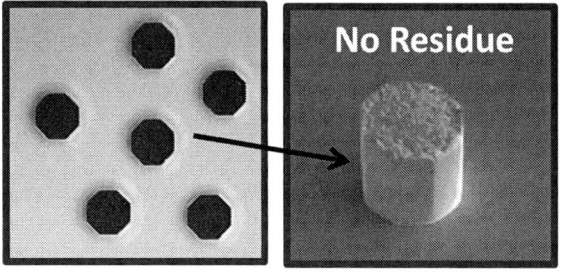

Figure 5. Images illustrating cleaning performance comparing an immersion process to a CoatsClean process after processing 150 wafers. [Top] Optical micrograph (left) and SEM image (right) showing resist residue on wafer after processing 150 wafers in a single immersion tank. Degradation of the active component leads to poor cleaning performance after an extended time or number of wafers processed. [Bottom] Optical micrograph (left) and SEM image (right) showing complete resist removal with no residue after processing 150 wafers in a CoatsClean process. The use of fresh, unused chemical on every wafer leads to consistent wafer-to-wafer cleaning performance regardless of the number of wafers processed.

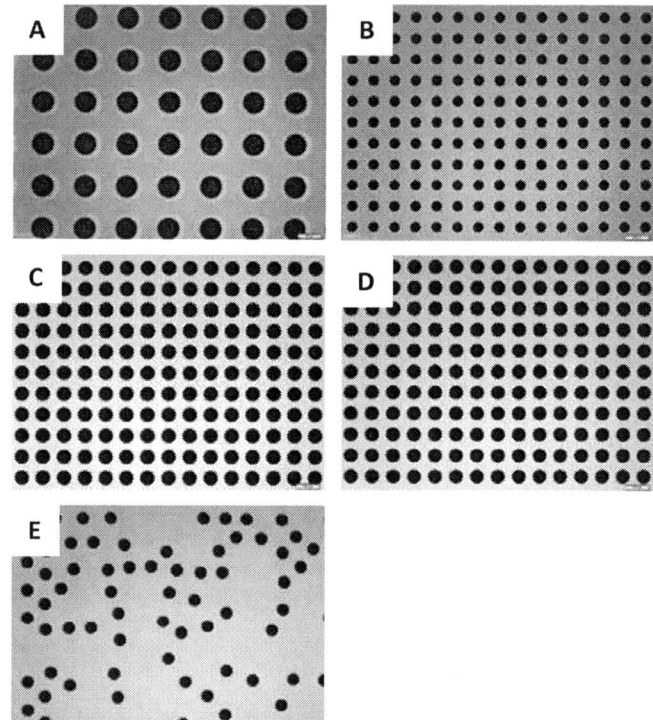

Figure 7. Optical micrographs illustrating complete Cu field metal etch for wafers shown in Figure 6. (A) 200mm wafer with 80 µm thick Asahi CX-8040 dry film resist with leadfree SnAg solder. (B) 200mm wafer with 80 µm thick Asahi CX-8040 dry film resist with Cu pillar with leadfree SnAg solder cap. (C) 200 mm wafer with 53 µm thick AZ 4620 spin-on resist with eutectic SnPb solder. (D) 200 mm wafer with 53 µm thick AZ 4620 spin-on resist with leadfree SnAg solder. (E) 300mm wafer with 120 µm thick TOK 50120 dry film resist with leadfree SnAg solder.

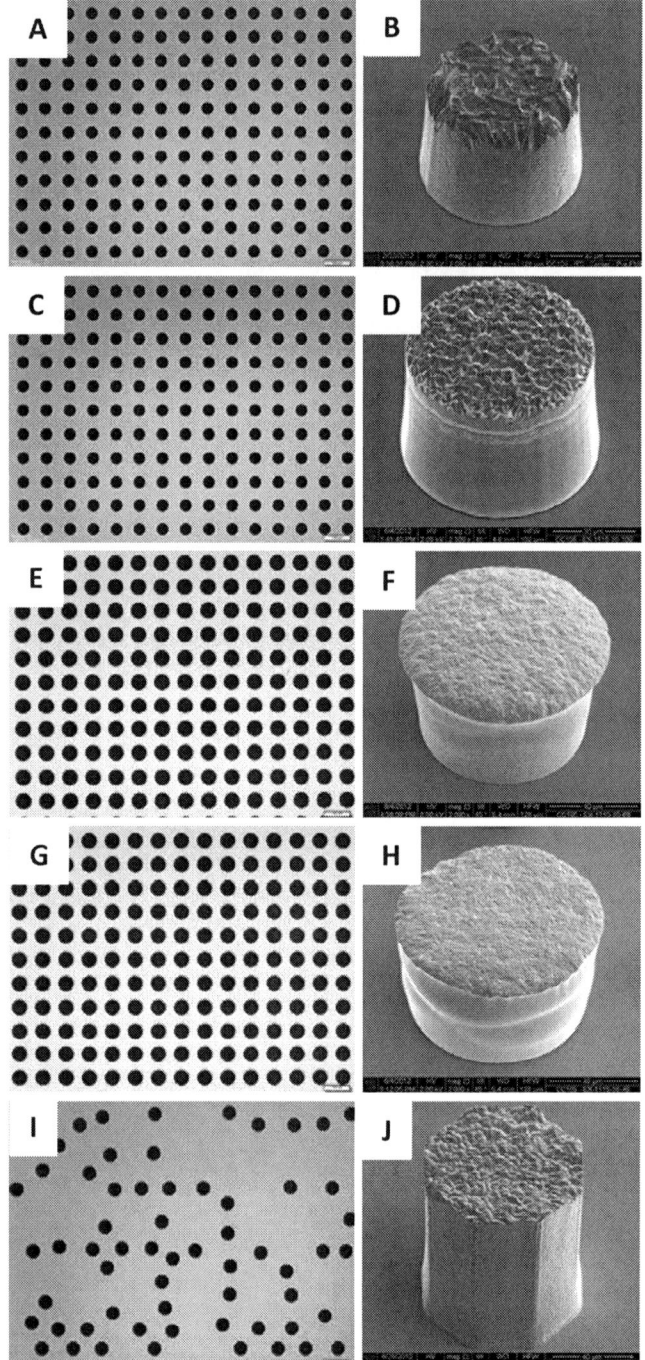

Figure 6. Optical micrographs and SEM images illustrating complete resist removal for bumped wafers with different photoresists and solder types. (A) and (B) 200mm wafer with 80 µm thick Asahi CX-8040 dry film resist with leadfree SnAg solder. (C) and (D) 200mm wafer with 80 µm thick Asahi CX-8040 dry film resist with Cu pillar with leadfree SnAg solder cap. (E) and (F) 200 mm wafer with 53 µm thick AZ 4620 spin-on resist with eutectic SnPb solder. (H) and (I) 200 mm wafer with 53 µm thick AZ 4620 spin-on resist with leadfree SnAg solder. (J) and (K) 300mm wafer with 120 µm thick TOK 50120 dry film resist with leadfree SnAg solder.

SILICONE AND CLEANING SOLVENT COMPATIBILITY

Michelle Velderrain and Danielle Peak
NuSil Technology, LLC
Carpinteria, CA
michelle@nusil.com

ABSTRACT

Silicone materials exhibit unique characteristics including chemical stability and a broad operating temperature range, allowing for preservation of mechanical properties when exposed to extreme conditions. These and other dynamic attributes of silicone help to provide multiple solutions to the complex challenges of electronic packaging and a wide variety of other applications. However, a major concern surrounding the use of silicones is the volatile component observed to outgas in extended exposure to high temperatures and low pressures (vacuum). This issue has sparked the design of new silicone materials to reduce the potential for contamination while maintaining silicone's valuable physical and chemical characteristics.

Volatile species may contaminate sensitive surrounding surfaces and equipment, making adhesion or soldering difficult in an upstream process. Wafer Level Packaging (WLP) requires various cleaning steps throughout the process. Silicones are becoming more popular for uses as adhesives, encapsulants and thermal management applications for their ability to absorb stress during thermal cycling and stability at elevated temperatures. This presentation will discuss the use of low volatility silicones to minimize outgassing of volatile components in conjunction with the need for cleaning solvents and their compatibility with silicones.

Key words: Low outgassing silicones, volatiles, contamination, solvent resistance

INTRODUCTION

Silicones are viable candidates for use as adhesives, encapsulants and interface materials due to their thermal stability, low modulus, and elastomeric (or rubber-like) properties, which can protect the components within the microelectronic package[1,2,3]. The electronic package may be exposed to a variety of different solvents by fabricators in the cleaning process. Figure 1 depicts the workings of a typical Flip Chip, in which silicone is a viable material choice for TIM-1 and TIM-2.

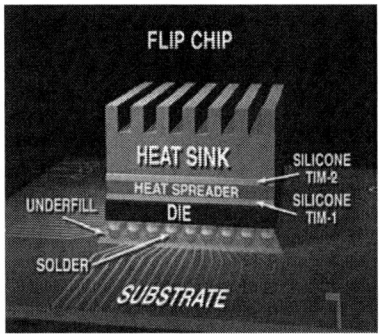

Figure 1. Thermally Conductive Silicones Used in a Flip Chip Assembly

Once the solvent evaporates, delamination and stress on the wire bonds can result—potentially bending and even shearing the bonds. If the solvent does not evaporate, it may cause outgassing (bubbles) once the equipment is in use. Another potential risk of using silicones is the re-condensation of outgassed components in closed systems, which can cause contamination or fogging of sensitive critical components in electro-optic or microelectronic mechanical systems (MEMS). Any extracted siloxanes remaining in the cured silicone can affect bonding or soldering at a downstream process. Reducing the potential for contamination from outgassing and swelling from solvents will require knowledge of the silicone's compatibility with the solvents utilized for cleansing.

SILICONES FOR MICROELECTRONIC PACKAGING

Several areas within and on a microelectronic assembly may need an elastomeric material to act as an adhesive, encapsulant or interface material. Thermal Interface Materials (TIMs) are typically polymeric materials that are able to absorb stress and transfer heat away from heat sensitive components. They can be electrically and/or thermally conductive, with a consistency ranging from that of a grease to that of a hard elastomer. A silicone can be used directly in contact with the die as a die attach or underfill, or as an encapsulant, also known as a "glob top" (See Figure 2). Silicone encapsulants work well to protect the entire microelectronic assembly from damage during transport, as well as from environmental conditions that can damage the components.

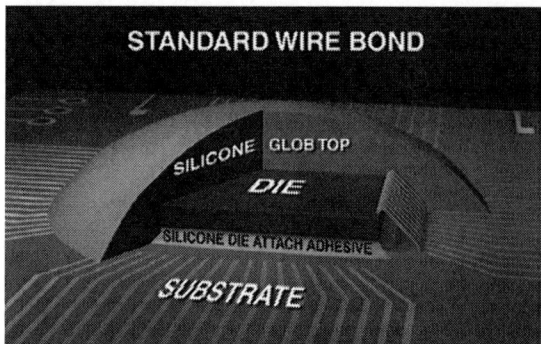

Figure 2. Silicone Materials in a Wire Bond

SILICONE POLYMER CHEMISTRY

The primary structure of many silicone materials used in microelectronics is the silicone polymer, which can account for up to 70 percent of the total formulation. Polymers are linear structures with alternating silicon-oxygen atoms that make up the polymer backbone. The large bond lengths and bond angles of the Si-O-Si bonds yield a large amount of free volume[4], giving silicone the ability to be designed to exhibit properties it does not inherently possess, such as optical clarity or thermal or electrical conductivity. Intrinsically, silicone possesses many unique characteristics due not only to its free volume but also to its unusually weak intermolecular forces. Most relevant to the microelectronic packaging industry are silicone's typical glass transition temperature of < -115°C, low shrinkage of < 1%, and dielectric strength of 500V/mil (0.001 inch), or 20 kV/mm.

For a given silicone polymer formulation, the number of monomeric, or singular, units dictates the chain length, which affects viscosity. The component commonly known as the R group, or the pendant group, will affect the chemical performance, and the "end blocker" typically contains the functional group that partakes in the crosslinking (curing) reaction. Silicone polymers, comprised of several repeating monomers, can be more technically referred to as poly (organo) siloxane polymers (See Figure 3).

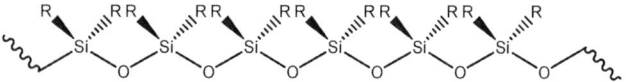

Figure 3. Diorganodisiloxane Polymer

The diorganodisiloxane structures in Figure 4 show that the R groups (methyl, trifluropropyl, phenyl, etc.) on the back bone can vary, allowing different types of organic groups to be incorporated as pendant groups within one polymer chain (R and R'). Varying the pendant groups is a common way to design silicones to suit an application's needs.

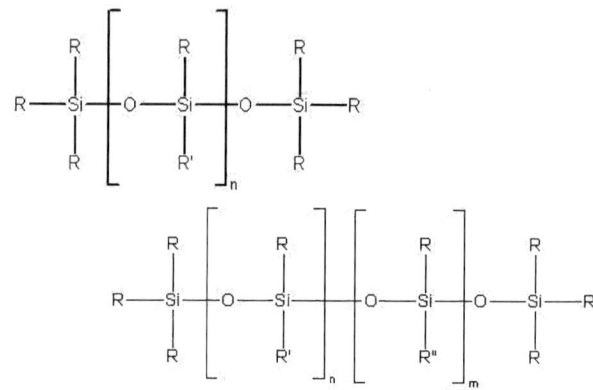

Figure 4. Generic Polymer and Generic Co-polymer

The organic pendant groups that will be evaluated for this discussion are methyl and trifluoropropylmethyl substituents[5]

Fluorosilicones are based on trifluoropropyl methyl polysiloxane polymers and historically used for applications that require fuel or hydrocarbon resistance. The trifluoropropyl group contributes a slight polarity to the polymer, resulting in swell resistance to common solvents, gasoline and jet fuels (See Figure 5).

Figure 5. Vinyl-terminated Polytrifluoropropylmethylsiloxane

Dimethylpolysiloxanes (PDMS) are the most common silicone polymers used industrially. Figure 6 depicts their chemical structure. These polymers are typically the most cost-effective to produce and generally yield physical properties desired in silicones for microelectronic applications. The silicones most often used in this industry are elastomers and gels.

Figure 6. Vinyl-terminated Polydimethysiloxane

Silicone Polymer Reinforcement

When silicone polymers alone are crosslinked together, the cured material is typically referred to as a "gel" since the cured silicone has minimal tensile and tear strength properties. Gels are also soft and have very low modulus. Silicone polymers can be reinforced by adding reinforcing fillers, such as silica, and/or silicone resins, that increase the elastic properties of the cured silicone.

Silica reinforces the cured silicone polymers through van der Waals forces and hydrogen bonding between hydroxyl groups on the silica surface and siloxane backbone of polymer. These weak interactions allow the cured silicone to absorb stresses and increase the viscosity of uncured silicone through the same polymer-filler interaction[6]. The silica is typically added to the polymer and treated with organosilicones to make it more compatible with the polymer, which subsequently stabilizes the viscosity to a certain degree. Silica-reinforced silicones are translucent and typically have non-Newtonian flow characteristics wherein the viscosity will decrease when shear is applied.

Silicone resins are highly branched polyorganosiloxanes. Silicone resins can reinforce the silicone polymer through more complex crosslinked architecture once cured, which helps distribute applied stress as well as combat entanglement of resin and polymer molecules. Of course, van der Waals forces and hydrogen bonding also play a role in stress response. Resin-reinforced silicones are typically transparent and have Newtonian flow characteristics wherein the viscosity is not greatly affected by the applied shear.

Silicone Polymerization

A silicone polymer is manufactured in several steps. Initially, a silicone polymer is produced via Ring Opening Polymerization (ROP). The process begins with polyorganosiloxane cyclics reacting with a chain terminating species, or "end blockers," in the presence of an acid or base initiator as shown in Figure 7.

The product of this polymerization reaction is a mixture of various molecular weight molecules, including cyclical and linear polymers of varying lengths; concentrations of each species are based on a thermodynamic equilibrium. When analyzed using gel permeation chromatography, a bimodal distribution can be seen, as displayed in Figure 8. A small peak represents cyclics wherein very short chained, low molecular weight polymers can be observed, and a larger peak represents larger molecular weight polymers.

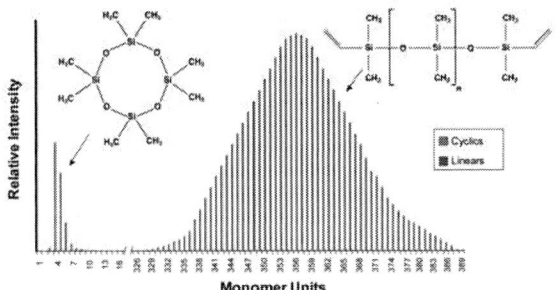

Figure 8. Molecular Weight Distribution of Final ROP Reaction Products of PDMS

The cyclics, often a low molecular weight siloxane species, can outgas under high temperatures and low pressures, which can lead to fogging and deposition of silicone oil that reduces adhesion and dimensional changes. Cyclics are volatile because they do not crosslink into the cured polymer matrix, unlike the linear polymer chains, for elastomeric properties. Removal of these cyclics, while keeping the longer-chained, higher molecular weight, linear polymers, will prevent or reduce the aforementioned problems. The use of low out gassing silicones in electronic packages can greatly reduce the risk of contamination, and thus the need for reworking or cleaning.

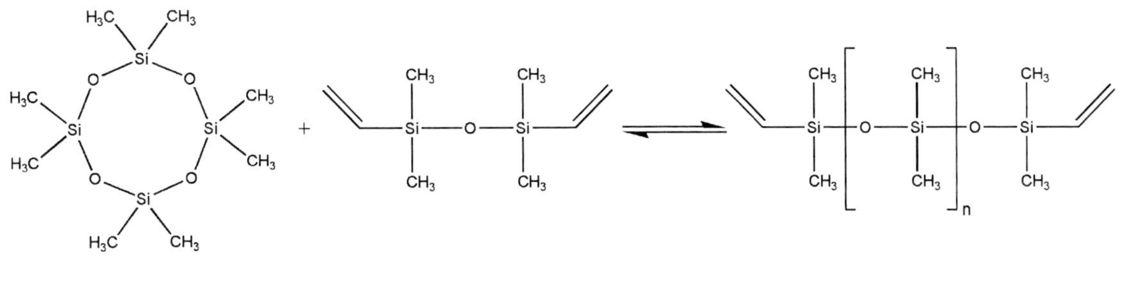

Octamethylcyclotetrasiloxane   Divinyltetramethyldisiloxane   Vinyl-endblocked polydimethylsiloxane

Figure 7. Basic Ring Opening Polymerization (ROP) Reaction for a Vinyl-terminated Polydimethylsiloxane

LOW OUTGASSING SILICONES

Low outgassing silicones are processed to meet specifications outlined in NASA SP-R-0022A and ESA PSS-014-702, with a maximum allowable Total Mass Loss (TML) of 1% and a Collected Volatile Condensable Material (CVCM) of less than 0.1%. However, certain electronic applications, namely space applications, may require even lower maximum levels of outgassing based on risk assessment. The newly developed *Ultra Low Outgassing[TM]* version exceeds typical American Society for Testing and Materials (ASTM) E595[7] requirements, achieving ≤0.1 % TML and ≤0.01% CVCM, and displays improvement in outgassing kinetics and volatile components as tested per ASTM E-1559. An *Ultra Low Outgassing[TM]* material keeps processing time down as no additional conditioning is required to achieve the desired outgassing values, nor are the physical properties altered in the conditioning. In addition, if device contamination is based on cumulative contaminant levels from all materials within the device's vicinity, using one material with exceptionally low levels of outgassing could allow use of other materials with higher outgassing levels in the same vicinity.

ASTMs E595 and E1559[8] provide standards for testing all silicone adhesives for extraterrestrial use. The E595 test measures weight loss directly under vacuum conditions. In the E595 test, each material sample undergoes preconditioning, conducted at 50% relative humidity and ambient atmosphere for twenty-four hours. The sample is weighed and loaded into a compartment within a test stand, which is then heated to 125°C at less than 5 x 10^{-5} torr, for 24 hours. Any volatile components of the sample will outgas in these conditions, escaping through an exit port and, if condensable at 25°C, condensing on a collector plate maintained at that temperature. The samples are post-conditioned in 50% relative humidity and ambient atmosphere for a 24-hour minimum. The collector plate and samples are then weighed again to determine the percentage of weight change, determining TML% and CVCM%. Standard criteria for low outgassing materials limit materials' TML to 1.0% and CVCM to 0.10%. To adhere to these requirements, NuSil Technology performs this as a standard, lot-to-lot test for low outgassing materials.

ASTM E1559

OSI laboratories conducted ASTM E1559 experiments and provided test reports (See Figure 9). The isothermal outgassing test apparatus is explained in detail by Garret *et al.* and will only be discussed here briefly.[8,9,10] The material sample can range from 0.5 g to 10 g and is placed in a temperature-controlled effusion cell in a vacuum chamber. All samples are preconditioned in accordance with ASTM E595 unless otherwise specified.

Outgassing flux leaving the effusion cell orifice condenses on four Quartz Crystal Microbalances (QCMs) that are controlled at selected temperatures. The QCMs and effusion cell are surrounded by liquid nitrogen shrouds to ensure the molecular flux impinging on the QCMs is due only to the sample in the effusion cell. The TML and outgassing rate from the sample are determined as functions of time from the mass deposited on an 80 K QCM and normalized with respect to the initial mass of the sample.

The amount of condensable outgassing species, VCM, is measured as a function of time from the mass collected on the 298 K QCM. After the outgassing test is complete, the QCMs are then heated to 398 K at a rate of 1K/min. As the QCM heats, the deposited material evaporates. The species that evaporate can be analyzed by a quadruple mass spectrometer to quantitatively determine the species observed.

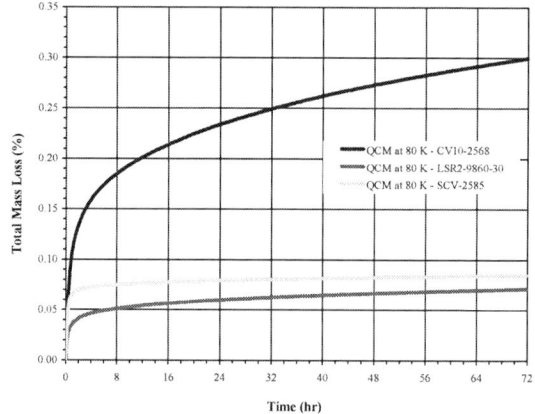

Figure 9. Total Mass Loss of CV10-2568, developmental Ultra Low OutgassingTM material, and SCV-2585 as a function of time

SOLVENT RESISTANT EVALUATION

Silicones are versatile systems that can be optimized to have chemical characteristics such as solvent resistance, and can also be processed so as to remove unwanted siloxanes created from the polymerization process. According to the general rule of "like dissolve like"[11], the more chemically similar the solute (in this case, silicone rubber) is to the solvent, the more the silicone will absorb the solvent into its cured matrix. The more dissimilar the silicone and solvent are chemically, the less solvent the cured silicone will absorb and the less swelling it will experience. Polarity is the chemical property most responsible for how similar the solute and solvents are. Organic solvents may have a polar functional group such as alcohol (-OH) or a halogen (-F, -Cl, or -Br) to increase the ability to dissolve substances comprised of similar functional groups.

Once the solvent evaporates, the subsequent volume loss undergone by the silicone can impact the package in several ways, including delamination of the silicone from the substrate surface. This particular consequence can be a positive effect if the silicone needs to be removed from the electronic package, such as for cleaning, but otherwise it is a hindrance to the application. It is important to note that the stresses associated with adhesive shrinkage from solvent evaporation can be considerable when compared to expansion and contraction due to thermal cycling, depending on the amount of swelling and the modulus of the silicone.

Siloxanes removed from a cured silicone are typically referred to as "extractables" and are measured by the difference in sample weight before and after immersion in the solvent. A low outgassing silicone that has been processed to remove more volatile siloxanes should theoretically have lower extractables if extracted with a solvent in which the silicone is soluble.

Materials

Solvents (or solvent families) commonly used in the microelectronics industry for cleaning were chosen (Table 1) to be evaluated for their compatibility with various silicone formulations. The silicone formulations were chosen based on polymer chemical composition and %TMLS %TMLs and %CVCMs on the cured material. Table 2 displays the abbreviations given to the silicone systems for the study, as well as lists the siicones' physical properties. The polymers used for DMR and DMRLO are composed of PDMS and reinforced with methylfunctional silicone resins; however, one was more highly processed for removal of the volatile species and is considered "space grade" (DMRLO). These PDMS formulations were compared to an experimental 100% Mol fluorosilicone and a commercialized 50% Mol fluorosilicone. All of the silicone formulations used in this study are platinum-catalyzed addition cure silicones, which were cured according to their design specifications. Note that the materials chosen may not be specifically for electronic packaging; however, their results can be used as guidelines when considering electronic packaging materials with similar chemistry and processing modes.

Experimental

Percent swell (Ref. ASTM D-471), which measures change in volume based on specific gravity, was measured to evaluate the differences before and after exposure to solvent. The method and calculations were adapted from ASTM D471-79[18] wherein the higher the percent change in volume, the more solvent the silicone was absorbing.

Percent change in thickness was also measured before and after submersion. The thickness was measured again after a 48-hour dry out period to evaluate how much solvent would evaporate in ambient conditions after 48 hours, as well as to verify if there were any further dimensional changes.

Table 1. Solvents chosen to be evaluated for their effects on various silicones

Solvent	General Description and Use
Isopropyl alcohol (IPA)	Polar solvent commonly used to wipe down a surface before applying silicone.
Specialty Fluoro Based Solution [12]	Fluorinated ether solvents are polar compared to hydrocarbon solvents. Common family of solvents used for cleaning microelectronics.
Terpene [13]	Limonene/Ester mixture. Aromatic product of citrus fruits and replacement for CFC.
Hexane	Hydrocarbon solvent with no double bonds.
H_2O	Water

Table 2. Silicone formulations chosen for evaluation before and after contact with various solvents

Product	Mix Ratio	Durometer	Tensile psi	Elongation %	ASTM E595 %TMLs/%CVCMs
DMR[14]	10 to 1	51A	1400	109	0.80%/0.25%
DMRLO[15]	10 to 1	51A	1050	142	0.05%/0.01%
F 100%[16]	1 part	45A	209	82	Not tested
F 50%[17]	1 to 1	27A	774	444	Not tested

Extractables data was obtained from measuring the change in weight before and after solvent immersion after 48 hours. Extractables are independent of the siloxane volatility since they are comprised of molecular species not crosslinked into the cured elastomer and which are soluble in certain solvents.

Calculations
a. Percent Swell (% V)

$$\% \, Swell = \frac{(M3 - M4) - (M1 - M2) \times 100}{M1 \quad M2}$$

M1 = initial weight of sample in air
M2 = initial weight of sample in water
M3 = weight of sample in air after immersion
M4 = weight of sample in water after immersion

b. Δ % Thickness (%ΔT)

$\% \, \Delta \, Thickness = \{(T2 - T1)/T1\] \times 100$

T1 = Original sample thickness
T2 = Final sample thickness

c. % Extractables

$\% \, Extractables = [(W1 - W2)/W1] \times 100$

W1 = original silicone film weight
W2 = extracted silicone film weight

Sample Preparation
1. Silicones were cured per standard cure schedule.
2. Sample size = 1" x 1" x 0.025".
3. Volume of the solvents was 50 ml in 2 oz glass container.
4. The samples were submersed for 1 hour, 6 hours and 48 hours and kept at 25°C.
5. Specific gravity, extractables and thickness were measured immediately after submersion (after excess solvent was allowed to drip off).

RESULTS
Percent Swell
The results are reported after 48 hours in order to demonstrate the worst case scenario that occurred when the cleaning solvents used were exposed to the types of silicones tested.

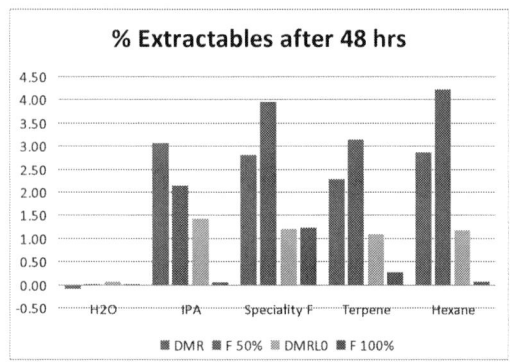

Figure 10. Comparative Percent Extractables After 48 Hours

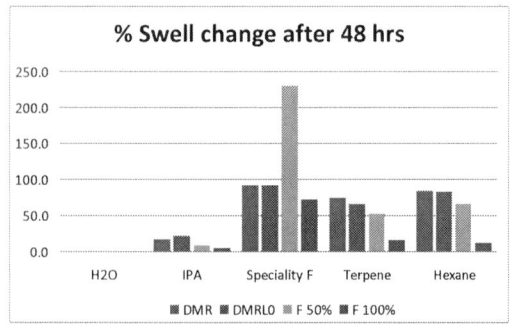

Figure 11. Comparative Percent Swell After 48 hours

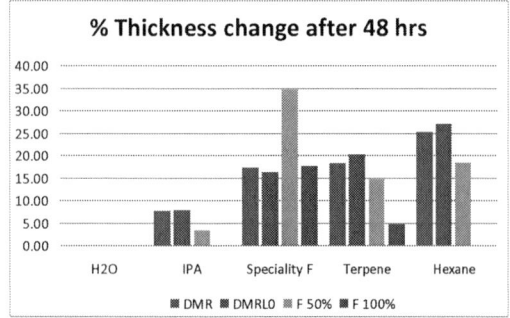

Figure 12. Comparative Percent Thickness After 48 Hours

DISCUSSION

Extractables
Overall, the product with the most extractables after solvent exposure was the F 50% silicone elastomer, which was highly soluble to all solvents tested except for IPA and water. This performance could be due to, among other factors, F 50%'s low durometer relative to the other materials tested. By contrast, The F 100% silicone had the lowest % extractables overall. This was attributed to its low solubility in the solvents used.

The standard PDMS silicone elastomer (DMR), exhibited the second highest overall extractables content and the highest when exposed to IPA. The low outgassing version, DMRLO, had a significantly lower amount of extractables than DMR, demonstrating that a more thorough removal of the volatiles from the silicone reduces potential contamination of surrounding areas.

Solvent Compatibility
There were no significant differences in results between 1 and 48 hours for percent swell (% V) and percent change in thickness (% Δ T). While the % Δ T does not take into account the specific gravity of solvent or silicone in the dimensional change after submersion, it does give a more accurate idea of the effect of solvent on silicone materials based on dimensional change.

All of the silicones and solvents used contained some methyl (-CH3) groups, which means all of them exhibited some solvent absorption (excluding absorption of water). Fluorosilicones performed very well against IPA and hexane even though IPA was polar and hexane, non-polar. Based on the positive performance of the fluorosilicones in both of these solvents, it was deduced that hexane is much less polar than the fluorosilicones, and that IPA must be significantly more polar.

The PDMS based silicones, DMR and DMRLO, followed the general trends based on chemical composition, wherein they swelled the most in hexane and experienced less swelling as the polarity of the solvents increased. The F 100% displayed the highest overall solvent resistance except in regard to the Specialty Fluoro cleaning solution. Both the DMR and DMRLO performed similarly based on % ΔT.

CONCLUSION
Not all silicones are the same with respect to their response to solvent cleaning processes. Fundamentals of silicone manufacturing allow silicones to exhibit different combinations of chemical characteristics that can react differently to various solvents. The change in thickness did not show that extractables had a significant effect on the volume occupied by the silicone. The extractables results indicate that residues can be deposited onto surrounding surfaces. To prevent problems with adhesion or other adverse effects from these residues it is recommended to perform a wipe down on these surfaces or use a low outgassing material.

By understanding how the electronic package is affected by different solvents, the compatible silicone/solvent system can be selected.

REFERENCES

[1]. H. Yu, S.G Mhaisalkar, E.H. Wong, J.F.J.M. Caers, " Evolution of Mechanical Properties and Cure Stresses for Non-Conductive Adhesives Used for Flip Chip Assemblies".

[2]. R.Viswanath, V.Wakharkar, A.Watwe, V.Lebonheur, 'Thermal Performance Challenges from Silicon to Systems'.

[3] R.Mahajan, C. Chiu, R.Prasher, "Thermal Interface Materials: A Brief Review of Design Characteristics and Materials", Electronics Cooling Feb 2004.

[4]. W. Noll, Chemistry and Technology of Silicones, Academic Press, New York, 1968.

[5] B.Riegler, S.Bruner, R.Thomaier, 'Low Outgassing Materials for Electro-Optic and Electronic Systems', IMAPS Conference on Device Packaging March 2005.

[6]. Mark, James E. and Burak Erman, Rubberlike Elasticity A Molecular Primer, John Wiley and Sons, 1988.

[7]. ASTM E595, "Standard Test Method for Total Mass Loss and Collected Condensable Materials from Outgassing in a Vacuum Environment."

[8]. ASTM E1559, "Standard Test Method for Contamination Outgassing Characteristics of Spacecraft Materials."

[9]. J.W. Garrett, A.P.M. Glassford, and J. M. Steakley, "ASTM E 1559 Method for Measuring Material Outgassing/Deposition Kinetics", Journal of the IEST, pp. 19-28, Jan/Feb 1995

[10]. A. P. M. Glassford and J. W. Garrett, "Characterization of Contamination Generation Characteristics of Satellite Materials", Final Report WRDC-TR-89-4114, Jun 82 - Aug 89

11. L.G Wade J., Organic Chemistry, Prentice-Hall Inc., New Jersey, 1987.

[12]. Super Clean Part # MCC-SPRG by Micro Care

[13]. Citrus Burst ™ 7

[14]. NuSil Technology R-2615

[15]. NuSil Technology CV-2500

[16]. NuSil Technology Experimental fluorosilicone

[17]. NuSil Technology CF2-3521

[18]. ASTM D471-79, "Effect of Liquids"

TSV PROCESS VARIATIONS FOR 2.5D AND 3D SEMICONDUCTOR PACKAGIG

Vern Solberg
Invensas Corporation
San Jose, California USA
vernsolberg@att.net

ABSTACT

Industry roadmaps point toward the use of through-silicon-via (TSV) for a broad range of semiconductor packaging applications. Although TSV is touted as the 'next big thing' in semiconductor packaging and the technology has the potential to revolutionize component density, many believe that it is currently limited to a very homogeneous family of products; MEMS, memory and image sensors.

The market driver for semiconductor package technology is to increase product functionality and performance without increasing product size. Vertically configured TSV package technology will address this issue for a broad number of enterprise products but the capabilities and methodologies for processing and joining wafers and die elements currently vary a great deal between suppliers.

This paper explores three basic approaches to TSV formation, via-first, via-middle and via-last:

- Via-first integration forms very small via holes in the wafer during front-end processing.
- Via-middle process is performed following the semiconductors fabrication process and may include surface redistribution
- Via-last ablation and via fill is executed from the backside of the finished wafer or after wafer fabrication for joining individual die elements to one another (Die-on-Die).

Key words: 2.5D, 3D, semiconductor packaging, through silicon via, TSV, die stack, wafer stack.

INTRODUCTION

Currently, die-stack package assembly using conventional wire-bond methodology has enabled developer's greater latitude in bringing heterogeneous semiconductor elements into system level package configurations. The use of wire-bond processing as the exclusive means of interconnection, however, is somewhat restrictive. The ideal scenario will adapt die elements that are progressively smaller than the one above so that the die attach and wire-bond operation can be performed in two stages.

This 'tiered' configuration allows all die to be sequentially attached on top of one another in a single operation leaving the edge of die elements accessible for the subsequent wire-bond operation. On the other hand, wire-bond processing three or more vertically configured die on the substrate carrier requires significant surface area to complete the interface (**Figure 1**).

Figure 1. Wire-bond die stack package

The SDRAM die elements are especially difficult to stack and terminate due to their center positioned wire-bond pad sites. This factor has further complicated the die stacking process and because of the excessively long and unbalanced wire-bond interface, signal speed is significantly degraded (**Figure 2**).

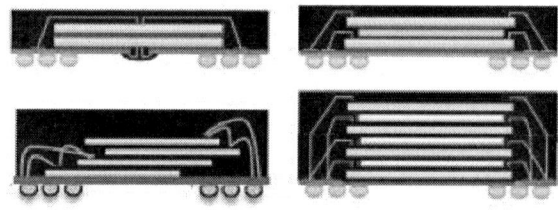

Figure 2. Stacking same size die.

The overall package height can be a critical roadblock for a number of applications as well. Even though the die elements can be made very thin, the accumulated stack-up height generated by the added spacer and wire-bond loop may not be acceptable.

Product developers are looking to the semiconductor package assembly specialist for a packaging process that can enable maximum efficiency for the next generations of high performance applications. The challenge is to develop a packaging technology that can meet the density and performance requirement for the next generation of electronic products. To achieve these goals the industry must consider adopting more innovative semiconductor package solutions.

THROUGH SILICON VIA PROCESS

A great deal of development has occurred in establishing an efficient and low cost method for providing a reliable interface between the top and bottom surface of the silicon based semiconductor. Providing very small plated or filled via holes through the contact sites initially designed for wire-bond processing is proving to be a very practical solutions. Recent advances in through-silicon-via (TSV) technology have enabled a more efficient die-to-die and die-to-substrate interface. In regard to via-hole ablation in silicon material, both Laser and Deep Reactive Ion Etching (DRIE) processes are commonly employed. Before implementing either of the ablation processes, however, a decision must be made in choosing which method will be the most practical for the anticipated volume.

In planning the resources needed for setting up a new TSV operation, users must consider that there will be a significant investment in the equipment and processes for via ablation; mask and coating systems. Also, precision imaging systems, alloy plating and chemical etching systems must be established. There will be a substantial waste treatment capacity as well to meet any regional environmental restrictions.

Through silicon via ablation process variations:

- Via First
- Via Middle
- Via Last

Via-first is more commonly used for 2.5D interposer applications. The TSV holes are ablated into the basic silicon wafer prior coating imaging and plating the copper circuit features required for interconnecting die elements mounted onto the surface. Although via-first process can be used before developing active semiconductor elements, the process in more commonly adopted for substrate interposer applications. The illustration furnished in **Figure 3** is typical of how a number of users have utilized via-first interposers for interconnecting uncased die elements.

Figure 3. System-in-Package using TSV Interposer.
(Example source: ST Micro)

The basic 2.5D interposer may include functional elements as well. It is not uncommon to integrate passive features on the surface of the substrate and, although preparation of the substrate is more complex, companies have also embedded very thin active semiconductor elements within shallow cavities formed in the structure.

Via-middle integration is most commonly applied to die elements that will be vertically joined. The small via holes will be ablated and plated in the wafer following front-end transistor formation and local interconnect processes.

In this process, via-holes are first ablated (using laser or plasma processes) through the pad features originally provided for wire-bond processing. Wire-bond pads are typically no less than 100µm pitch. Although some companies are able to provide 5µm vias, the via-hole diameter will more often range between 10 and 25µm. When the bond pitch is less than 100µm the fabricator may utilize the smaller diameter via holes or modify the hole pattern to be offset to a new location using a metallization process.

Via-last hole formation and plating processes are performed from the backside surface following wafer thinning. The via-last process will typically require significantly larger via-holes than the via-first and via-middle variations (**Figure 4**).

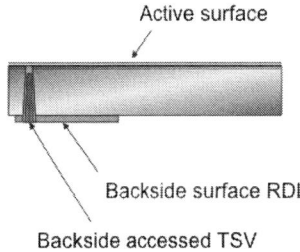

Figure 4. Via-last hole ablation and plating.

Companies' providing via-last process capability state that the diameter of via holes can range between 20µm and 50µm to depths of 20µm to 200µm.

Most TSV applications will be processed while the die elements remain in the wafer level format. Although small volume applications and developing process may adapt smaller wafer sizes, the wafer size for high volume TSV applications will range between 200mm and 300mm diameter. The wafer thickness before thinning can range between 700µm to 800µm. The following furnishes an overview of the two primary methods used for hole formation in silicon.

LASER ABLATION PROCESS

The laser process is preferred by companies with low to medium volume production because it avoids the need for a number of lithographic and coating steps normally required for alternative ablation methods. With laser processing, via locations, diameter, shape and depth are digitally programmed. This factor simplifies the implementation by enabling faster setup and product changeover. The manufacturers of the laser systems developed for TSV claim that the process can be employed for both front-end and back-end fabrication and allow for either via-first, via-middle or via-last hole ablation.

The actual ablation rate is determined by the laser repetition rate and the speed at which individual via-holes can be formed. The number of laser pulses and the pulse energy dictate via depth while the beam size governs via diameter. Overall, the short-pulse laser process is said to produce a narrower beam that enables 10μm diameter via-hole formation. These systems can operate at what is considered a very high speed for laser ablation in silicon, up to 10,000 via sites per second (depending on the via hole diameter and depth).

PLASMA ABLATION (DRIE) PROCESS

Forming high aspect ratio via holes can be achieved using a process described as 'inductively coupled plasma' (ICP) ablation. The ICP methodology is commonly referred to as the 'Bosch Process'. In preparation for the plasma ablation processes, a dedicated system will be required to apply a high-viscosity photo-resist material onto the semiconductor wafer surface. The coating selected must be able to withstand the aggressiveness and relatively long duration of time required for the ablation process. The via-hole pattern is typically imaged and developed using a glass photo-mask stepper process or, if available, pattern ablated with a numerically controlled 'direct-laser-imaging' process.

The ablation process uses SF_6 plasma to progressively etch away a thin layer of the exposed silicon via sites. Each dry etch sequence is followed by a C_4F_8 plasma deposition that furnishes a very thin fluorocarbon polymer passivation layer onto the via-hole sidewalls. Passivation on the sidewalls of these via features protects them from any further horizontal chemical etching. The process is repeated until the via-hole reaches its specified depth. The actual etch rate is dependent on via-hole diameter and depth. The smaller the via diameter, the higher the speed for ablation. A via diameter in the range of 10 to 25μm, for example, will ablate the silicon material at approximately 9 microns per minute. The TSV examples shown in **Figure 5** are 10μm diameter.

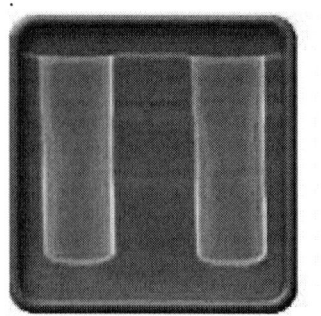

Figure 5. Plasma ablated and Cu filled TSV
(Example source: ALLVIA, Inc.)

In regard to the overall performance of the DRIE process, via-hole wall formation is relatively smooth and uniform. Furthermore, the ablations of all via-holes on the wafer are formed simultaneously. Via-holes must be uniform in diameter in order to control the plasma ablation rate. If larger diameter vias are required in select areas it will require a secondary masking and ablation process.

Selecting which TSV ablation process will be most practical will depend a great deal on economics and throughput criteria. Laser systems developed for via-hole ablation, although relatively slower than chemical ablation, are numerically controlled, requiring little, if any, complex tooling or masking operations. With either process, via-holes are commonly ablated from the active side of the wafer to a depth that is slightly less than the overall thickness of the silicon material.

VIA FILLING

Following via-hole formation the holes are filled with a conductive polymer or plated closed with copper using an electroplating process. A number of companies are furnishing electrically conductive polymer products for via filling, however, a majority of TSV users have adopted a copper plating process to fill the tiny vias.

Copper via plating process

The copper via filling process is achieved through a number of alloy deposition steps that begin with applying a thin adhesion layer to the wafers surface and via features using an RF magnetron sputtering process. This is followed by a metal-organic compound deposition to provide a conformal, continuous, and low resistivity Cu seed layer. Electroplating is finally performed using a copper sulfamate plating solution. The copper electro-plating process is complete in a relatively short time, however, because electroplating of the via features results in a thick Cu layer (~10-20μm) over the active surface of the wafer. The example shown in **Figure 6** is typical of the solid copper TSV connecting the RDL on the active surface of the silicon to a land pattern on the backside surface of the thinned wafer.

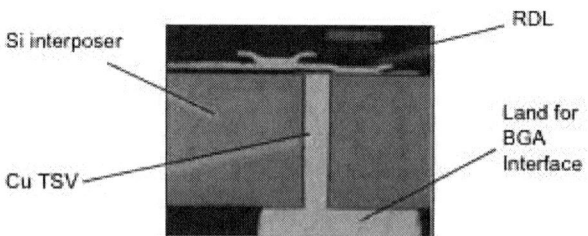

Figure 6. Post-process TSV interface.
(Example source: Fraunhofer IZM)

The active side of the wafer is commonly plainarized to ensure that the solid copper via topography is uniform and level with the wafer surfaces. Even though the chemical via ablation process requires specialized equipment, it has proven to be most efficient for high volume TSV ablation applications.

WAFER PLANRIZATION and THINNING
Four primary methods are available for wafer thinning: mechanical grinding, chemical-mechanical polishing, wet etching and atmospheric downstream plasma and dry chemical etching. Because of its higher thinning rate, mechanical grinding is currently the most common technique for wafer thinning.

Mechanical grinding systems typically use a two-step process that includes coarse grinding with a thinning rate of about 5 microns per second followed by fine-grinding at a rate of only 1 micron per second). When completed, the plated copper blind via features that were initially processed from the active topside surface of the wafer will be exposed on the backside for subsequent wafer-to-wafer joining. Although significantly slower than mechanical grinding, dry-etching is an alternate method for wafer thinning. One successful dry-etch process uses a Ar/CF_4 plasma to thin the surface. It's stated by the developer that the silicon material is removed at a rate of about 20 microns per minute. Another dry-etch process uses fluorine or chlorine-containing plasmas to reduce wafer thickness, however, some experts' state that the surface finish may not be as uniform as that provided by the Ar/CF_4 plasma process.

Wet chemical etching is another common thinning technique. Wet chem. thinning is accomplished by directing a thin stream of an etching agent over the surface of the rotating wafer. The etching chemistry developed for silicon wafer thinning is a combination of Hydrofluoric acid (HF) and Nitric acid (HNO_3). Etch rates can be modified by adjusting the percentages of the chemical elements. The etch rate for silicon wafers will depend on the spin rate and the flow of the etching agent across the wafer surface. The target etch rate for wet chemical spin etching of silicon is about 10 μm per minute. In preparation for wet etching, the active surface of the wafer is protected, either by applying a mask coating or by use of special chucks designed to protect the surface during the process.

Many of the second tier service providers have brought wafer-thinning capability inside, however, developing a high yielding thinning procedure for TSV can be challenging. Key concerns for handling the thin and rather brittle silicon wafer includes breakage, edge chipping, damage that may occur during the grinding process and wafer bow and warp conditions. Specialized fixtures and wafer transfer systems are necessary to individually handle and stack the thinned wafers. This includes wafer cassettes specially designed to handle very thin wafers, robotic end effectors, pre-aligners and various process modules.

WAFER JOINING PROCESS
The intermetallic joining process for wafers will require a rather aggressive cleaning process to ensure that all surfaces are free of contaminates and foreign particles. To reduce physical contact throughout the joining process, the thinned wafers are transferred from their cassette magazines using robotic vacuum chuck fixtures. The vacuum chucks are designed to precisely align and bring the wafers into contact to one another. To prevent damage to the wafers during this process the vacuum or pressure must be tightly controlled between the chuck and the surface of the wafer.

Three primary wafer-to-wafer bonding methodologies currently being used for 3D TSV interconnects are: 1) fusion (or molecular) bonding, 2) adhesion thermo-compression bonding and 3) metal-metal thermo-compression bonding. There are also alternative joining variations that may employ conductive polymer bonding, oxide bonding or a fusion-metal bonding process. Each of these methods has advantages and disadvantages and although there is no standard process or universal recipe for TSV wafer preparation and joining, a number of companies have directed most of their resources toward some form of metal-to-metal joining.

Many in the industry prefer the direct Cu-to-Cu bonding process for TSV joining. This is because the process simultaneously forms both a mechanical and electrical connection between wafer layers. Compared to solder or conductive polymer, this joining method ensures low electrical resistivity and it provides a greater resistance to the affects of electro-migration. The bonding process, however, does require very high temperature, high pressure, and a relatively long time under heat and pressure to complete the joining process. The Cu-to-Cu bonding process involves heating the bonded wafers to 350-400°C for 30 to 120 minutes while under pressure.

Fusion bonding
Fusion wafer bonding is considered by some to be a more efficient method for metal-to-metal via joining. This joining process is a two-stage procedure that begins with the initial precise alignment and a room temperature pre-bonding of the wafers. The system selected for performing this task must be able to maintain precise wafer-to-wafer alignment accuracy. Following room temperature pre-bond, the wafer is exposed to a high temperature annealing process. This joining process can be significantly enhanced with the deposition of a thin layer of tin-alloy onto the exposed copper TSV features. When the stacked wafers are heated to approximately 400°C, the tin alloy completely diffuses into the apposing copper TSV features to form a stable Cu-Sn-Cu (Cu3Sn) intermetallic at the TSV interface (**Figure 7**).

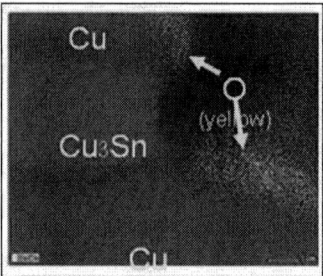

Figure 7. Cu3Sn intermetallic interface. *(Invensas).*

There are a number of key process controls that will require close monitoring: Wafer-to-wafer alignment required for the fusion metal process is +/- 150nm. Metal oxide, if allowed to form, can prevent adequate bond formation and excessive heat and bonding force can result in non uniform bond strength between the wafer layers.

For either Cu-to-Cu process variations, control of surface roughness and oxidation is important to allow the apposing metal surfaces to come into intimate contact, especially for diffusion bonding. Where the basic stages for Cu joining are in the initial alignment, stacking and bond operations, the metal-fusion-bond process will require an aggressive plasma activation process prior to the stack and join procedure.

3D TSV DIE STACKING

Wireless electronics continues to dominate a wide number of market segments. Due to the insatiable consumer demand for complex applications, smart phones and other personal electronic products have an increasing need for data storage capability. The e-book and small format computing products are heavily dependent on data storage as well. The primary driver for meeting the demand for greater data capacity is to furnish higher density memory packaging. Many experts agree that increasing memory storage at the device level is the most efficient solution. In addition to significantly increasing memory capacity, adoption of TSV die-stack methodology will play a significant role in minimizing PCB surface area. In addition, the shorter electrical interfaces between the die elements will likely result in reduced power consumption, faster signal propagation and a reduction in noise and cross-talk.

Due to the process complexities and semiconductor yield concerns associated with wafer-to-wafer joining, many companies are stacking individual die elements. Although the TSV process is performed at the wafer level, the singulated die elements are easier to test. Even though wafer fabrication processes for memory has a relatively high yield, the pretesting of individual die elements before joining ensures that every die within the stack are 'known good'. Several methodologies have been developed for direct joining of the TSV configured memory die elements. Progress in this area has accelerated through the cooperation and joint development programs between a number of government, industry and technical universities. In addition to these joint development programs, many package

assembly service providers have independently developed proprietary processes for TSV. Memory semiconductors in particular, are ideal for direct vertical joining because many of the I/O (pins) can be connected in parallel (**Figure 8**).

Figure 8. 4Gb DRAM, 512Mbx8 die stack memory. *(Example source: Mitsubishi)*

Mounting the basic TSV stacked die assembly to a common organic based PCB platform will not be practical because the via sites are generally too close to each other. Two solutions commonly utilized to accommodate PCB attachment are; adopting a fan-out silicon-based (2.5D) interposer surface or, re-distribution on the lowest die in the stack. The surface redistribution process requires a pattern plating process on the lower die to re-direct the TSV sites to a wider pitch array contact pattern for accommodating conventional circuit board routing techniques.

FAN-OUT INTERPOSER DEVELOPMENT

If the die outline cannot provide the surface area to accommodate a suitable contact array for SMT assembly, the fan-out silicon based interposer may become necessary. One process developed for the silicon based fan-out interposers first applies a polyimide passivation coating onto a bare wafer. To enable the copper filling of the TSV features and the redistribution of the TSV sites to a wider spaced contact pattern, the wafer is subjected to a series of metallization processes. It is common practice to first sputter coat a metal adhesion layer onto the wafers surface to provide the required conductivity for the electro-plating process. Adhesion-promoting metals include: nickel (Ni), molybdenum (Mo), chrome (Cr), tungsten(W), and titanium (Ti). Next, a thin 'seed layer' of copper is deposited onto the adhesion metal.

Following the silicon ablation and copper deposition process to form the TSV features, a photo-imageable mask coating is applied over the copper seed layer surface and developed to define the circuit pattern for subsequent copper plating processes. The finished Cu circuit pattern on the wafer can be as small as 15µm in width with a space separating the conductors as narrow as 10µm. Ideally, the final array contact pitch will not be less than 400µm.

When the electroplating is complete the copper conductors connecting the small TSV features to the newly defined contact sites (typically arranged in a column and row array format). The copper conductors are finally over-coated with a photo-imaged resist material leaving the copper contact sites free of the resist coating for the eventual solder bumping or solder ball attachment process.

STACKED DIE TO INTERPOSER ASSEMBLY
Although stacked wafer joining has been on all of the major semiconductor industry roadmaps for many years, user companies have been reluctant to endorse the concept for all but a few applications. As noted, TSV technology remains most suited to a homogeneous family of products. Industry roadmaps, however, continue to point toward the eventual use of TSV in developing high performance system-in-package products. Broad implementation of TSV, however, has been slow. This is due in part to the significant costs associated with implementing high volume capability for TSV. The process for joining the semiconductor elements to the silicon wafer interposer may employ any of the methods noted above for the die stacking process. The example shown in **Figure 9** represents die-stack-to-wafer assembly using a tin-alloy based coating to complete the copper-to-copper TSV interface.

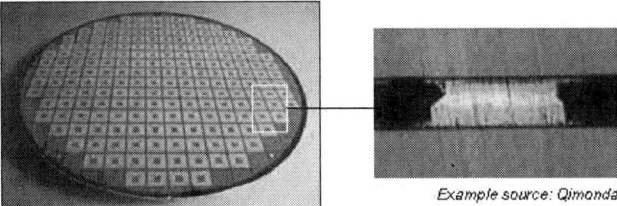

Example source: IBM

Figure 9. TSV stacked die to wafer bonding

Following the joining process the entire wafer assembly is commonly over-molded with polymer composition similar to that used in single die packaging.

While all units remain in the wafer format, solder bumps or preformed solder ball contacts are applied to the array contact pattern on the lower surface. The solder bump process is the most economical. For this process, a precise volume of solder paste is deposited onto each contact site and heated to the liquidus stage to form a near-spherical contact profile. Applying preformed solder spheres to the wafer is more complex. The most common approach prints or deposits a tacky flux to each contact site. Solder alloy spheres are then deposited onto the flux and reflow soldered to complete the joining process.

The mold and contact forming operation is followed by singulation, typically using a dicing saw. The example detailed in **Figure 10** represents a singulated 3D stacked memory semiconductor package with TSV technology mounted onto a fan-out TSV silicon interposer.

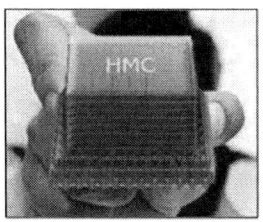

Figure 10. Pre-joined TSV memory die mounted onto an array contact configured silicon interposer.
(Example source: Samsung/Micron Technology)

Even though a great deal of progress has been made in process refinement for memory applications, the industry will still need guidance and a degree of standardization. Package outline and contact layout standards will be required to better enable end users to procure TSV memory products from multiple sources.

SUMMARY AND CONCLUSIONS
The industry recognizes that the major challenge for today's electronic products is the control of overall system costs and reducing power consumption. Because volumetric area for many electronic applications is somewhat limited, die-stack using through-silicon-via technology would seem to hold an attractive solution. The advantage is significant. Direct joining of one die element to another provides electrical paths between devices that are significantly shorter. The shorter interface will contribute lowering resistance and inductance. Adoption of TSV technology also furnishes the potential for maximizing package functionality while minimizing overall package height and a reduction of surface area needed for mounting.

Foundries considering joining this evolution may find that they will utilize only a limited number of existing resources because there are numerous process steps that are well outside the normal front-end process flow for semiconductor wafers. Whether TSV capability remains within the semiconductor foundry or outsourced to a specialist, companies will need to make some critical monetary decisions. In addition, there are a number of processes and methodologies that are considered proprietary and may require licensing agreements and additional fees for there use. Investing in already proven processes, however, may help to accelerate implementation because a licensing agreement may include assistance in selecting the best equipment set and provide technical training.

There is also general consensus that thermal management is one of the most critical issues, especially when combining logic and memory in the vertically stacked configuration. In effect, the power dissipation of the logic die (typically an application processor or a digital baseband semiconductor) can transfer any generated heat into the memory elements directly above. In the low to high-end 3G platforms the power dissipation of the bottom die can range between one Watt and three Watts.

REFERENCE:

1. 2011/2012 International Technology Roadmap for Semiconductors (ITRS)
2. Rodin, A., *'High Throughput Low CoO Industrial Laser Drilling Tool'*, XSil Ltd. White paper.
3. Kim, D., *'Evaluation for UV Laser Dicing Process and its Reliability for Various Designs of Stack Chip Scale Package'* , Amkor Technology Korea, Inc., White paper.
4. Bauer, T., *'High Density Through Wafer Via Technology'*, NSTI-Nanotech 2007.
5. Reif, R., *'3-D Interconnects Using Cu Wafer Bonding : Technology and Applications'*, IEEE ESTC 2010.
6. Matthias, T., *'Thin wafer processing and chip stacking for 3D Integration'*, IEEE ESTC 2010.

SINGLE SIDED WET ETCHING FOR TEXTURING, THINNING AND PACKAGING APPLICATIONS

Ricardo I. Fuentes, Ph.D.
Materials and Technologies Corp., (MATECH)
Wappingers Falls, NY
info@matech.com

ABSTRACT

The advancement of wafer thinning and surface treatment technologies has been somewhat limited by the methods commonly employed; namely immersion, spin, or spray etching. All these technologies suffer from inherent transport characteristics that result in non-uniformity, limitations in the minimum thickness attainable, and lack of process flexibility.

We will report on results produced by a new technology that has been gaining momentum over the last few years: Linear Scanning. It is inherently single-sided and more uniform as well as cost effective and flexible.

LinearScan etching technology exposes only the process side of the substrate to a thin line of flowing chemicals and scans in a continuous, repeating sequence as many times as necessary to achieve the desired results. The process time is similar to that required by the same chemistry with a conventional method.

The non-process side and edges of the wafer are protected from the etchant by a shaped gas flow making it a truly singled process. Without the dynamics involved in spinning, spraying or immersion techniques, this technology is suitable for very thin wafers, fragile materials, or delicate structures such as those used in 3D packaging and solar applications. The non-process side can be bumped, taped, patterned or otherwise structured and is not affected by the process. Common Linear Scan applications include texturing, thinning, stress relief, among others.

Keywords: Texturing, Chemical Thinning, Etching, Stress Relief, WaveEtch, Linear Scan.

INTRODUCTION

Even though, in its essence, wet etching is a simple process; i.e. the removal of material during the interaction between a liquid and a solid substrate, it can involve a series of complicated steps which combined effect leads to the resulting structure or surface. It remains popular in the semiconductor, solar, packaging, optoelectronic, as well as in many other industries because it is often the fastest and most cost-effective way to remove material[1] selectively or across an entire surface.

Thinner packages, higher power densities, and the ever increasing functionality of systems-in-a-pack (SiP) are driving the need for more robust, lower cost, higher yield thinning technologies[2]. When the substrate can be wetted on both sides, immersion is a common choice for etching and thinning. If the substrate can only be exposed to the chemicals on one side, spin or spray etching become reasonable candidates, but all these technologies have their shortcomings, such as radial and transport-induced non-uniformities[3]. Also, conventional technologies often result in undesirable exposure of the non-process side to residual liquid or vapors.

The need to alleviate the shortcomings of conventional etching technologies, as well as the increased power and flexibility that the control of some of the kinetic variables during etching brings to wet processing, make LinearScan etching a very exciting and powerful new technology. Surface texturing is an excellent case in which the LinearScan added control yields engineered surfaces in a faster, more cost effective manner.

New device manufacturing techniques have become more demanding, which often requires the use of single-wafer, single-sided processing. In addition, thinner, denser packages, the push for increased functionality and decreased cost are driving this trend[4].

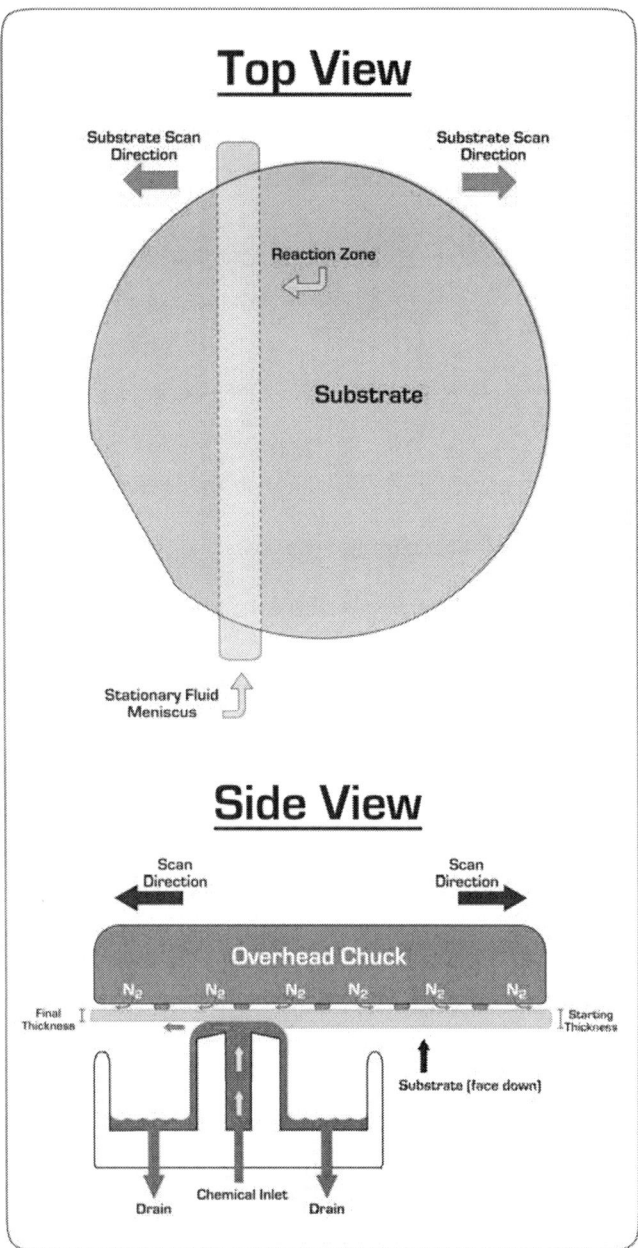

Figure 1. Schematic representation of the LinearScan process depicting the substrate material being removed by the etching process (held process side down), as well as the orthogonal paths of the reactants and the byproducts. The shaped flow of gas (DynamicConfinement), preventing the encroachment of fluid and vapors onto the non-process side (top), is also shown. Note the removal of material (etching) as the wafer is being gently scanned over a narrow pool of chemicals, and the different paths the reaction byproducts take to avoid interference with the supply of fresh reactants to the reaction zone. The top view illustrates the wafer being scanned repeatedly in alternating directions, as necessary, over the fluid meniscus showing the narrow reaction zone where thinning is in progress (wafer is being etched from below).

The new technology WaveEtch™ LinearScan™ etching provides high uniformity as well as true single-sidedness on ultra-thin, large substrates[5,6]. It addresses the main shortcomings of conventional wet processing by providing a consistent and uniform supply of chemicals throughout the liquid-solid interface while making available an orthogonal path for the byproducts, such as gases and vapors. Exposure of every surface element to the same chemical and transport environment makes the process intrinsically uniform (Figure 1). The solid-liquid interface (boundary layer) is not subject to speed gradients, convection, or other transport-related gradients that may cause variations in its thickness and its concomitant impact on uniformity. The system eliminates virtually all transport-related and centrosymmetrical non-uniformities, which plague spin/spray and immersion processes. Reactants enter the reaction zone through the bottom of the pool, while the byproducts exit in a plane parallel to the substrate surface; this delays solution saturation, extends bath life, and insures a consistent supply of fresh chemicals to the surface. The substrate is not immersed, but merely put in contact with the top of the pool's meniscus, as also illustrated in Figure 1.

Most chemistries used with these processes do not require surfactants and are used in smaller volumes at lower flow rates, allowing for more efficient chemical usage[7].Together, these features lower chemical usage and its associated purchase and disposal costs, as well as often easing environmental regulatory compliance, resulting in overall production and costs-of-ownership reduction.

LinearScan etching processes are size- and shape-independent. Since all areas are exposed to the same chemical and transport environment, the size and shape of the substrate are largely irrelevant. A process developed for a given substrate geometry, can be readily used for another substrate geometry, thus making product process migration effortless and cost effective. These systems naturally accommodate odd, noncircular, thick shapes, and structures larger than 300mm.

TEXTURING

LinearScan technology allows increased control over events that influence surface development which is crucial in the formation of textured or engineered surfaces. The separate paths that reactants and by-products (including gaseous byproducts) take during linear scan etching as well as the control over the time chemicals are in contact with the surface between scans, allow for a variety of surfaces to be engineered. Textured surfaces are required in applications such as solar cells and backside metallization to achieve lower reflectivity and improved adhesion, respectively.

Gas evolution during etching plays an important role in surface development, as well as "bubble" nucleation and detachment. LinearScan allows for control of some of the parameters that determine such bubble formation and the impact that the size and spatial distribution of their gas-solid interfaces has on the formation of the surface during etching. Parameters such as scan speed and surface fluid velocity determine the thickness of the boundary layer and the consequent transport characteristics of the process.

Acidic textures are oftentimes desirable because they are faster to achieve and produce complex non-faceted surfaces.

THINNING AND PACKAGING

Wafer thinning is one of the native applications of the LinearScan etching technology. It allows for thinning of mounted or un-mounted, taped or un-taped substrates with superior uniformity, no edge damage, and without requiring any form of backside protection. Substrate assemblies, at any point in the packaging process, of virtually any thickness, structure, and size are all compatible with the LinearScan etching process.

LinearScan etching systems are particularly well suited to handle and process very thin substrates. The unique process

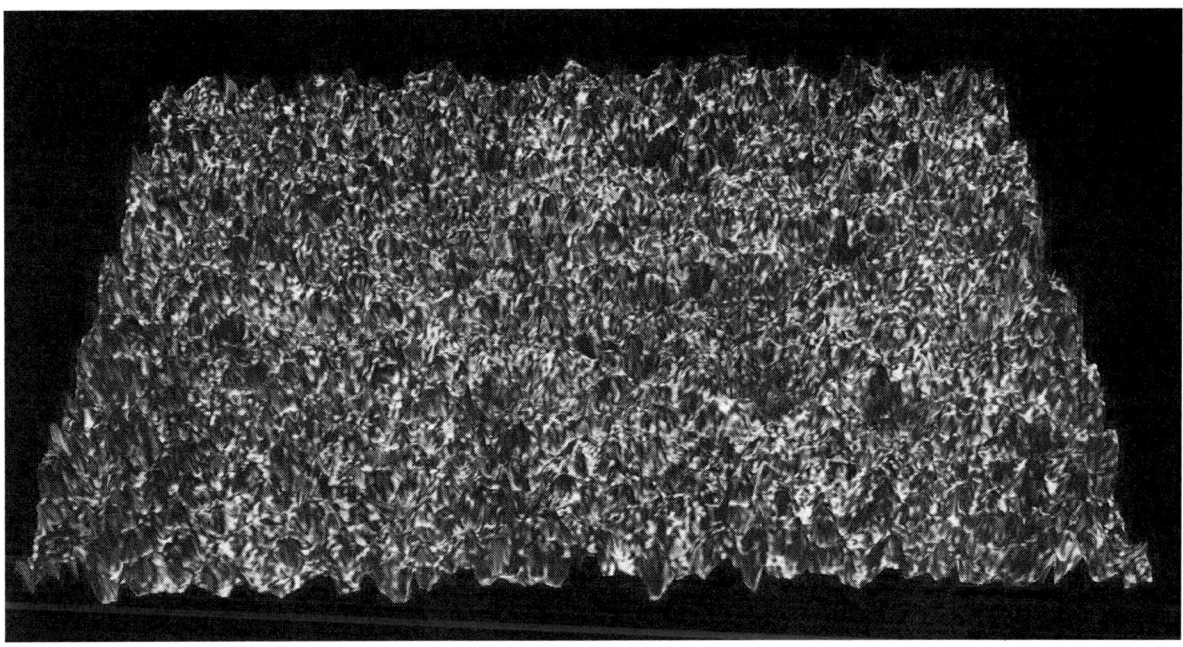

Figure 2. Confocal microscope rendered image of a LinearScan textured surface with an acidic solution viewed at 50X magnification. Note the complex and convoluted nature of the surface, and the absence of smooth flat facets. Sa=0.65 μm, Sq=0.81 μm, Sz=7.52 μm (average, RMS, and maximum roughness depth areal parameters respectively, as per ISO 25178). Such surfaces offer an increased surface area for mechanical interlocking of deposited films, as often required in backside metallization.

LinearScan produces acidic textures of a wide range of scales and on different materials. Figure 2 shows the image of a silicon textured surface for a backside metallization application. The peak-to-valley scale is approximately 7 μm although the attainable range by the LinearScan acidic texturing processes is from sub-micron to tens of microns.

These surfaces are produced in a few seconds to a couple of minutes depending on the amount of material to be removed and the desired final roughness. It is important to notice the complex and non-faceted nature of the surfaces which is a desirable morphology for applications such as backside metallization. The convoluted nature of the surface allows for a very large interfacial area between the metal film and the substrate material resulting in considerable "interlocking" and consequentially increased adhesion strength. The technology is equally capable of producing faceted surfaces by alkaline processes [7].

is carried out with no violent spinning, no need for lateral confinement by pins or other hard devices that may damage the wafer's edge, and no dynamic loading due to high rotational speeds. In the absence of hydrodynamic edge effects, the edges of the wafers are free of edge sharpness and the formation of other features common in spin/spray etch systems that significantly weaken the substrates [8].

The ability to use virtually any chemistry to interact with any substrate material enables the systems to process any material of interest. In addition to packaging applications, the systems are being used to etch or thin InP, Ge, GaAs, Si, polysilicon, glass, and quartz, among others. Substrates of odd shapes and within a large range of size and thickness are being processed. This method provides a new way to do wet thinning for packaging applications in a more precise, efficient, and environmentally friendly manner.

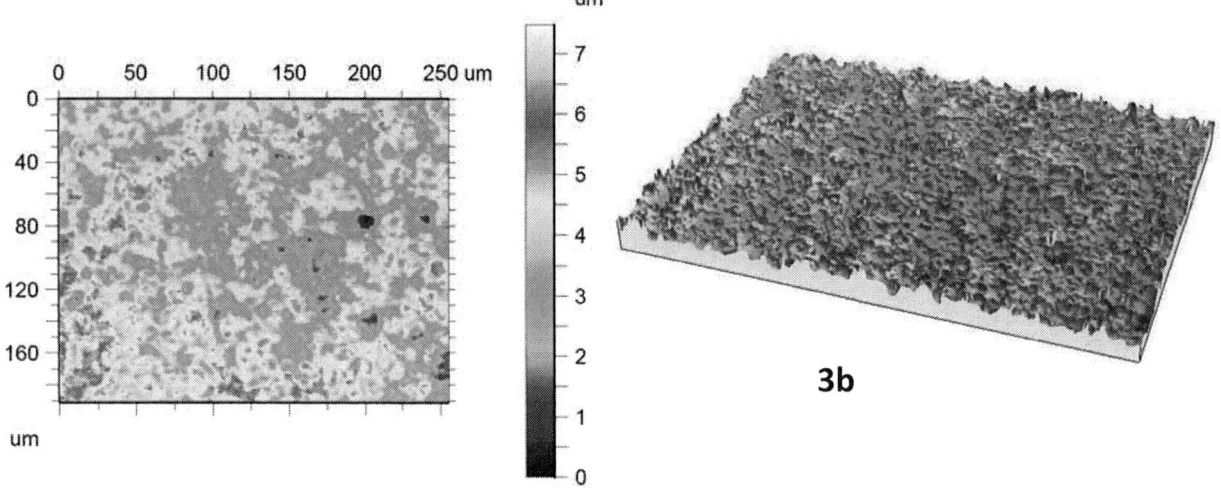

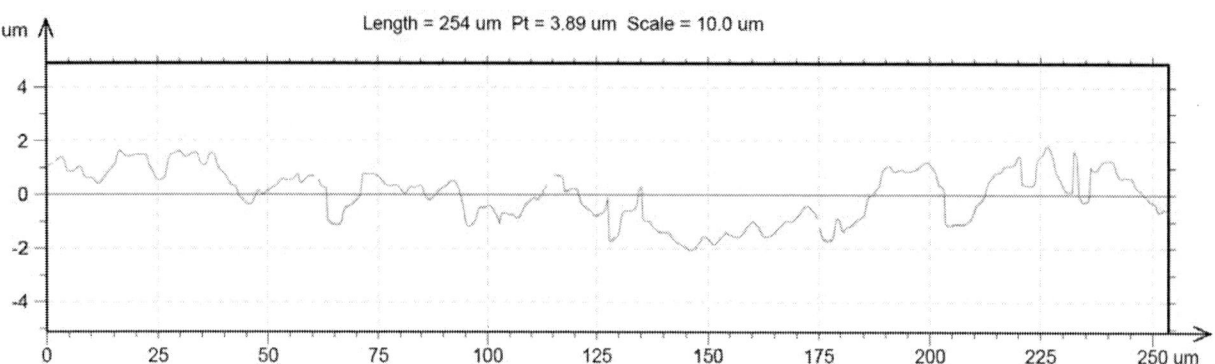

Figure 3a, b, c. Top-left to right clockwise. Confocal microscope map (3a), image (3b), and profile (3c) of the same surface (50X) depicted in Figure 2. Ra=0.62 μm, Rq=0.74 μm, Rt=3.23 μm (average, RMS, and maximum roughness depth line parameters respectively, as per ISO 4287). The additional control that the LinearScan process exerts over the solid-liquid interface allows the production of surfaces like this in fast and cost effective manner. Note the consistency between the linear (Ra and Rq) and their corresponding areal parameters (Sa and Sq).

ACKNOWLEDGEMENTS
WaveEtch™, LinearScan™, and DynamicConfinement™ are trademarks of Materials and Technologies Corp. (MATECH)

Ricardo I. Fuentes is the founder and president of Materials and Technologies, Corp., (MATECH), 22 Bill Horton Way, Wappingers Falls, NY 12590, United States. Phone (845) 463-2799, e-mail info@matech.com.

REFERENCES
[1] Marc J. Madou, "Fundamentals of microfabrication: the science of miniaturization" 2002 CRC Press, LLC, p. 110.

[2] Robert Castellano, "Wafer & Device Packaging and Interconnect" September/October 2010 p.10.

[3] Ricardo I. Fuentes, "Intrinsically Uniform Single-sided Wafer Thinning", Levitronix User Conference, 2011.

[4] IWLPC, 6th Annual International Wafer-Level Packaging Conference proceedings, October 27-30, 2009, Santa Clara, CA.

[5] Ricardo I. Fuentes, "Extending Process Flexibility for Single-wafer Etch," Solid State Technology, July 2007.

[6] Ricardo I. Fuentes, "Single-sided Wafer Thinning and Handling," Chip Scale Review, Jan/Feb 2011.

[7] Ricardo I. Fuentes, "Single-sided Wafer Thinning for 3D Integration," Annual International Wafer-Level Packaging Conference proceedings, 2010, Santa Clara, CA.

[8] G. Coletti, C.J.J. Tool and L. J. Geerligs, "MECHANICAL STRENGTH OF SILICON WAFERS AND ITS MODELLING," 15th Workshop on Crystalline

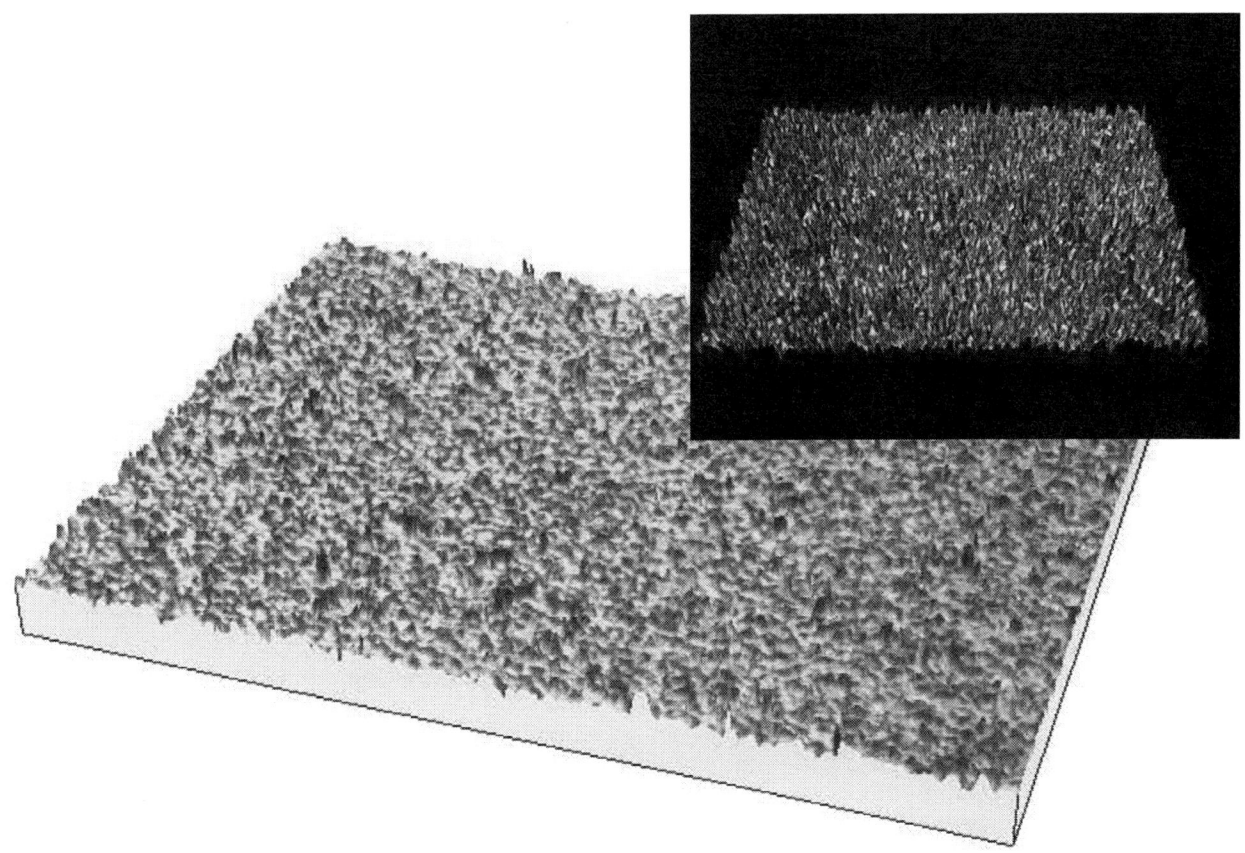

Figure 4. Acidic textures, shown in this confocal microscope image at lower magnification (20X) can be easily achieved with the LinearScan etching technology. Ra=1.82 µm, Rq=2.25 µm, Rt=13.4 µm, Sa=2.03 µm, Sq=2.62 µm, Sz=51.9 µm (average, RMS, and maximum roughness depth, line and areal parameters, respectively, as per ISO 4287 and 25178). Roughness levels achievable with this technology can range from the above to sub-micron levels. Top-right corner shows a rendering of the same X-Y-Z data.

Silicon Solar Cells & Modules: Materials and Processes,
Vail Colorado, USA, 7-10 August, 2005.

DEPOSITION PROCESSES FOR COMPETITIVE THROUGH SILICON VIA INTERPOSER FOR 3D

Cyprian Uzoh, Rezwana Sharna, Pejman Monajemi, Michael Newman,
Charles Woychik and Terrence Caskey
Invensas Corporation
San Jose, CA, USA
cuzoh@invensas.com

ABSTRACT

The 3D-IC and related technologies enables higher device and package bandwidth, improved device performance at lower power consumption, reduced form factor, and the potential for lower cost for applications in logic and memory integration including heterogeneous technologies in a single package. Invensas Corporation is pursuing many unique enabling solutions to address many of the most difficult 3D-IC fabrication challenges. The convention methods for the design and implementation of Through Silicon Via (TSV) interposer is reviewed. These various process modules include TSV etch, TSV fill, chemical mechanical polishing (CMP), BEOL, Bonding backside silicon removal processes, via reveal, passivation, wiring and bumping. These technologies were developed using 200mm and 300mm foundry equipments ranging from chemical vapor deposition (CVD), physical vapor deposition (PVD), deep reactive ion etching (DRIE), electrochemical deposition (ECD), silicon grinding, CMP, temporary bonding and de-bonding and various cleaning steps. Efforts on optimizing some of the most expensive processing steps, especially the metal filling in deep multi-micron via cavities is discussed. The difficulties of depositing void-free deep via cavities and the differences between conventional BEOL and newer TSV plating chemistries, cost reduction methodologies will be presented. Keywords: tsv, drie, ecd, beol. bottom up, cavity fill efficiency, plating planarization factor, gapfill anisotropic index, additives

INTRODUCTION

Silicon interposer for 2.5D and 3D IC stacking has emerged as a viable solution for higher interconnect density, smaller from factor, significant reduction in power consumption and major increase in bandwidth. The Interposer provides flexible and high density wiring platforms, from which component and system designers can explore the potential benefits of 3D memory, low latency logic-memory pairing and full logic partitioning [1-6]. In the hybrid memory cube, thousands of TSV interconnections provide 15 times higher bandwidth and 70% less power compared to DDR3 in one-tenth silicon real estate [2]. 3D stacking of memory above processor fulfills the stringent bandwidth, scalability, and power requirements, particularly for wide-IO mobile DRAM [3]. 2.5D TSV interposer for SerDes application in

field programmable gate array is reported in [4] for multi-Terabit-per-second die-to-die bandwidth.

Integration challenges associated with adoption of TSV technology are the following:

Deep via cavity formation by reactive ion etching technologies routinely performed using Borsch process is typical. The long etching times required to etch the deep vias (30 to 300 microns) results in low process thruput of less than 4wph and high COO. Etching defect such as undercut and inadequate via sidewall taper, excessive via sidewall roughness can pose subsequent problems for the barrier, seed coating step and the plating step.

The of coating continuous barrier and seed layer on walls of deep cavities with aspect ratio often greater than 5 presents its own sets of difficulties. Barrier and seed layer coating in micron scale submicron BEOL technologies and TSV features bears great similarity. The composition and the texture of the barrier exerts a great influence on the properties of the seed coated thereon. Both needs to be continuous, however 3 to 5nm of good barrier materials and seed have been shown to be effective in BEOL applications. However, thin copper seed layer 3-5nm may be problematic if portions of the thin seed is oxidized to form a discontinuous seed. In practice, for TSV applications, depending on via depth and diameter, thicker barrier layers 100 to 300nm, and thicker seeds (200 to 3000nm) are required. The thicker barrier and seed reduces process thruput < 10 wph, depending on tool configuration.

Copper electrodeposition (ECD) for submicron gapfill in damascene features is well established in the industry, thanks to the pioneering works of C. Uzoh and his colleagues in the early 1990's (7,8). It is of note that the same fundamental principles that governs the coating of void-free copper in submicron and micron scale damascene features applies to deep TSV vias with depths order of magnitude higher than BEOL features.

This paper compares gapfill processes in moderate to very high aspect ratio BEOL features to gap fill in 25 and 100 micron TSV features with aspect ratios between 4 to 8 using conventional copper BEOL chemistry and without prewetting process, using only direct current plating mode. This paper also present results from an experimental plating chemistry with a highly efficient bottom-up growth profile within the via cavities.

EXPERIMENTS

The gapfill wafer test vehicle was built using the conventional via middle process flow. Figure 1 shows the cross section Scanning Electron Microscope (SEM) view of the 15μm diameter and 100μm deep TSV's after DRIE step. The etch recipe was optimized to obtain best roughness at the bottom of via while minimizing necking at the top (below 0.25μm). The etch profile is made slightly tapered (about 1°) to enhance PVD coverage without compromising the roughness.

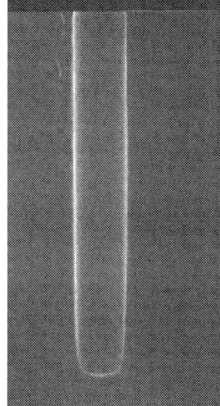

Figure 1. Cross section view of 15μm via

Post-etch vapor-HF clean further removes the etch residues and improves roughness, as shown in Figure 2 for a 15μm TSV before and after clean.

Orthosilicate (TEOS) liner isolates silicon sidewalls from the consequent metallization. Sidewall coverage is critical to electrical performance. TEOS films applied by plasma enhanced CVD provide excellent via coverage.

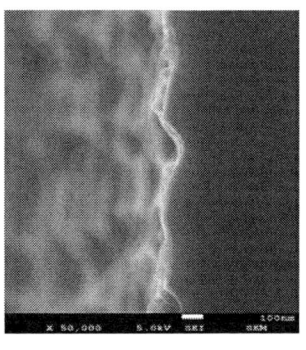
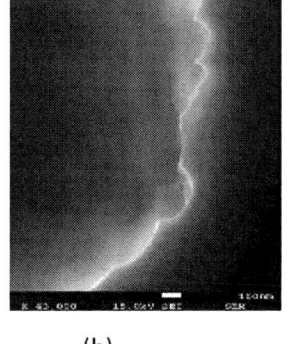

(a)   (b)

Figure 2. Cross section SEM view of etched TSV: a) sidewall residue after DRIE, b) after vapor HF cleaning

For barrier layer 200nm tantalum (Ta) was coated to provide continuous and thick enough barrier at via sharp corners Fig.3 To reduce copper agglomeration within the via cavities, the substrate temperature was kept below 80°C also, the seed layer deposition was split into multiple small coating steps.

The seed layer coated via were plated in various low sulfuric acid based BEOL copper plating solution using Enthone, Inc. Viaform™ additive package to evaluate BEOL bath limitations in coating deep TSV cavities. Also, wafers from the same lot were plated in experimental acidic methanesulfonate chemistry with a three additive systems. Finally the results of the various coatings were compared to earlier work on plated damascene features and shallower TSV structures.

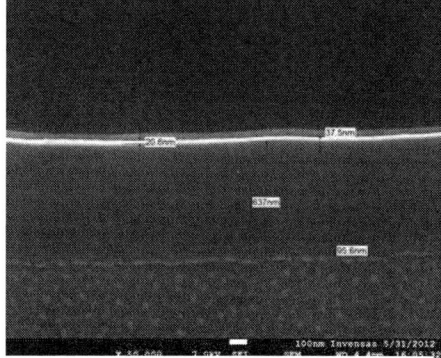

Figure 3. Cross section of bottom of via after seed deposition

After the via deposition process, 2D images of the plated substrates were obtained using Dage XD7600NT Xray Inspection System to quickly evaluate defects within the plated cavities. A high resolution 3D Xray microscope Xradia - Micro XCT-200 3D was used to examine non-destructive 3D images of arrays of the plated vias. Selected plated vias of interest were cross section using SEM for more detailed analysis of the nature and dimensions of the via defects.

The standard initial BEOL plating solution contained 40g/L copper ions, 10g/L sulfuric acid and 50ppm/L of chloride ions. The organic additives system was Viaform™ (Suppressor, Accelarator and Leveler) were manufactured by Enthone, Inc. The seeded wafer were plated at room temperature without any prewetting operations using only DC current without any pulse reversal . The experimental current densities were between 0.5 to 20mA/cm^2 after the plating steps, the substrates were rinsed in DI and dried with a nitrogen gun.

RESULTS AND DISCUSSION

The fundamental electrochemical origin of preferential bottom-up growth in chip metallization in damascene cavities is well established.(7-13). The role of the chloride ion polarization, preferential additives adsorption, (suppressor, accelerator, and lever) in electrochemical mechanical deposition (ECMD) (14) in which bottom-up growth in large cavities are greatly enhanced at the expense of the nominal expected conformal coating in large cavities have further enhanced the body of knowledge in this field. Figure 4 are FIB sections of damascene vias plated with two different current densities A and B. from the same low acid BEOL chemistry.

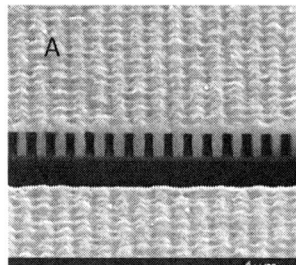 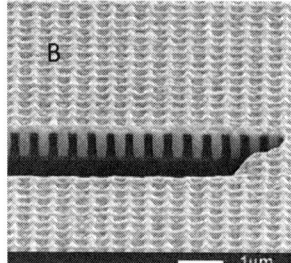

Figure 4. FIB section of vias plated at two current densities A and B respectively

Both vias are void-free, substrates A exhibits a convex growth front, while that of B is concave (dish shaped). In electrochemical suppression of overplating in submicron vias Uzoh et. al eliminated overplating by combining the convex and concave growth fronts during metal coating to eliminate overplating on the submicron features. With similar approach very high aspect ratio (16 to 18) BEOL features have been successfully filled with copper Fig5. For these plating formulations any significant increase in leveler concentration was shown to produce seams in the via cavities.

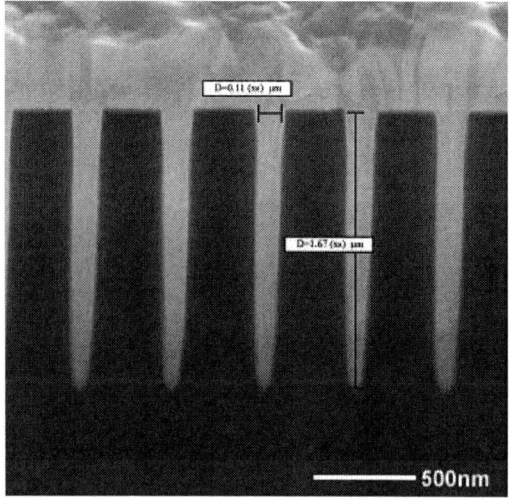

Figure 5. FIB section of vias plated very high aspect ratio damascene cavity

Similarly the addition of leveling agents in plating baths suppresses electrochemical planarization phenomenon (14).

Unlike traditional damascene scale features, plating formulation without leveling agents produces very large continuous seams in TSV features.

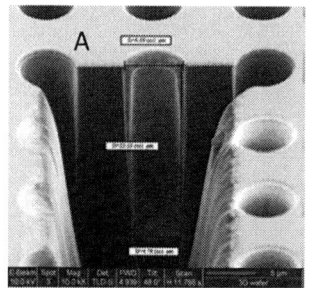 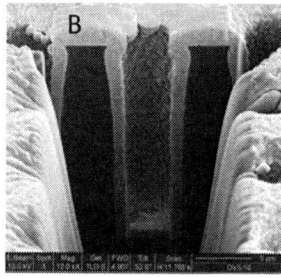

Figure 6. FIB section of TSV a) before plating b) after plating.

Figure 6A shows cross sections of TSV via (6 μm x 24 μm) after the barrier and seed deposition and Fig6B after the substrate was plated at 10mA/cm^2 for 900s in a plating bath containing Enthone Viaform™ BEOL additives without leveling agents. Unlike BEOL features, for TSV features, increasing leveler concentration promotes bottom-up growth. Apart from leveling additives, mass transport plays a critical role in the evolution of void free gap filling in TSV scale structures.

For void-free gap fill, current distribution within TSV cavities is controlled and dominated by kinetics and mass transport. The plating additives are typically formulated in minute amounts (in ppm ranges) their flux inside the via cavity is mostly transport limited. For deep TSV channels, the additives are transported solely by diffusion.

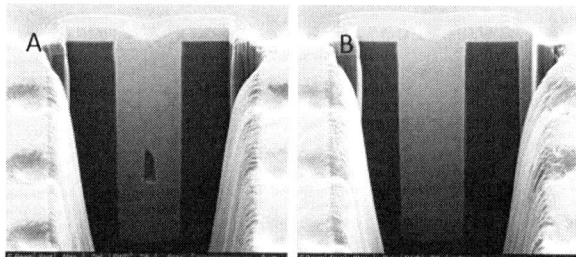

Figure 7. Cross section SEM view of 4:1 aspect ratio via A) plated with reduced agitation B) plated with enhanced agitation.

Also copper ions concentration at the bottom of the TSV cavities is controlled by mass transport limitation. As a result, higher agitation promoted void-free filling. This is demonstrated from the results of two substrates plated at 10mA/cm^2 for 900s in the same bath. Higher agitation Fig 7B sample produced void-free copper as compared to Fig 7A the substrate plated with reduced agitation the occluded a near midvia void.

For deeper TSV cavities such as those shown in Fig1, nominal copper ion concentration of 40g/l was found to produce voided vias at reasonable plating current densities. Increasing the concentration of copper ions in the BEOL bath formulary to 75g/L, boosted copper transport to the bottom of the blind structures. Depending on details of the plating process, void-free vias could be obtained with BEOL additives. To study copper growth evlution within the vias,

76

various TSV substrates were plated with varying amount of copper charge. The cross sections of the coated substrates were examined in electron microscope. Figure 9 is the SEM or partially filled deep via after plating an equivalent charge of 5.67μm of copper on the substrate. In Fig8 the measured copper thickness in the field 4.8 μm, the field seed layer thickness of 2μm. From the above, the electrodeposited copper at the top of the substrate is 2.8μm, which is about 49% of the total plated copper charge. This implies that 51% of the plated metal resides within the via cavities. Within the via cavity, the bottom of the via front or the height of the plated metal is 41μm. The ratio of the height of the metal in the via to the field plated copper thickness is the vertical plating planarization factor (ppf). The plating anisotropy factor is a measure of growth anisotropy within the via cavity and is an index of the efficiency of bottom-up growth or lack of it.

Figure 9. Completely filled TSV structures.

The plating of additional copper charge on the partial filled substrate of Fig 8, using the same plating bath produced a fully filled TSV structure of Fig 9.

Similar experimentation using methane sulfonate plating chemistry with high copper ion loading and new three additive system is shown in Fig. 10.

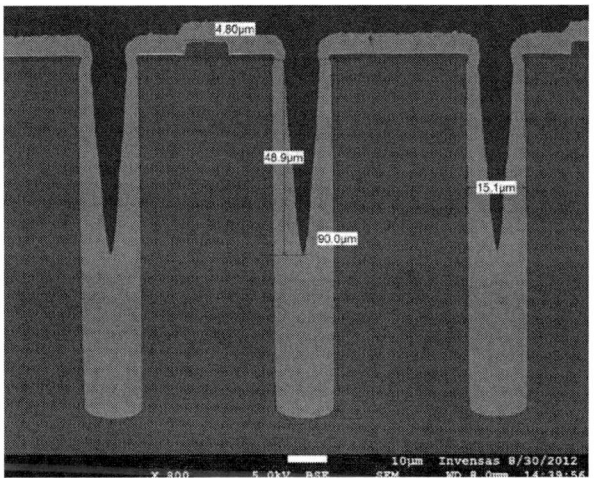

Figure 8. SEM or partially filled deep TSV cavity

The planarization factor for Fig 8 process is (41/2.8) ~ 14.6. From the above analysis, the via coating process can be described by a cavity plating efficiency of 51%, and a planarization factor of 14.6. The cavity plating efficiency informed that only 51% of the total plated metal is coated with the via cavity, while the planarization factor foretells the propensity for voided vias.

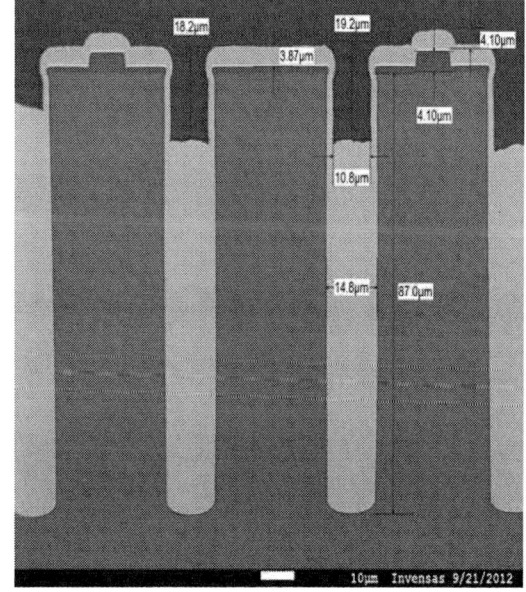

Figure 10. Partially filled TSV with new chemistry

Close examination of the electron microscope image of the cross sections, for a plating charge of 5.46μm and 2μm copper seed, the field copper thickness is 3.87μm. From the above, the electroplated copper in the field is about 1.87 μm (3.87-2). This implies that the equivalent of 3.59μm (5.46-1.87) of the plated copper resides within the TSV via cavity, yielding a cavity plating efficiency of 66% . The height of the plated copper within the via cavity is about 73μm, using a plated field thickness of 1.87 μm, the resulting plating planarization factor (ppf) for the partially filled TSV is about 39. Both the high cavity filling efficiency and the high planarization factor suggests a very efficient and effective cavity plating process. By any measure, the cavity filling efficiency of the new process is superior to that of obtained from modified conventional BEOL type chemistry.

Also, of interest is the planar copper growth profile within the via cavity of Fig 10. Planar growth profile are least likely to occlude a void. Of singular importance is the profile of the sidewall of the via cavities of figures 8 and 10. The growth profile within the via cavity with the modified BEOL chemistry is superconformal, here there is ample evidence of lateral growth front. The simultaneous combination of vertical and lateral growth front results in a conical sidewall structure in the unfilled portion of the via. For the bottom-up growth evolution of Fig10, after initial sidewall copper coating of 2μm, further lateral growth was effectively suppresses, in favor of highly accelerated vertical growth. The gapfill cavity anisotropic index at the growth front (ratio of Cu height/Cu sidewall thickness) is (73μm /2μm) is 36.5 While the gapfill anisotropic index for the superconformal evolution is about (41.1μm /7.5μm) ~ 5.5. These estimates assumes that the copper seed thickness on the sidewall and bottom of the TSV are negligible compared to the plated copper Fig 3. The gapfill anisotropic index is a direct estimate of void occlusion in plated cavities. In conformal deposition, this value is 1. For superconformal coating the index >1, and for bottom up the index >> 1. In practice, it may be preferable to measure the plating planarization factor and the gapfil anisotropic index at the or near 50% of the height of the via cavity. For 90μm deep TSV, the mid via height is 45μm and the gapfill anisotropic index is 6 (45μm/7.5μm) and 22.5 (45μm/2μm) for the BEOL chemistry and methane sulfonate bath respectively. For the sulfonate bath, the planar growth front profile subsisted throughout the cavity gap filling as shown in Fig 11, where additional copper charge was coated on the substrate.

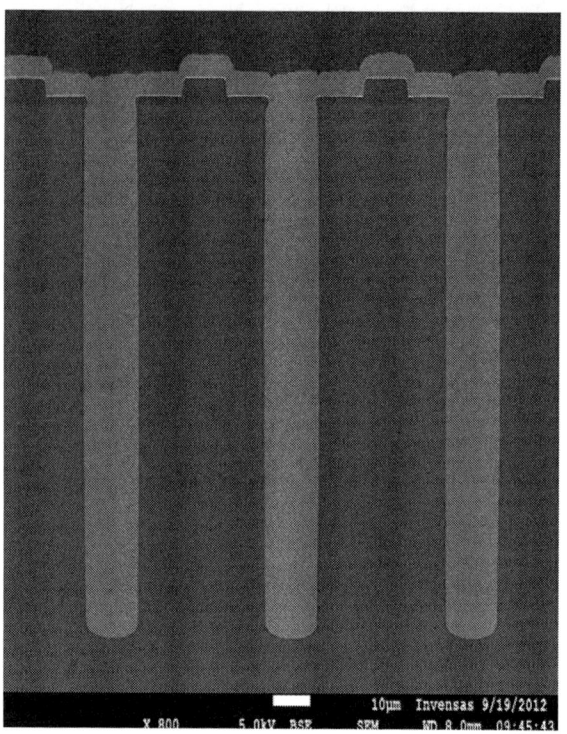

Figure 11. Copper coating with planar growth front.

A high resolution 2D and 3D Xray microscope image of wafer plated with BEOL chemistry is shown in Fig 12. Close inspection of the images from the top fig 12A and 3D images Fig12B from the wafer backside confirms the absence of voids within the via cavities.

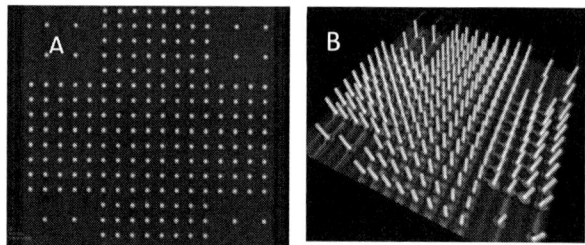

Figure 12. Top down 2D X-ray image and 3D image of void-free TSV plated from modified BEOL chemistry

In general, highly efficient cavity filling processes exhibits a high filling efficiency typically >0.5, a high plating planarization factor and a high gapfill anisotropic index. Plating chemistries and processes with high gapfill anisotropic indices will result in TSV with thinner overburden and significantly reduced TSV plating time and associated costs. The thinner metal overburden improves CMP thruput is a welcomed bonus.

CONCLUSION

We demonstrated void free copper electrodeposition in deep TSV cavities using BEOL and non-BEOL chemistries without pre-wetting process using DC without pulsing or pulse reversal methods. Conventional BEOL copper plating chemistry is effective in plating TSV cavities below 50 micron. Deeper TSV cavities will require significant increase in copper ions content of the plating chemistry. Plating chemistries and processes with high gapfill anisotropic indices are more efficient for plating TSV structures.

ACKNOWLEDGEMENTS
The authors would like to express their gratitude to Rajesh Katkar, Dileep Goyal and Michael Huynh of Invensas Corp. for their contributions to the characterization efforts.

REFERENCES
[1] M. Jackson, "A silicon interposer-based 2.5D-IC design flow, going 3D by evolution rather than by revolution", *2011 3D Architectures for Semiconductor Integration and Packaging Conference.*

[2] J. Jeddeloh, B. Keeth, "Hybrid memory cube new DRAM architecture increases density and performance", Proc VLSI Conf, pp. 87-88, 2012.1.

[3] J-S. Kim et al., A 1.2V 12.8GB/s 2Gb Mobile Wide-I/O DRAM with 4X—128 I/Os Using TSV-Based Stacking, ISSCC 2011, pp. 496 – 498

[4] N. Kim, D. Wu, K. Dongwook, A. Rahman, P. Wu, Interposer design optimization for high frequency signal transmission in passive and active interposer using through silicon via, ECTC 2011, pp. 1160-67.

[5] P. Vivet, "A three-layer 3D-IC stack including wide-IO and 3D NoC practical design perspective", *Proc architectures for semiconductor integration and packaging,* 2011.

[6] W. C. Chiou *et al*, "An ultra-thin interposer utilizing 3D TSV technology", *Proc VLSI Conf,* pp. 107-108, 2012.

[7] IBM Corp. 1997 Annual Report Page 18

[8] P .C. Andricacos, C. Uzoh, J. O. Ducovic, J. Horkans and H. Deligianni IBM J. of Res. Dev. 42, 567 (1998)

[9] J. J. Kelly, A. C. West "Copper Deposition in the Presence of Poly(ethylene glycol) I, J of Electrochem. Soc Vol 145 (1998) pp3472-3476 and pp 3477-3481.

[10] T. P. Moffat, J. E Bonevich, etc. "Superconformal Electrodeposition of Copper in 500-90nm Features" J. of the Electrochem. Soc. Vol 147, No12 (2000) pp.4524-4535.

[11] D. Josell, D. Wheeler. W. H. Huber, J. E. Bonevich and T. P. Moffat, " A Simple Equation for Predicting Superconformal Electrodepositionin Submicron Trenches" J. of the Electrochem. Soc. Vol 148, (2001) pp. C767-C773

[12] David Roha and Uziel Landau, "Transport Controlled Leveling Plating Additives: Steady-State Model for Blocking Additives," J Electrochem. Soc. 137 (3) pp. 824-834.

[13] USPTO 6,709, 562B " Method of Making Electroplated Interconnection Structures on Interated Circuit Chips"

[14] B. M. Basol. S. Erdemli, C. E. Uzoh and T. Wang. "Planarization Efficiency of Electrochemical Mechanical Deposition and Its Dependence on Process

Parameters" J of Electrochem. Soc.2006, Vol 153, issue 3 pp. C176-C181.

BONDING AND CONTACTING OF VERTICALLY INTEGRATED 3-D MICROSCANNERS

M. Wiemer, J. Frömel, C. Jia
System Packaging Department, Fraunhofer Institute for Electronic Nanosystems (ENAS)
Chemnitz, Germany
maik.wiemer@enas.fraunhofer.de

S. Bargiel, M. Barański, N. Passilly, C. Gorecki
Micro Nano Sciences and Systems department, FEMTO-ST Institute UMR CNRS 6174
Besançon, France
sylwester.bargiel@femto-st.fr

ABSTRACT

In this work we describe the bonding and contacting of a micromachined vertically integrated 3-D microscanner, which is a key-component for a number of scanning imaging microsystems, such as confocal microscopes on-chip or optical coherence tomography (OCT) probes for micro endoscopy. The 3-D microscanner is composed of two electrostatic silicon MEMS micro actuators which are vertically aligned and bonded with glass and ceramic components to create hermetically sealed cavity for scanning microlenses.

We demonstrate the 3-D stacking technology, based on sequential multi-level anodic bonding, successfully tested with deeply structured silicon/glass wafers. In addition, the through wafer vias (TWV) technology is employed to establish electrical connection from the top to the bottom through a stack of two SOI wafers, one glass wafer and one ceramic wafer. Presented methods are of general importance for various silicon-based vertically integrated devices.

Key words: MOEMS; optical micro scanner; assembly technology; vertical integration; bonding; glass micro lens

INTRODUCTION

The lack of miniaturization in conventional 3-D display and imaging systems limits their application fields and imaging capabilities. Responding to strong consumer demand for ultra-compact devices and global passion for "greener", more power-efficient products, marketplace demands are also pulling 3-D integration into the mainstream.

Firstly, micro-opto-electromechanical systems (MOEMS) technology combining MEMS and micro-optics is well suited for manipulating light. A number of different ways can be envisioned to scanning, steering, or modulating the light beams. This technology allows a large array of micromechanical light manipulators to be batch-fabricated at low cost. A number of MOEMS display and imaging products and technology demonstrators have been developed for defense, aerospace and medical markets in the form of miniature devices for projection displays, imaging devices, barcode readers, and scanners.

Secondly, the use of the 3rd dimension by employing multi-wafer integration, stacking and interconnecting several functional wafers based on disparate technologies, enables the creation of truly 3-D devices that are smaller, thinner and lighter in weight than existing devices.

The main goal of this work is to design, develop and validate experimentally a fully integrated prototype of vertically integrated micro optical scanner, suitable for a wide number of imaging systems such as the confocal microscopes on-chip or OCT probes.

One of the challenges of this project is the proposed 3-D packaging that combines several dies vertically by using multi-wafer technology. This approach offers the possibility to fabricate complex MOEMS device that consists of vertically stacked building blocks (microlens in glass, Si MEMS micro actuator, beam splitter, detector) and effective integration of heterogeneous technologies in a minimum space.

This paper describes the methods of bonding of two silicon MEMS micro actuators (X-Y scanner, Z-scanner) with other glass and ceramic components of 3-D microscanner as well as their simultaneous electrical connections by vertical through wafer technology.

DESIGN AND BONDING CONCEPT
Design
The design of the 3-D microscanner relies on vertical integration of five silicon, glass and ceramic building blocks, which are mechanically and electrically connected on the wafer level (Fig. 1). The silicon components are electrostatic X-Y and Z micro actuators, described elsewhere [1, 2]. Two glass scanning microlenses are integrated onto movable platform of these micro actuators to provide well controlled deflection of laser beam, whereas one glass microlens is monolithically integrated within bottom lid for focusing purposes. The integration of

microlenses is realized by glass frit bonding using a paste, which melts at relative low temperature (420°C).

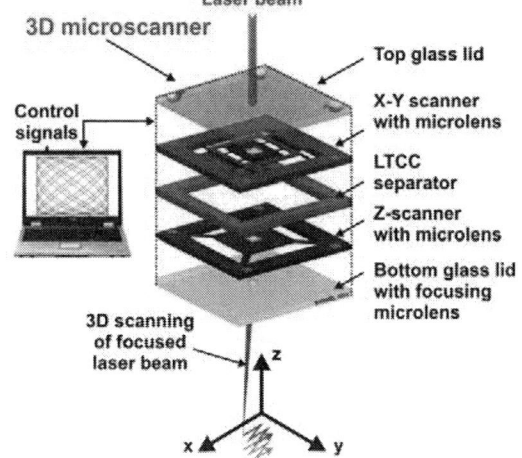

Figure 1: Conceptual drawing of the vertically integrated 3-D microscanner

Glass substrates encapsulate also the whole device from both sides while ensuring the optical transparency through the stack. Due to optical reasons, the distance between scanning microlenses has to be well controlled within tolerances of a few micrometers. This is achieved by use of mechanical spacer made of low temperature co-fired ceramic (LTCC), which can be anodically bonded to silicon due to matched coefficient of thermal expansion.

In order to drive electrostatic micro actuators, sandwiched between glass and ceramic components, the technology of through wafer vias (TWV) is applied to create electrical connections from the pads, located on the top glass lid, through the stack of two SOI wafers, one glass wafer and one ceramic wafer. The cross-sectional view of individual chip is presented in Fig. 2. The vias on different levels of the stack are connected during sequential multi-level anodic bonding forming pressure-contacts between Cr/Au pads.

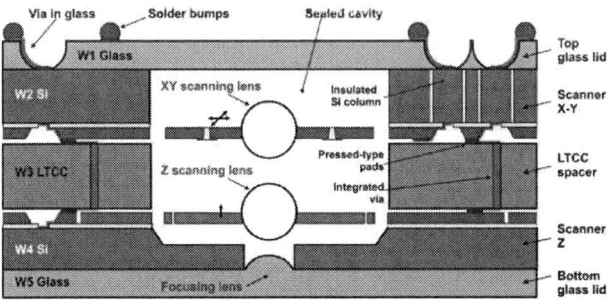

Figure 2: Cross-sectional view of 3-D microscanner

Bonding concept

In order to assemble the whole MEMS scanner seven different substrates have to be bonded. Based on this stack three different bonding technologies are necessary. These are Si-Si direct bonding to fabricate the customized SOI substrates for scanner etching, the glass-Si anodic bonding to protect the scanner against the environment and the LTCC-Si bonding to mechanically and electrically join the

X-Y scanner with the Z scanner. Furthermore the LTCC has the task to create the space between both scanners to allow free movement of the micro lenses. Because the anodic bonding can only be realized between Si and glass or Si and LTCC, respectively the bonding sequence has been chosen in way that Si and glass or LTCC are used alternatively to each other. The bonding sequence starts with fabrication of SOI wafer. These substrates are used to etch the X-Y scanner as well as the Z scanner. For this direct bonding a wet and plasma pretreatment, a vacuum bonding process and an annealing step are applied. The pretreatment consist of RCA1, RCA2 and again RCA1 followed by low pressure oxygen plasma. The bonding is done at low pressure (< 1x10-4 mbar) using a standard bonding equipment. For the annealing step the parameters are 800°C, 6h in nitrogen in a horizontal furnace. At the beginning of the bonding process a glass ball lens has to be integrated into the Z scanner. Next wafer 4 (Z-scanner) and wafer 3 (LTCC spacer) are bonded anodically, schematically shown in figure 3.

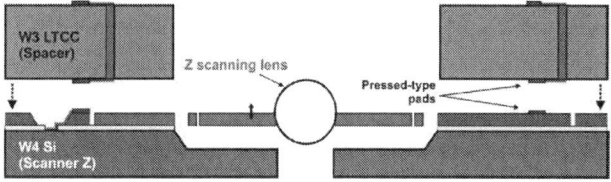

Figure 3: Anodic bonding between wafer 4 (SOI substrate) and wafer 3 (LTCC)

This first stack must be bonded to the X-Y scanner based on an anodic bonding process also. In this case the interface consists again of LTCC and Si.

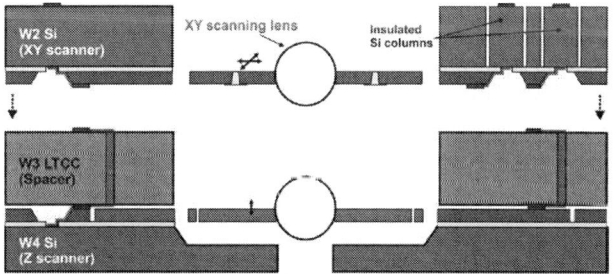

Figure 4: Anodic bonding between first stack (wafer 4 and wafer 3) and wafer 2 (X-Y scanner)

Afterwards a second lens has to be integrated into the X-Y scanner. To close the system a cap lid has to be bonded on top. The cap lid is formed from borosilicate glass and can be anodically bonded to the scanner stack. The glass wafer contains isotropic etched holes for contacting the underlying layers.

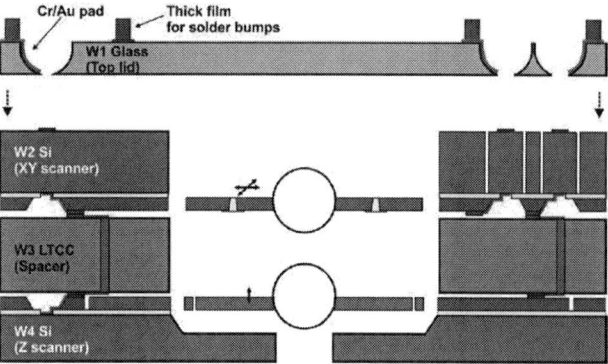

Figure 5: Anodic bonding between the scanner stack and the top glass lid.

The last step to realize a hermetic sealing a bottom lid made of glass must be bonded on the bottom side of the Z scanner. That can also be done anodically.

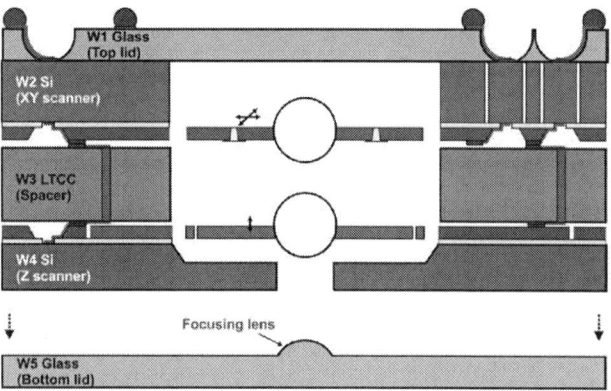

Figure 6: The whole wafer stack after bonding the last glass bottom lid.

Because all the necessary contacts (GND, driving voltage 4x - X-Y scanner and 1x - Z scanner) should be on the top side of the top lid, vertical through silicon/glass/LTCC vias have to be integrated. During the wafer bonding the contacts on each wafer must be connected to the next level. The contacts between the wafers two, three and four are formed by gold thermo compression bonding during the anodic bonding processes of the wafers stacks. The used chromium/gold layer thickness is 20/200nm thick. The contacts in the holes in the top glass lid are sputtered directly after the wafer bonding by applying again chromium/gold 20/1000 nm thick.

This bonding concept was chosen because we have to integrate three different materials. Glass is necessary for the optical system. This is a reason why other materials are necessary which are compatible to the used glass and which give us the opportunity to form the scanners as well as the spacer. The applied LTCC material has a similar CTE to the glass and Si and is also anodically bondable. Si on the other hand is well known for the structuring of the X-Y and Z scanner. Concluding the anodic bonding, as a very simple bonding technology, is especially fitted to assemble the prefabricated components into an optical system. In addition, anodic bonding does not need additional interface

materials therefore it does not increase the mechanical stress of the whole stack.

RESULTS
Anodic Bonding Si- Glass
Anodic bonding is an essential and important tool for the assembly of multi-wafer micro optical systems. In order to join different functional modules of the micro optical system together, the components are arranged in such a way that silicon and glass (or anodically bondable ceramic) substrates are placed alternatively in the assembly sequence. [3] In case of need, dummy silicon or glass (or anodically bondable ceramic) wafer is inserted between two neighboring components, so that the appearance of identical material on different sides of a bonding interface can be avoided. Sequential anodic bonding steps are then performed to assemble the modules into a functional measurement system. Before assembly, the thickness and surface finish of each module are adjusted in case the wafers can withstand such process according to requirement by using grinding and polishing processes.

For the assembly of a multi-layer Si-glass (or ceramic) stack, a simultaneous contacting and bonding process is developed. In this method, the silicon and glass (or ceramic) components are alternatively and sequentially stacked above each other. External electric field is then applied on the wafer stack to initial bonding. Because anodic bonding requires the silicon side of a bonding interface to be connected with anode and the glass (or ceramic) side with cathode, the wafers that will bring bonding interfaces that is reverse to the overall external electric field must be bonded together first, as is depicted in Fig. 7. In this case, two glass wafers and two silicon wafers are required to be bonded together. The first and the third bonding interfaces are coincide with the general external electric field and could by joined together by normal anodic bonding process. But the second interface between the Si-01 and middle glass wafer is just opposite to the external electric field and could not be bonded by the same process. Consequently, these two wafers must be bonded first, after that, this component is bonded simultaneously with the top and bottom most wafers in the second bonding step. In order to ensure strong enough bonding strength, the process temperature of the second bonding step may be increased slightly. For example, if the Si-01 and glass intermediate component is bonded at 360°C and 800V, then the second bonding step can be performed at 400°C and 600V.

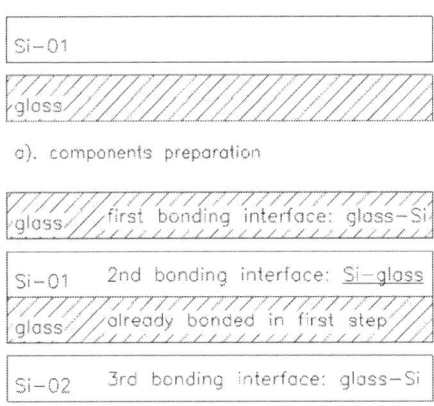

a). components preparation

b). simultaneous contact and bonding

Figure 7: A schematic view that shows how multilayer wafer stack can be bonded together step by step.

Fig. 8 and 9 shows some typical multiple-layers sandwich structures that are fabricated by this method. With such low temperature vertical stacking technology, one can integrate both glass lens and metal conductors between the bonding interfaces.

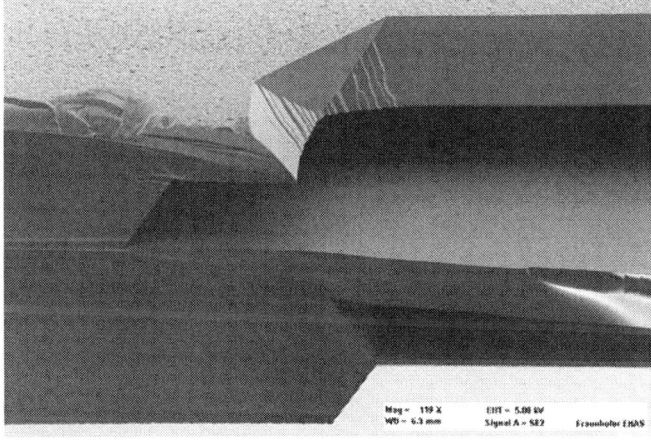

Figure 8: Si-Glass-Si triple-layer water stack realized by sequential bonding of three patterned substrates

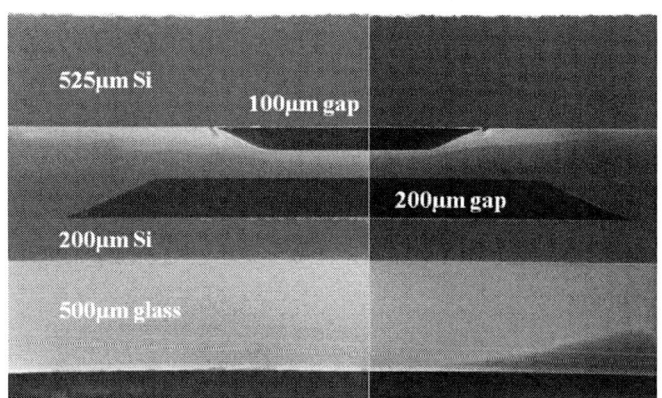

Figure 9: Si-Glass-Si-Glass 4 layer wafer stack realized by simultaneous contact and bonding of 4 patterned wafers

If a microsystem composes only three or four components, one can join the components together either step by step or simultaneously. However, if the amount of component excesses five, sequential bonding will be more inefficient. In order to form more complex micro optical system, a hierarchic assembly procedure is proposed and demonstrated. To do this, the wafers are first arranged into module groups. Suitable elementary bonding processes are then used to build the individual functional modules. For example, for the formation of X-Y and Z scanners, direct bonding is chosen because it can provide the desired assembly accuracy. For the preparation of the top electrode with optical window, anodic bonding is selected because glass wafers are used in this module. Once a functional module is ready, surface damages on them are removed by using grinding and polishing tools, so that this module is again eligible for the next bonding operation. Following that, suitable bonding methods, such as glass frit bonding, thermal compression bonding etc., are used to establish reliable connections between the modules. Finally, hermetic sealing of the complete system is realized by carrying out the last bonding step under vacuum condition. Through this hierarchic assembly method, the 7 constitutive components of the 3-D microscanner can be successfully joined together with satisfactory bonding yield-rate.

During multi wafer bonding process, the quality of each bonding step is recorded by an infrared microscope. Fig. 10 shows the results of the first and second bonding steps. Satisfactory yield-rate is achieved by using this method.

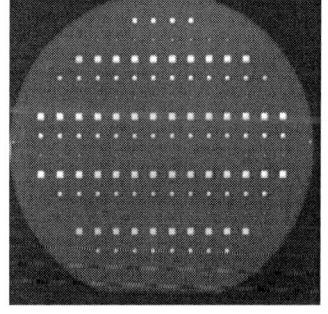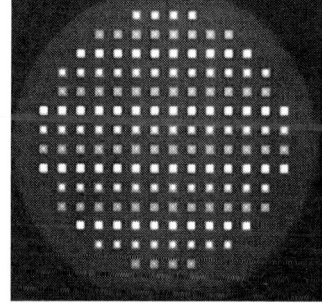

Figure 10: Bonding quality inspection by using infrared camera. Left: first bonding; right: second bonding.

After multi-wafer stacking and bonding, the quality of the overall bonding is inspected by using a scanning acoustic microscope to get an idea on bonding yield-rate. Then the bonded wafer stack is diced into 5mm size square chips. Fig. 11 shows the evaluation results. Both investigations indicate that reliable bonding has been established.

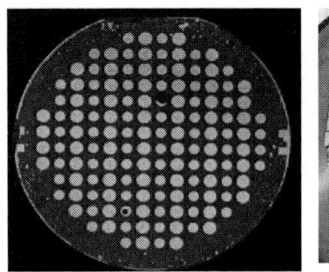 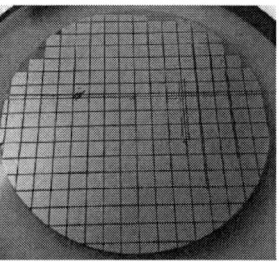

Figure 11: Bonding quality inspection. Left: scanning acoustic photo; right: dicing experiment.

Anodic Bonding Si- LTCC

LTCC is composed from glass and ceramic filler material. The CTE is determined by the particle size and particle form. The sintering temperature of this LTCC is adjusted around 800°C to 900°C, which is below the melting temperature of low resistance conductors, such as Ag and Au. These materials can be used for LTCC via filling. Cavities and interlayer connecting vias are created by pin punching the green sheet, and then the vias and conductors are screen printed, and finally laminated and fired. This is an IVH (Interstitial Via Hole) structure, which allows easy processing of the vias and unrestricted positioning of the vias. IVH also offers flexibility to the design of the internal layer patterns. Figure 12 is a SEM cross-section photo of a LTCC substrate sample.

The LTCC that is being used here has been recently developed. With a CTE of 3.4 ppm/°C it is not only matched to the CTE of silicon but its special composition also allows it to be anodically bonded to silicon just as typical borosilicate glass wafers. [4]

Figure 12: SEM cross-section photo of a LTCC multilayer substrate (6 layers) with integrated vias. [4]

For the bond strength the temperature and the voltage are of great importance. Therefore we varied the temperature and the voltage to evaluate their influence on the fracture toughness. For this purpose we applied similar bonding sequence like for borosilicate glass. The temperature has been changed between 300°C and 400°C and for voltage we used two different values, 400V and 800 V, respectively. As the result we can see that the temperature has the higher influence. At 400°C the fracture toughness is comparable to borosilicate glass. With decreasing temperature the fracture

toughness decreases faster than during the Si-glass bonding (see fig. 13). For the application fracture toughness higher than 0.6 MPa√m is acceptable. From experimental results can be concluded that at bonding conditions of 400°C and 400V this value can be achieved.

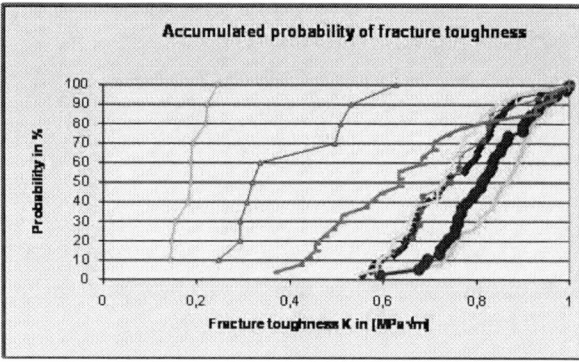

- ◆ LTCC 800V 400°C
- ✳ LTCC 400V 400°C
- △ LTCC 400V 350°C
- ✕ LTCC 400V 300°C
- ⋯ Borofloat 800V 400°C
- ● Borofloat 400V 400°C
- Borofloat 400V 350°C
- Borofloat 400V 300°C

Figure 13: Comparison of fracture toughness of LTCC/Si and Glass/Si bonds at different bonding parameters.

The bonding yield, defined by a value higher than 0.6 MPa√m, is more than 90%. The distribution of fracture toughness over a 4" wafer can be seen in figure 14.

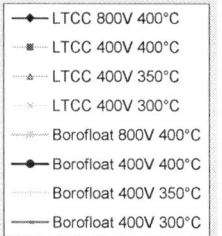

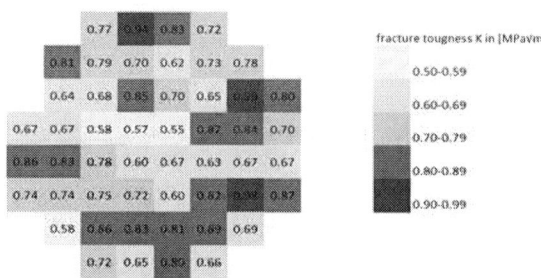

Figure 14: Distribution of fracture toughness of a 4" LTCC/Si wafer stack bonded at 400°C and 400V.

In order to prevent a short cut though the vias, integrated into the LTCC between the top electrode and the Si substrate, an additional glass spacer between the LTCC and the top electrode is used during the bonding process. The electrical contact formation between the gold layers on both substrates can be realized in parallel with the described bonding process. Designing a contact area of 100μm x 100μm a contact resistance smaller than 3.5 ohm could be realized [5]

Thermo compression bonding of gold contacts

The contacts between the wafers W2 and W4 and the ceramic wafer W3 are formed by gold thermo compression bonding during the anodic bonding processes. The used chromium/gold layer thickness is 20/200nm thick, deposited by sputtering. Pre tests for gold/gold contact formation have shown that a temperature of at least 300°C are necessary to form a low ohmic contact. The thermo compression bonding was done in a SUSS wafer bonder at 400°C for 30 minutes at a tool pressure of 7 MPa. Before annealing we measured the electrical resistance between both wafers (middle wafer and top or bottom counter wafer) showing values between 2 and 4.8 ohm with a mean value of 3.5 ohm having a contact area of 100µm x100µm. After an annealing step at 300°C for 3h the contact resistance decreased to a mean value of 3.2 ohm (see figure 15).

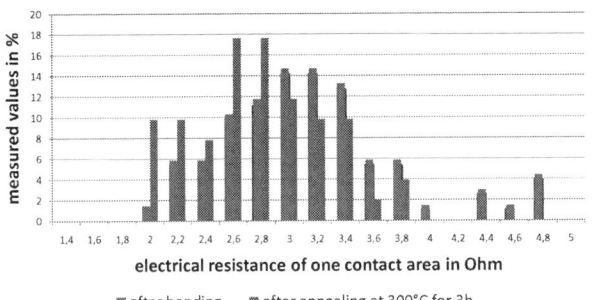

Figure 15: Contact resistance measurements after gold/gold thermo compression bonding

The reliability of the contacts has been tested by thermal shock cycles (100 cycles, -40°C to 120°C, 30min-1min-30min). To compare the results we measured the pull strength before and after shock loading. The results have shown that the shock loading has no significant influence to the fracture force.

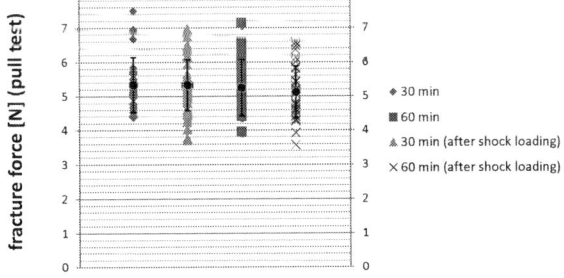

Figure 16: Distribution of fracture force (pull test) before and after shock loading in dependence of bonding time (30 min and 60 min)

The results show that there is no obvious impact of bond time and shock loading to the fracture force of the gold/gold thermo compression contact.

Micro lens integration
The bonding of discrete micro optical components on micro actuators at wafer level is a big challenge. The microlenses have to be individually bonded to prefabricated micro actuator wafer in a reliable way without causing stress-

induced deformation of fragile structures of micro actuators while keeping their optical performances (e.g. focal length). If intermediate bonding material is concerned, its application on the bonding area has to be well controlled ensuring required relative distance of one scanning microlens to the other. Moreover, the full compatibility of bonding material with anodic bonding process temperature is necessary.

In this work, a low melting-point glass frit bonding process is proposed and successfully demonstrated. First, several droplets of glass frit paste (FX-11-0366, FERRO) is applied on the side walls of each lens holder that is made on a silicon wafer. Glass ball microlens with diameter of 300µm ± 1µm (N-BK7 glass, Tg= 557°C, Edmund Optics) is then mounted in the microlens holder and is stuck by the glass frit. The lens-holder unit is then annealed at 120°C for 15min to drive out organic solvent, and at 420°C the glass frit is melted to join the lens and the silicon carrier. Fig. 17a shows a SEM photo of the suspended 15-µm-thin silicon platform with bonded microlens.

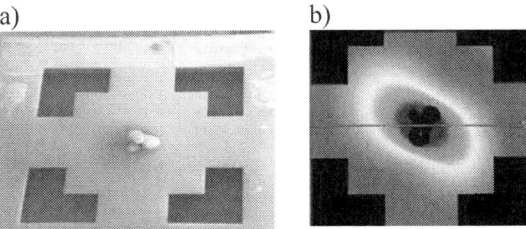

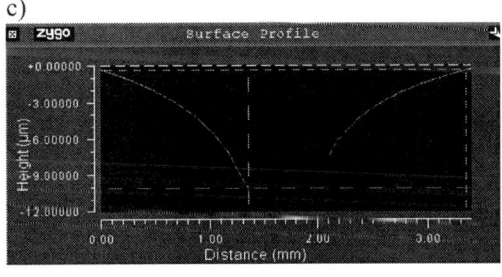

Figure 17: SEM photo of glass ball lens mounted on the movable platform and the static deformation of the Z-scanning platform caused by thermo-mechanical stress.

After annealing, the reliability of this integration method is inspected by ultrasonic agitation in water, by centrifugal experiment and by brutal force separation. All these tests validate the effectiveness of the bonding process. The connection between the microlens and the silicon carrier is so strong that even with a sharp tweezer, it is difficult to separate the mounted lens forcefully from the holder, which indicates that this lens bonding method is reliable enough for future operation in labor environment.

Fig. 17b presents the static deformation of the suspended platform as a result of thermo-mechanical stress on the glass frit / silicon interface caused by slight mismatch in the thermal expansion coefficients of silicon and glass frit. The measured deformation is asymmetrical due to unsymmetrical application of glass frit material. As shown in Fig. 17c, the maximum deformation at the center of the 15μm thick membrane is about 9μm. This initial deformation can possibly be compensated by using counter electrostatic force.

ACKNOWLEDGEMENTS

The work was performed in the frame of DWST-DIS, founded by Project Inter Carnot-Fraunhofer (PICF 2010).

REFERENCES

[1] Laszczyk K. et al, "A two directional electrostatic comb-drive X-Y microstage for MOEMS applications", Sensors and Actuators, A Physical, Vol.163, Issue 1, (2010), 255-265

[2] S. Bargiel et al, "Electrostatically driven optical Z-axis scanner with thermally bonded glass microlens", Procedia Engineering (Eurosensors XXIV, September 5-8, 2010, Linz, Austria), Vol. 5, pp. 762-765

[3] Jia C, et al, A Hierarchic Bonding Procedure for the Assembly of Micro Confocal Microscope, Proc. 3rd IEEE International Workshop on Low Temperature Bonding for 3D Integration, Tokyo, May 22-23, 2012

[4] Mori M., Okada. A., Fukushi H., Tanaka S., Esashi M., "Hermetic seal using anodically-bondable LTCC substrates", 23th Conference on Electronics Packaging Electronics, Yokohama, 2009 Mar. 11-13, pp. 51-52, proceedings.

[5] Froemel, J.; Haubold, M.; Wiemer, M.; Gessner, T.: Thermo compression bonding with gold interfaces. Smart System Integration 2009, Brussels (Belgium), 2009 Mar 10-11; proceedings (ISBN978-3-89838-616-6)

MEMS HERMETICITY AND RELIABLILITY TESTING TODAY

Michael Shillinger
Innovative Micro Technology
Santa Barbara, California
mjs@imtmems.com

ABSTRACT

Successful wafer-level packaged MEMS devices need hermeticity and reliability testing to prevent failure from particles, contamination, moisture, among other factors. Unfortunately, legacy hermeticity test protocols fail when applied to the measurement of micron-sized volume of cavities, where MEMS devices typically reside. Recently, higher resolution and more accurate test methods have evolved in market. When combined with standard reliability tests, MEMS manufacturers can ensure they ship high quality products.

Reliability is a critical parameter to take into consideration; if a product does not deliver the design performance during its expected lifetime it cannot be commercialized. Reliability in MEMS devices is particularly important because failure can be time consuming and costly.

Assurance of reliability involves intelligent design, reliable test strategies, and accurate testing techniques. Selecting the right test(s) depends on the MEMS device and how the device needs to perform in the field. Engineers cannot design reliable MEMS without first understanding the factors that can lead to failure and how to test for these factors.

INTRODUCTION

MEMS encompass many different types of devices and applications as illustrated in Table 1. Because of the diversity of the MEMS market, certain performance criteria that are very important in one MEMS device may not matter at all in another.

Table 1: Classification of MEMS Devices

Characteristics	Examples	Hermeticity Tests	Reliability Tests
No moving parts	Accelerometers Pressure sensors Planar light circuts Free space optics	Membrane deflection, Q measurement; CHLD; Radio isotope	Shock and vibration Burn-in, ESD sensitivity, Thermal cycling, thermal shock, electrial stress
Moving parts with no impacting surfaces	Gyros Resonators Fabry Perot filters WSS	Same as Class A	Class A tests performed and charge accumulation
Moving parts and impacting surfaces	Relays Valvues Pumps Cell sorting chips RF switches DC DSL cwitchoc	Samo ao Claoo A	Class B tests performed and stiction testing

This paper will describe multiple different hermeticty and reliability tests available today. The hermeticy tests covered are: membrane deflection, radio isotope, cumulative helium leak detection (CHLD), electrical resistance, Q measurement, and pressurized steam testing. The reliability tests that will be discussed are: shock and vibration, lid torque strength, burn-in, electrostatic discharge and electrical stress sensitivity testing, thermal cycling, thermal shock, trapped moisture detection, thin film corrosion, storage life, highly accelerated life testing (HALT), charged accumulation, and stiction testing.

HERMETICITY TESTING

Membrane deflection is accurate for finding both fine and gross leaks. In the fabrication of the lid wafer, cavities are etched which form a very thin membrane on the top surface of the lid wafer itself. Typically these membranes are on the order of 30 microns thick. After the lid wafer is bonded to the device wafer, the membrane deflection is measured with an optical tester. With vertical resolution in angstroms, a small deflection of the lid membrane indicates pressure or vacuum in the package. The membrane in the lid is convex if package is pressurized and concave if the package is in vacuum. Figure 1 shows a Wyko interferometer image of a deflected lid membrane that indicates hermeticity.

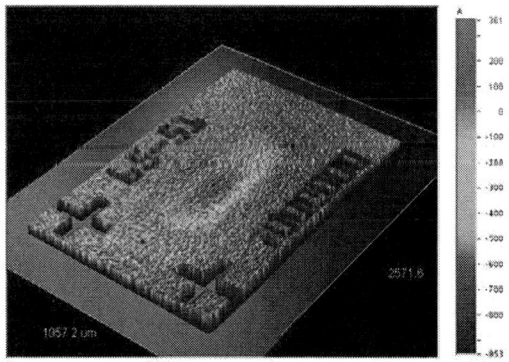

Figure 1: Wyko Image of a Deflected Lid

This fast and automated wafer-level test is non-destructive, and currently in use at IMT for production screening. When applied to hermeticity testing, any small deflection on the lid membrane will indicate hermeticity.

Radio isotope or radioactive decay testing is highly sensitive. Packaged chips are placed in a chamber which contains pressurized, radioactive Krypton-85, and dry nitrogen. After a period of time, the chips are removed from the chamber and measured for Kr-85 atoms using a scintillator. Any number of atomic particles present inside the package would emit alpha radiation, indicating that the package is not hermetic. While this methodology will detect both fine and gross leaks, it is more costly because the test is performed at die level. Although this is extremely precise test of hermeticity, it is not widely used because the use of radioactive material requires an Atomic Energy License.

Cumulative Helium Leak Detection (CHLD) is the only hermetic device testing method currently approved by the military, Mil-STD-750 Method 1071. This die level test is more sensitive than traditional fine and gross leak testing used for vacuum device packaging by four orders of magnitude. The CHLD method can detect leaks less than 1×10^{-14} ATM-cc/sec. Besides helium, this test can also detect krypton, fluorocarbons, and argon, utilizing an integrated mass spectrometer. Testing can be performed by using Inficon's Pernicka Series Systems (Figure 2 courtesy of Inficon). This system uses a specially designed cryo-pump, which accumulates all of the helium that escapes through a leak path during the test. A change in the helium signal as a ratio to a standard is observed, which determines if the package is hermetic or not.

Figure 2: Inficon's Pernicka Series Tool

Another effective way to confirm hermeticity is to employ integrated thermistors to record Resistance-hot (R-hot) vs. resistance-cold (R-cold) measurements. (Figure 3). After completion of wafer bonding and pad reveal, Innovative Micro Technology (IMT), a MEMS foundry located in Santa Barbara, California, uses an automatic probe station to pass current through the thermistors to measure resistance.

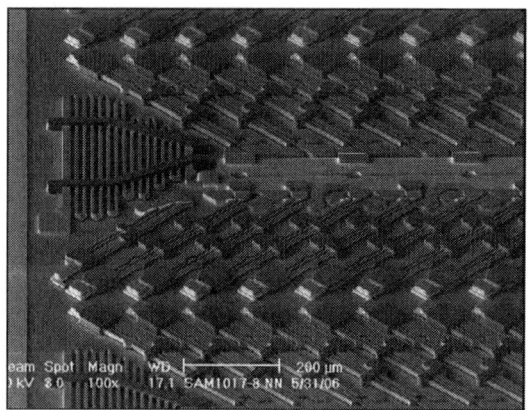

Figure 3: MEMS Switch with on Board Thermistor

Whether the bonding is with vacuum or partial pressure, the change in resistance compared to the baseline resistance measurement of the thermistor indicates hermeticity. Screening for high resistance would indicate that high vacuum is present. Because this is a wafer-level electrical test that allows 100% testing of every die, wafer maps can be created to identify good and bad dies (Figure 4). This simplifies sorting immensely. The low cost and speed of this automated testing is well-suited for production.

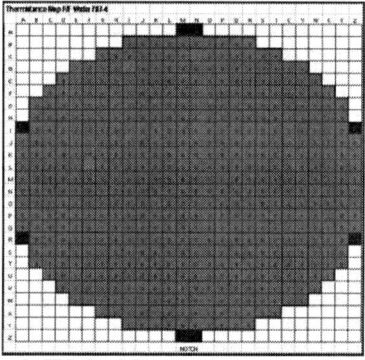

Figure 4: Wafer Map Showing 3 Failed Dies For Hermeticity

Hermeticity can also be quantified using mechanical Q measurements of MEMS resonators. Hermetic bonding achieves vacuum levels on the order of 1 mTorr inside the wafer level packaging. At that point, the viscous drag of gas on the resonator device inside the package is eliminated. The Q is strongly dependent on the pressure; therefore, high vacuum results in higher Q. Excitation of the resonance can be driven by external transducers or using integrated electrostatic forces. Sensing of the resonance amplitude can also be done externally (Doppler Vibrometry) or using integrated piezo-resistive devices. This test is ideally matched for measuring resonators, since the devices require no additional test structures or electrical I/Os. This test performed at wafer level, making it production friendly.

Lastly, pressurized steam is also used for hermeticity testing. Packaged devices are placed in an autoclave chamber. The chamber is then filled with high temperature pressurized steam (130 °C, 2.7 ATM, and 100% RH) for a prescribed time. If the bonding is not hermetic, the pressurized steam will penetrate through the small crevasses in the bondlines. Devices are, then de-lidded and optically inspected for moisture or corrosion inside the packaged device. This testing would be performed during development and for audit testing in volume production.

RELIABILITY TESTING
Shock and vibration testing pushes the device until failure (Figure 5). Afterwards, failure analysis procedures are employed to determine the root cause.

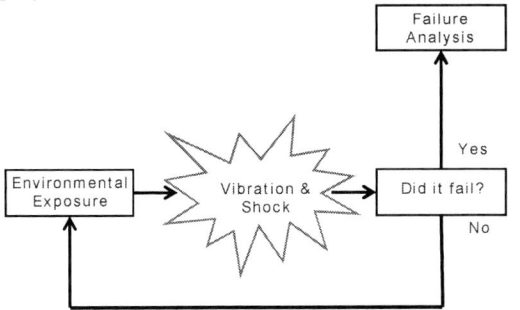

Figure 5: Shock and Vibration Testing Steps

In shock testing, MEMS devices are attached to a durable metal fixture which is then attached to a movable shock table. The shock table has specific height settings for producing certain G-level shocks. The various height settings are used to produce increasing shock loads until device failure. Vibration tables are also employed and dies are subjected to a variety of amplitudes and frequencies. Devices are visually inspected and/or electronically actuated to determine when device failure occurs.

The lid bond strength of the wafer-level packaged (WLP) MEMS device needs to be characterized during development and audited during production. In this test, torsional forces are applied to the lid of the device through a specially machined wrench driven by a torque watch. The force at which the bond fails is recorded and compared to the historical data. If the lid fails are lower torque than the specification, failure analysis is performed. Figure 6 is a histogram of audited lid torque strength from a production MEMS program at IMT.

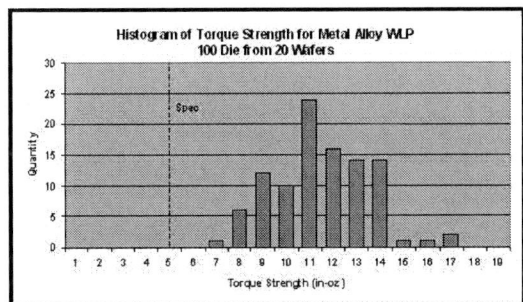

Figure 6: Lid Torque Strength Histogram

Burn in testing is necessary to separate parts that exhibit infantile failures from the general population of product being shipped. In this test the devices are active and

placed in an oven (80 – 120°C) for the duration of 8 to 24 hours. The device must function electrically during and after test to qualify for shipment. Infantile failure data is collected in a bathtub plot to ensure that the infantile parts are not shipped (Figure 7).

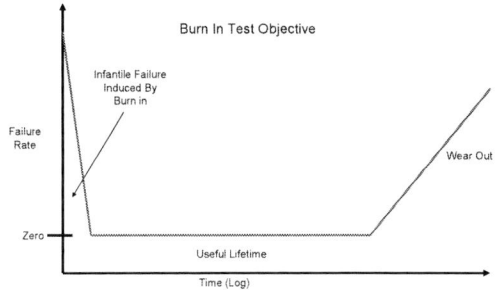

Figure 7: Bathtub Plot for Burn In Testing

Electrostatic discharge (ESD) and electrical overstress (EOS) damage in some MEMS devices have been identified as new failure modes. Previously, these failure modes have been mischaracterized as stiction, attributing to the mechanical nature and functionality of these systems. Typical ESD testing is performed using an IMCS Model 2500 Tester with improved Human Body Model (HBM) waveform characteristics. It has now been determined that some types of MEMS devices can be damaged at 250 volts, which is very low considering the HBM. HBM testing is dictated by JEDEC standard 22-A-114-B. EOS testing is typically performed using a Pragmatic Instruments Model 2414A Waveform Generator, the voltage is increased in 10 Volt increments until the device fails. The voltage level where failure occurs is judged to be acceptable or non-acceptable depending on the device type. If the voltage level is unacceptable, failure analysis is conducted which can ultimately lead to device re-design. The Machine Model test circuit consists of charging up a 200 pF capacitor to a certain voltage then discharging this capacitor directly into the device being tested through a 500 nH inductor with no series resistor. If excessive device failure occurs for either HBM or EOS reasons the work environment and/or machines need to be characterized and ESD sources need to be systemically eliminated.

Any MEMS device that has movable parts needs to go through thermal cycling and thermal shock testing. Thermal shock testing consists of rapidly exposing the device to one temperature cycle between -60°C and 150°C, and then cooled to room temperature (Figure 8a).

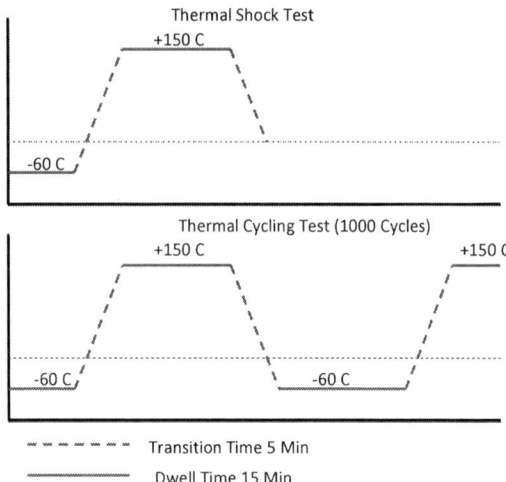

Transition Time 5 Min
Dwell Time 15 Min

Figure 8a & b: Thermal Shock and Thermal

Cycling Test Parameters

Extensive inspection and electrical measurements are conducted to test for changes from benchmark values. Typically thermal shock testing is employed to initially understand the potential device damage from mismatched material TECs. If changes in device performance are deemed unacceptable after this test failure analysis is conducted, then the root cause determinations are made using 8D analysis tools. If the changes are acceptable, thermal cycling tests are performed, where the devices are subjected to cycles from -60°C to 150°C, ramping in 5 minutes with 15 minute dwell times at the high temperature (Figure 8b). This cycle is repeated up to 1000 times, measurements are usually taken after 100, 200, 400, 700, and 1000 cycles. Blue-M thermal shock and cycling ovens are used (Model WAP-109-D) with full data capture. These tests are compliant to MIL-STD-202, Method 107.

If a MEMS device contains metal films, trapped moisture inside of the package could cause destructive corrosion. To prevent this, trapped moisture testing and thin film corrosion testing are required. The specifics of these tests need to be considered depending on the metal films. For instance a different methodology would be used to ensure highly corrosive metals, such as copper or silver, are intact rather than more inert metals (e.g., gold or platinum). These tests are typically performed in a cyclic corrosion chamber with contents varying from salt fog, high humidity, or solution spray to full immersion. After testing, parts are characterized using microscopes (optical and SEM). Device performance is compared before and after these tests to ensure no change.

Storage life testing is necessary to determine the effect of time and temperature for thermally activated failure mechanisms in MEMS devices. These tests are performed in high temperature storage chamber, typically for 1000 hours, without electrical stress being applied. Conditions varying with device type and range from 85°C

to 300°C. The visual condition and device performance are characterized based on JEDEC Standard JESD22-A103, performance prior to and after the test are compared.

Highly Accelerated Life Testing (HALT) is used to identify potential design problems not typically encountered. HALT uses the step stress method to define the operating/destruction limitations of a MEMS device. The complete HALT program is a much faster and more economical test approach to determine design flaws. HALT is comprised of five individual step stress tests. The first three tests define the temperature and vibration parameters. The fourth test is a rapid thermal cycling from -100°C to 200°C. The fifth test is a combination of rapid thermal cycling as well as vibration step stress. This test is the most practical test used to validate a product that can withstand such environmental exposure. One type of chamber used for HALT testing is shown in Figure 9.

Figure 9: HALT Hanse Chamber

Charge accumulation testing in a device, such as a MEMS switch, is the characterization of pull in voltage over actuation cycles. Devices which accumulate charge have ever increasing pull in voltages. If pull in voltages exceed the level that the drive-ASIC can produce, the switch will not function. Essentially this test is looking for degradation as a function of charge accumulation in dielectric films. It has been reported for MEMS switches, for every five volt reduction in pull in voltage a lifetime increase of a factor of ten is achievable. Figure 10 indicates an increase in pull-in voltage as a function of cycles therefore charge accumulation, while Figure 11 indicates no increase in pull-in voltage and therefore no charge accumulation in the device which results in extended lifetime.

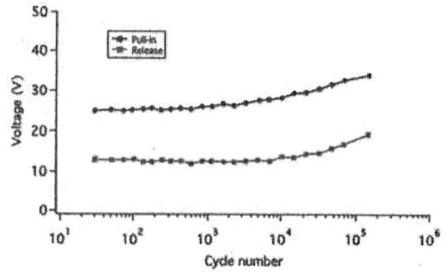

Figure 10: Graph Showing Charge Accumulation

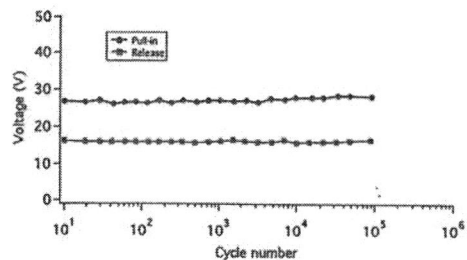

Figure 11: Graph Showing No Charge Accumulation

Stiction is a phenomenon where the moveable elements in a MEMS device come in contact with stationary elements and are held in place permanently due to Van der Waals force. Because many MEMS devices, such as accelerometers, require moving parts to function, stiction can be destructive. The dynamic event causing the elements to touch is generally applied shock. The higher the G-load of the applied shock, the higher frequency stiction events can occur. There is a relatively simple test set up to characterize what the G-level limits are (Figure 12).

MEMS Stiction Test Setup

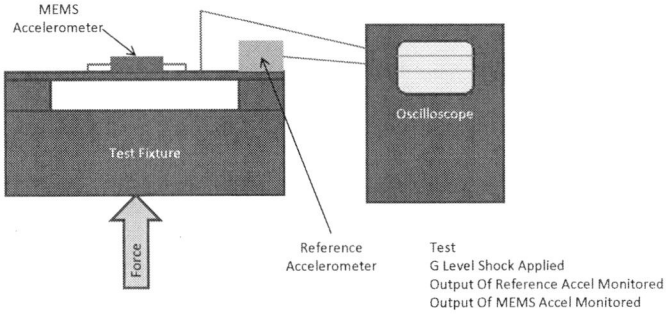

Figure 12: Stiction Test Set Up

In this tester ever increasing G-loads are applied to determine the limit for a particular design. Once this characterization is complete, a determination is made relative to the acceptability of the critical shock value. If the failure shock value is unacceptable device redesign is required. This can be accomplished by changing the geometry of the suspension members. A 2X safety factor is generally the goal, meaning the actual shock limit of the device is two times more than the documented specification.

CONCLUSION
MEMS technology is relatively new and highly diversified; therefore, there are no design standards to follow. Unlike integrated circuits, most MEMS devices need protection from the ambient environment; thus, WLP is critical for successful commercialization. To that end, many tests have been developed to ensure that these devices are hermetically packaged. Additionally,

reliability is not guaranteed at the completion of the design and fabrication process; therefore, reliability testing is necessary to ensure that the devices will function in the real world over their specified lifetime.. Some of the traditional reliability tests developed for the semiconductor industry also apply to MEMS technology, but many do not. Specialized reliability testing has been developed specifically for MEMS devices, such as stiction testing and hermeticity testing. These testing methods are crucial to successful deployment of MEMS technology.

ACKNOWLEDGEMENTS
The author would like to thank the following individuals for their contribution to this paper: Aria Birang, Ted Chi, and Craig Trautman.

RELIABILITY OF TSV AND WAFER-LEVEL BONDING FOR A 3D INTEGRABLE SOI BASED MEMS APPLICATION

Torleif André Tollefsen[1,2,*], Maaike M. Visser Taklo[1]
[1]SINTEF ICT, Instrumentation
Oslo, Norway
[2]Vestfold University College, Dep. of Micro and Nano Systems Technology
Borre, Norway
*Torleif.Tollefsen@sintef.no

Thor Bakke[3], Nicolas Lietaer[3]
[3]SINTEF ICT, Microsystems and Nanotechnology
Oslo, Norway

Per Dalsjø[4], Jakob Gakkestad[4]
[4]Norwegian Defence Research Establishment (FFI)
Kjeller, Norway

ABSTRACT

The reliability of a 3D integrable miniaturized wafer-level encapsulated MEMS acceleration switch has been studied. The through silicon vias (TSV) and the benzocyclobutene (BCB) bonding process used for the encapsulation were examined regarding the survival of various environmental stress tests. Sample groups using different metallization schemes (NiCr or TiW as adhesion layer/ diffusion barrier for the Au metallization) and different parameters for the bonding (temperature and environment) were fabricated. Samples were exposed to low temperature storage (LTS) at -55 °C, high temperature storage (HTS) at +230 °C and thermal shock cycling (TSC) between -55 and 125 °C. Four-point resistance measurements of daisy chain structures, die shear tests, and cross-sectioning were done to characterize the TSVs and the bond properties associated with the different sample groups and tests.

The initial bond strength was high for all sample groups (above 24 MPa). LTS and TSC reduced the bond strength of NiCr/Au metallized samples (more than 50% after LTS), while it remained unchanged for TiW/Au metallized samples. Lowering the bonding temperature from 250 °C to 200 °C increased the bond strength by 10 to 20 %, assumedly due to less residual stress. The electrical resistance of the TSVs with NiCr/Au metallization increased after HTS, while it remained low both after HTS and TSC for the TiW/Au samples. The demonstrated TSV technology with TiW/Au metallization and encapsulation by BCB bonding proves to be well suited for 3D MEMS packaging for harsh environment applications.

Keywords: BCB, TSV, MEMS switch, harsh environment, wafer-level 3D integration.

INTRODUCTION

Wafer-level 3D integration – 3D stacking prior to singulation of wafers into individual chips – has become an increasingly active research topic [1]. It is an appreciably complex technology, but it has significant potential advantages, including: high density of interconnects and low electrical parasitics (ICs), lowered high-volume manufacturing cost and small form factor (number of devices or functionalities per unit chip area) [1]. By combining wafer-level 3D integration with wafer-level encapsulation, another inherent advantage can be obtained; protection of fragile components, e.g. micro-electro mechanical systems (MEMS), during the singulation process [2].

There exist numerous wafer-level bonding techniques that can be used for wafer-level encapsulation of MEMS. Direct bonding [3], anodic bonding [4], solder bonding [5], eutectic bonding [6], thermo-compression bonding [7], low-temperature melting glass bonding [8] and adhesive wafer bonding [9] are among the techniques that are used. Adhesive wafer bonding, which is non-hermetic, offers a robust and low-cost process [2]. Among the different alternatives for adhesive wafer bonding, benzocyclobutene (BCB) polymers, have attracted substantial attention because of their desirable properties [10]. BCBs are thermosetting polymers which can be used for low temperature bonding. They have a high tolerance for surface topography and strong, high yield wafer bonds can be achieved [10].

Vertical through silicon vias (TSVs) are beneficial for 3D integrated MEMS devices. TSVs allow to connect electrical structures enclosed in a MEMS cavity to the outside [11, 12], and at the same time a significant reduction in form factor can be obtained as compared to traditional lateral interconnection methods. Additionally, it opens up for flip-chip bonding instead of wire bonding, which contributes to further miniaturization.

An acceleration switch, a device that closes (or opens) a circuit above a certain acceleration threshold, can be used in safety and arming devices (SADs) in smart ammunition fuzes [2]. SADs ensure that the ammunition is safe during storage, handling and launching, and reliably arm the ammunition when required. Due to the limited volume in an ammunition fuze, a MEMS acceleration switch utilizing wafer-level 3D integration, TSVs and BCB bonding is an attractive solution.

In this work the reliability of a 3D integrable miniaturized wafer-level encapsulated MEMS acceleration switch has been studied. The acceleration switch shall be used as a SAD in ammunition subjected to a setback acceleration pulse with an amplitude exceeding 60 000 g (1 g = 9.81 m/s^2) and a centripetal acceleration that increases radially to 9000 g/mm. Importantly, the ammunition must withstand severe climatic conditions, e.g. a temperature range from -54 to 71 °C [13]. Sample groups using different metallization schemes (NiCr or TiW as adhesion layer/ diffusion barrier for Au metallization) and different parameters for the BCB bonding (temperature and environment) have been fabricated and their reliability has been tested and characterized in this study.

EXPERIMENTAL PROCEDURE
Fabrication
100 mm diameter SOI wafers were used as substrate material. The substrate had a 40 µm thick silicon device layer, separated from a 300 µm thick handle wafer by a 2 µm buried oxide (BOX) layer (a schematic cross-section of a sample is shown in Figure 1). 1.2 µm aluminium was sputtered on both sides of the wafers. The aluminium on the device layer side was patterned to serve as a hard mask for the via etch, while the aluminium on the backside served as an etch stop layer. Rectangular via holes, 7 x 70 µm^2, were used. This gave an TSV aspect ratio close to 50:1 (in one direction). The width of the via holes was kept relatively narrow (7 µm) which made the subsequent via filling faster. The TSVs were etched consecutively through all layers – device, BOX and substrate – using the aluminium hard mask. The etches were performed on an Alcatel AMS200 SE I-Productivity etch tool. The silicon layer etches (device and substrate) were performed with an optimized Bosch deep reactive ion etching (DRIE) process [14, 15]. The extremely high aspect ratio of the TSVs was only possible by changing the etching conditions at different stages during the etch [16]. Upon the completion of the via hole etch, the aluminium was stripped from the wafers by wet etching. A 1 µm thermal oxide was then grown to isolate the TSVs from the bulk silicon material. The vias were filled using four consecutive chemical vapor depositions of 1 µm undoped polysilicon. Between each deposition step the polysilicon was heavily doped with phosphorus using POCl$_3$ gas phase

doping. At this point the excess polysilicon deposited on the wafer surfaces was removed by reactive ion etching (RIE).

After completion of the TSV processing, the fabrication of the MEMS structures was started. The thermal oxide was stripped from the device layer side by wet etching, and a 100 nm polysilicon layer was deposited by chemical vapor deposition. This was done in order to protect the oxidized TSV sidewalls from etching during the subsequent release etch of the MEMS structures. Since the release etch was done using a single sided process, the polysilicon could at this point be removed from the backside of the wafers by RIE. A 2.6 µm thick HiPR6517 photoresist was then spin-coated and patterned on the device layer side to define the MEMS structures. This resist protected the TSVs from the subsequent device layer etch. The device layer etch was performed with a Bosch DRIE process, with the buried oxide acting as an etch stop layer. After removal of resist and polymer residues (from the DRIE etch) using O$_2$ plasma stripping, the sacrificial buried oxide below the movable silicon structures was etched away using a 1 hr HF vapor release etch at 35 °C. The thin polysilicon layer protecting the TSV sidewalls during the release etch was then removed by a short RIE etch. Two different combinations of backside and device layer side metallization were then sputtered: (i) 150 nm NiCr / 250 nm Au on the backside and 12 nm NiCr / 500 nm Au on the device layer side. (ii) 50 nm TiW / 300 nm Au on the backside and 12 nm TiW / 500 nm Au on the device layer side.

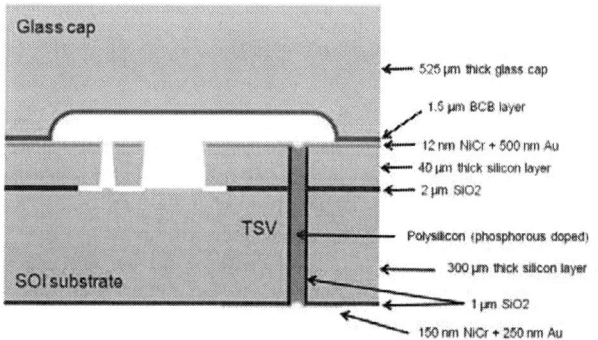

Figure 1. Schematic cross-section of tested samples with NiCr/Au metallization.

Glass cap wafers were then processed. 20 nm TiW and 1 µm Au was sputtered on both sides of the wafers followed by patterning and wet etching using KI/I$_2$ and H$_2$O$_2$. This provided an etch mask for the subsequent cavity etching. The cavity etching was performed at room temperature for 3 minutes in 49 % HF, producing cavity depths of 20 µm. The TiW / Au etch mask was then removed. The glass wafers were spray-coated on the side with cavities with 1.5 µm Cyclotene 3022-35 (BCB) using an airbrush pressurized with dry N$_2$. The wafers were then baked on a hotplate at 110 °C for 90 seconds in order to remove the solvents. A Suss BA6 bond aligner was then used to align the glass wafers with cavities to the MEMS wafers, thereby encapsulating the fragile MEMS structures. The actual bonding was performed in a Suss SB6 thermo-compression bonder. The bonding chamber was evacuated and the wafers were pre-heated to 150 °C for 5 minutes for dehydration. The wafers were brought together with a pressure of 300 mbar, heated to the final bonding temperature, and kept there for 60 minutes. Different bonding schemes were used to investigate the effect of temperature and environment on the bond integrity. An overview of the used process parameters is given in Table 1. The bonded wafers were cooled down prior to the removal from the bonder. After bonding, the bond pads on the backside of the device wafer were defined by patterning the NiCr/Au or TiW/Au layer using a 6 μm thick AZ4562 photoresist and wet etching. Finally, the wafers were diced, producing the completed MEMS acceleration switches.

Reliability testing

In order to accelerate the potential failure mechanisms of the investigated acceleration switches, the samples were subjected to various thermal treatments. An overview of the different testes groups can be found in Table 1. The most critical region for the TSVs in the investigated samples is at the edges of the polysilicon part on the device layer side. The actual shape is hard to predict as it depends on the scalloping of the DRIE etch and the wet etch of oxide. The shape may vary across the wafer and from wafer to wafer depending on the amount of over-etch experienced during etch back of the polysilicon. Failures caused by both stress and diffusion are likely to appear in this region.

Table 1. An overview of the process parameters that were used and the reliability tests that were performed. HTS, LTS and TSC are abbreviations for high temperature storage, low temperature storage, and thermal shock cycling.

Wafer Metallization (front and backside)	NiCr / Au	NiCr / Au	TiW / Au	TiW / Au
Bond temp (°C)	250	250	250	200
Atm (mbar)	500	Vacuum	500	500
# as bonded samples	13	13	13	10
# HTS samples (230 °C, 1000 hours)	9	9	9	-
# LTS samples (-55 °C, 1000 hours)	10	10	10	10
# TSC samples (-55 °C /+125 °C) 10→1000 cycles	29→16	29→16	29→16	20→10
# HTS+TSC samples	3	3	3	-

Selected sample groups were exposed to high temperature storage (HTS) to investigate the reliability of the TSVs. A test temperature significantly above the maximum operation temperature of 71 °C was selected in order to accelerate possible diffusion processes. The devices were stored for 1000 hours at 230 °C in a Binder laboratory oven.

The glass cap was bonded to the device using BCB at a temperature of 200 or 250 °C (see Table 1). Since the BCB glass transition temperature is above 350 °C, the bond is expected to retain its strength after HTS. The largest stresses in the bond layer are anticipated at the minimum specified

operation temperature of -55 °C. Low temperature storage (LTS) was therefore performed for selected sample groups. The samples were stored for 1000 hours at -55 °C in a Heraeus HT 7012S2 chamber.

To simulate large stresses in the device, thermal shock cycling (TSC) was performed according to MIL-STD-883G, method 1010.8, test condition B, i.e. a temperature range of -55 to +125 °C, 4 cycles per hour with 6 seconds between temperature extremes. The samples were evaluated after 10 and 1000 cycles. The TSC was performed in a Heraeus HT 7012S2 thermal cycling chamber.

Selected TSV samples were exposed to HTS followed by TSC.

Characterization

Certain samples included electrical test structures like daisy chains and Kelvin structures, making it possible to measure the electrical resistance of the TSVs. Four point electrical resistance measurements were performed on as-bonded, HTS, TSC, and HTS + TSC samples in order to investigate potential degradation of the TSVs and their metallization.

The die shear strength of as-bonded, LTS, and TSC samples was measured on a Dage 2400A shear tester with 50 kgf load cartridge in order to investigate potential degradation of the BCB bond and / or metallization. Eight samples from each group were tested using a test speed of 17 μm/s and a test height of 175 μm.

Cross-sectioning and SEM analysis were performed on as-bonded, HTS and TSC samples to see if there were changes in the TSVs or metallization after thermal exposure. The sample preparation was done on a Vion plasma FIB (focused ion beam).

All shear tested samples were inspected in an Olympus BX60 optical microscope in order to investigate the fracture surfaces. The fractures were classified as adhesive (e.g. at the interface between Au and BCB), cohesive (e.g. inside the BCB bulk), or as a combination of the above.

EXPERIMENTAL RESULTS
TSV metallization

The electrical resistance per TSV measured on daisy chains for samples with NiCr/Au and TiW/Au metallization are shown for as-bonded, HTS, TSC, and HTS+TSC samples in Figure 2 and Figure 3, respectively. The resistance remained low both after HTS, TSC and HTS+TSC for the samples with TiW/Au metallization. For samples with NiCr/Au metallization the resistance remained low after TSC, whereas there was a considerable increase after HTS.

In Figure 7 SEM cross-sections of an as-bonded and a HTS sample with NiCr/Au metallization are shown. From these images it can be seen that gaps appeared in the Au layer after HTS, probably resulting in the increased / (open) electrical resistance for the samples with NiCr/Au metallization.

SEM images of an as-bonded and a HTS sample with TiW/Au metallization are shown in Figure 8. These images show that for this metallization, the Au layer was continuous also after HTS, even though it appeared somewhat thinner.

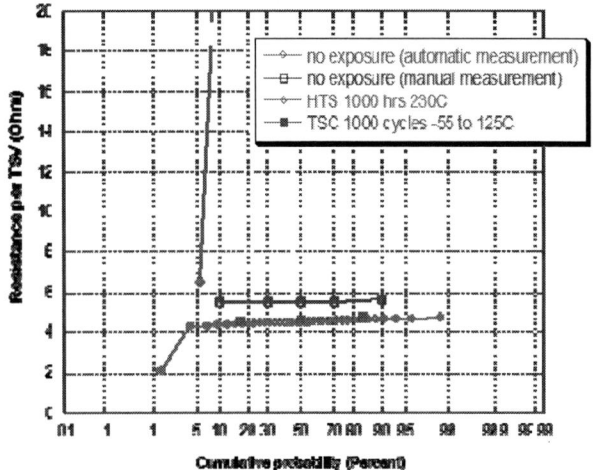

Figure 2. The electrical resistance per TSV measured on daisy chains for samples with NiCr/Au metallization for as-bonded, HTS, and TSC samples.

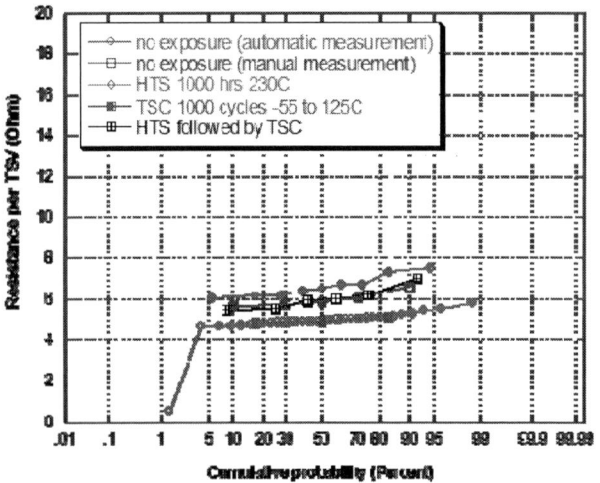

Figure 3. The electrical resistance per TSV measured on daisy chains for samples with TiW/Au metallization for as-bonded, HTS, TSC and HTS + TSC samples.

BCB encapsulation
The average die shear strength for the different sample groups is shown in Figure 4. It was satisfactory for the as-bonded samples, with an average above 25 MPa for all sample groups.

The average shear strength of the "NiCr-250-500" sample group (NiCr/Au metallization, bonded at 250 °C, at 500 mbar) decreased from 44 to 30 MPa after 10 TSC (10 cycles), and further down to 28 MPa after 1000 TSC (1000 cycles). After LTS the shear strength for this sample group

was reduced by almost 70 % to 15 MPa. Similar trends were seen for the NiCr-250-vacuum samples, but with less pronounced changes. The die shear strength of NiCr-250-500 samples is shown as a function of thermal treatment in Figure 5, supporting the observation of lowered shear strength as a function of thermal treatment.

Lowering the bonding temperature from 250 to 200 °C resulted in a 10 to 20 % increase in the average die shear strength. The average die shear strength of the "TiW-250-500" and "TiW-200-500" sample groups did not degrade significantly with thermal exposure. This is confirmed by Figure 6 which shows the die shear strength of the TiW-200-500 sample group as a function of thermal treatment.

The majority of fractures were considered as adhesive in the device / BCB and BCB / cap interfaces. Some cohesive fractures inside the BCB layer were also observed. There was no trend regarding which sample group having which fracture mode. Due to the complexity of the fractures it was difficult to determine the primary crack initiation.

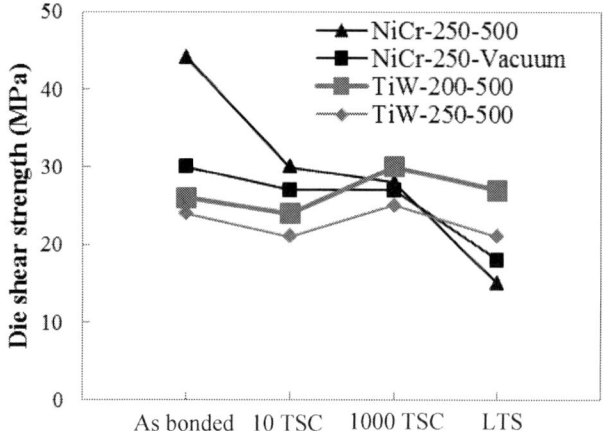

Figure 4. Average die shear strength of as-bonded, thermal shock cycled (TSC), and high temperature stored (HTS) samples.

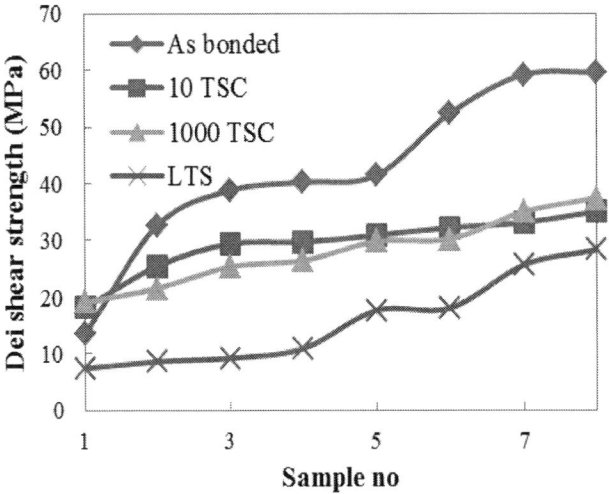

Figure 5. The die shear strength of NiCr-250-500 samples as a function of thermal treatment. Each point represents a single measurement value.

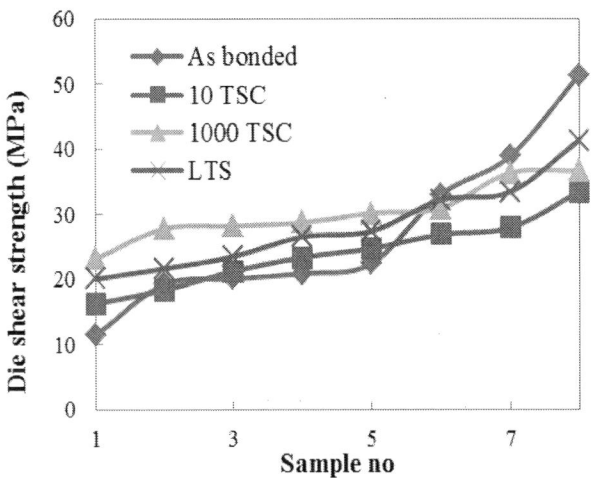

Figure 6. The die shear strength of TiW-200-500 samples as a function of thermal treatment. Each point represents a single measurement value.

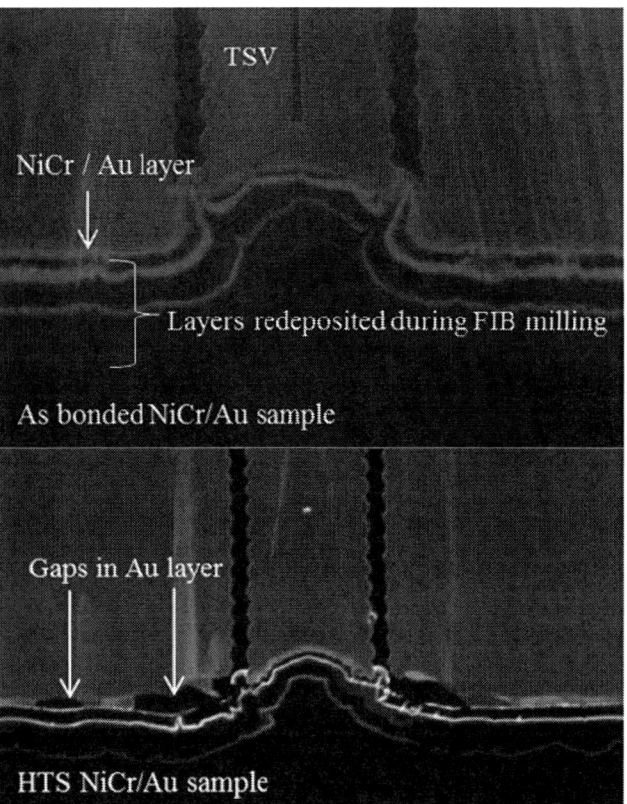

Figure 7. SEM cross-section of an as-bonded (upper) and a HTS (lower) sample with NiCr/Au metallization. Notice the gaps in the Au layer for the HTS sample, and the extra layers redeposited during FIB milling.

DISCUSSION
TSV metallization

Our results show that both samples using NiCr and samples using TiW as a diffusion barrier/ adhesion layer for the Au metallization initially had a low electrical resistance through the daisy chains with TSVs (as-bonded). This was also the case after TSC. However, only the electrical resistance of TiW/Au metallized samples remained low after HTS. For NiCr/Au metallized samples there was a substantial increase in the electrical resistance, or in some cases even a broken electrical connection, after HTS. From the cross-section images it was possible to see large gaps in the Au layer of the NiCr/Au metallized samples. The Au had probably diffused into the Si, leaving behind gaps in the conductor layer. Challenges associated with NiCr as a diffusion barrier for Au have also been reported earlier [17].

On the other hand, samples with TiW/Au metallization had a low electrical resistance through the TSVs even after combined HTS and TSC exposure. It was possible to see some degradation (thinning) of the Au layer, but the electrical resistance remained low. These results demonstrate that TiW is well suited as an adhesion layer/diffusion barrier for Au metallization for applications subjected to harsh environments.

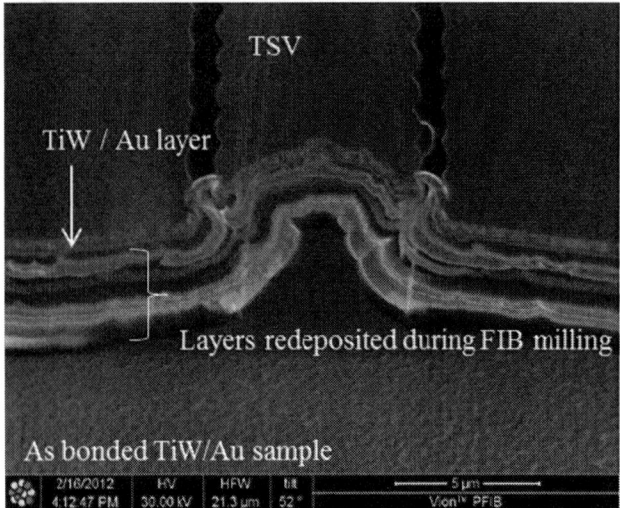

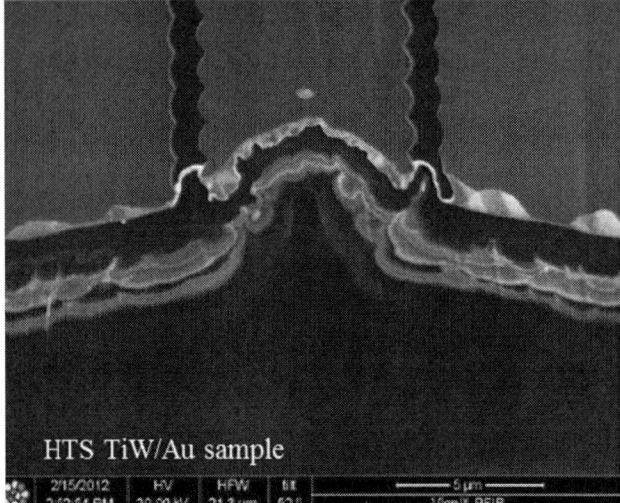

Figure 8. SEM cross-section of an as-bonded (upper) and a HTS (lower) sample with TiW/Au metallization. Notice the extra layers redeposited during FIB milling.

BCB encapsulation

It was observed that the initial (as-bonded) die shear strength of BCB bonded MEMS structures with NiCr/Au metallization are higher than that of samples with TiW/Au metallization. Since the majority of the fractures were adhesive device/BCB or adhesive BCB/cap fractures – i.e. most probably adhesive fractures between NiCr/Si or TiW/Si – NiCr probably has better adhesion to Si than TiW.

After thermal exposure, the die shear strength of NiCr/Au metallized samples was greatly reduced. However, even for the weakest group, the NiCr-250-500 samples subjected to LTS, the average die shear strength (15 MPa) is still well above the requirement of the MIL-STD-883H standard (6 MPa). Furthermore, the results indicate that NiCr/Au metallized samples bonded in vacuum are slightly more stable during thermal exposure than samples bonded in a 500 mbar atmosphere.

There was no significant indication of die shear strength degradation after thermal exposure of TiW/Au metallized

samples. The die shear strength of samples bonded at 200 °C was 10 to 20 % higher than that of samples bonded at 250 °C. This is probably due to the fact that those samples have a smaller amount of residual stresses due to the lower bonding temperature.

Importantly, these results show that a wafer-level 3D integrated MEMS acceleration switch utilizing TSV technology with the combination of TiW/Au metallization and BCB encapsulation is well suited for harsh environment applications.

CONCLUSIONS

The reliability of a 3D integrable miniaturized wafer-level encapsulated MEMS acceleration switch was studied. The investigated MEMS device used TSV technology and adhesive (BCB) wafer-level bonding for encapsulation.

TSVs using NiCr as a diffusion barrier/adhesion layer for Au metallization was shown not to be suitable for high temperature applications. However, by changing the NiCr layer to TiW, long time exposure at elevated temperatures was possible.

The initial die shear strength of BCB bonded samples with NiCr/Au metallization was higher than that of samples with TiW/Au metallization. Thermal exposure greatly reduced the die shear strength of the NiCr/Au metallized samples, while it was kept largely unchanged for the TiW/Au metallized samples. Lowering of the bonding temperature from 250 °C to 200 °C, increased the bond strength by 10 to 20 %, which is believed to be due to the reduced residual stresses.

The demonstrated TSV technology using TiW/Au metallization and BCB encapsulation is well suited for 3D MEMS packaging for harsh environment applications.

ACKNOWLEDGEMENTS

This work received financial support from the Research Council of Norway through the HTPEP project (contract No 193108/S60), the European ENIAC Joint Undertaking project No:120016 JEMSIP-3D, and the European Commission Framework Programme 7 within the project e-BRAINS (Project-no FP7- ICT- 257488).

REFERENCES

[1]   C. T. Tan, *et al.*, *Waferl Level 3-D ICs process technology*, in Integrated Circuits and Systems. Cambridge, Massachusetts: Springer, 2008.

[2]   N. Lietaer, *et al.*, "Wafer-level packaged MEMS switch with TSV," in *International Wafer Level Packaging Conference*, Santa Clara, CA, 2011.

[3]   K. Petersen, *et al.*, "Silicon fusion bonding for pressure sensors," in *IEEE Solid State Sensor and Actuator Workshop*, Hilton Head, USA, 1988, pp. 144-147.

[4]   H. Henmi, *et al.*, "Vacuum packaging for microsensors by glass silicon anadic bonding"

Sensors and Actuators a-Physical, vol. 43, pp. 243-248, May 1994.

[5] D. Sparks*, et al.*, "Wafer-to-wafer bonding of nonplanarized MEMS surfaces using solder," *Journal of Micromechanics and Microengineering,* vol. 11, pp. 630-634, Nov 2001.

[6] Y. T. Cheng*, et al.*, "Localized silicon fusion and eutectic bonding for MEMS fabrication and packaging," *Journal of Microelectromechanical Systems,* vol. 9, pp. 3-8, Mar 2000.

[7] G. S. Park*, et al.*, "Low-temperature silicon wafer-scale thermocompression bonding using electroplated gold layers in hermetic packaging," *Electrochemical and Solid State Letters,* vol. 8, pp. 330-332, 2005.

[8] S. A. Audet*, et al., Integrated sensor wafer-level packaging,* Proc. Transducers, Chicago, 1997, pp 287-289

[9] F. Niklaus*, et al.*, "Adhesive wafer bonding," *Journal of Applied Physics,* vol. 99, Feb 2006.

[10] C. H. Wang*, et al.*, "Chip scale studies of BCB based polymer bonding for MEMS packaging," in *58th Electronic Components & Technology Conference, Proceedings,* 2008, pp. 1869-1873.

[11] M. M. V. Taklo*, et al.*, "3D MEMS and IC integration," in *Materials and Technologies for 3-D Integration.* vol. 1112, 2009, pp. 211-220.

[12] N. Lietaer*, et al.*, "3D Interconnect Technologies for Advanced MEMS/NEMS Applications," in *ECS Transactions,* Vienna, Austria, 2009, pp. 87-95.

[13] Dep. of Defense, "MIL-STD-810G," in *Test method standard - Environmental engineering considerations and laboratory tests,* 2008.

[14] F. Lärmer and A. Schilp: Patents DE 4241045, US 5501893 and EP 625285

[15] A. Summanwar and N. Lietaer, "Etching Burried Oxide at the Bottom of High Aspect Ratio Structures," in *22nd Micromechanics and microsystems technology Europe workshop (MME),* Tønsberg, 2011, pp. 326-329.

[16] J. Hopkins*, et al.*, "The benefits of process parameter ramping during the plasma etching of high aspect ratio silicon structures," in *Materials Science of Microelectromechanical Systems.* vol. 546, 1999, pp. 63-68.

[17] M. M. V. Taklo*, et al.*, "Strong, high-yield and low-temperature thermocompression silicon wafer-level bonding with gold," *Journal of Micromechanics and Microengineering,* vol. 14, pp. 884-890, Jul 2004.

PAD LIFT FAILURE MODE INVESTIGATION
FOR WAFER LEVEL PACKAGE

Laurent Gay, Sebastien Gallois-Garreignot, Francois Guyader, Romain Brouillac, Pauline Boissiere
STMicroelectronics
Crolles, France
laurent.gay@st.com

ABSTRACT

The reliability of lead-free solder joints is a major concern of new packaging technology and assembly (Through Silicon Via (TSV), 3D IC integration...). For CMOS image sensors using TSV technology, 300µm balls are deposited by screen printing on the copper surface of the redistribution layer (RDL). Many tests, such as drop test, thermal cycling and total ball shear, are performed to ensure the structure's reliability. After experimental investigations and failure analysis, it has been shown that early and non standard failures occurred due to the pad lift phenomenon.

The first part of this paper aims to briefly describe TSV wafer level packaging technology as well as the reliability tests performed. In the second part, the failure mechanism of the pad lift will be described and characterized. In a few words, this failure is caused by the whole RDL plate pullout when the device is placed under mechanical or thermal loading. Various solutions (passivation enclosure, insulation/barrier layer influence, metallic layer undercut ...) are then proposed and evaluated to avoid such structural damage. Each solution has been thoroughly simulated and then compared to experimental trials. A discussion is proposed for each parameter.

Finally, based on numerical and experimental results, guidelines and recommendations in regards to design and process steps have been established in order to prevent from the pad lift mechanism and optimize mechanical robustness.

Key words: TSV (Through Silicon Via) - Pad lift failure mode - WLP - Total ball shear test - Simulation

INTRODUCTION

The reliability of lead-free solder joints is a major concern of new packaging technology like Through Silicon Vias (TSV) or 3D packages. Many Wafer Level Packaging (WLP) technologies have been already presented in term of structure and design [1]. The performance during thermal cycling is mainly affected by the mismatch of the coefficient of thermal expansion (CTE) between silicon die and plastic PCB board. To improve the reliability performance, one solution is to increase the flexibility of the structure in order to decrease the stress applied to the ball. The second rule is to enhance interface area of solder balls where fatigue failures occur [1].

For CMOS image sensor applications, a study was started to increase mechanical robustness of the solder joint. For this study, ball placement process is used to put 300µm balls directly on the copper surface of the redistribution layer (RDL). The original opening diameter of the passivation layer is set to 295µm. One solution to improve the strength of the assembly is to modify this opening to a larger value, 310µm for this study. This new design improves ball shear value but early failures were observed during reliability assessment (Drop Test and Thermal Cycling). On top of that, non standard failures were observed during total ball shear test for enlarged PSV mask (310µm). After investigation and characterization, it has been shown that both problems were linked and that pad lifts were occurring under mechanical or thermal loadings.

The aim of this study is to analyze pad lift mechanism. A primary guideline and recommendations to decrease chance of pad lifting during reliability test are also given.

ARCHITECTURE INVESTIGATED: TSV TECHNOLOGY

Through Silicon Via (TSV) is a technique to create 3D packages with high density of vias and short connections. Through-silicon vias replace wire bonding by creating vertical electrical connections through the chips. This technology is interesting for size reduction, performance improvement and cost reduction.

Details and explanations regarding the process flow of this technology have been already reported in a previous study [2]. Our application uses TSV for packaging of CMOS image sensors (CIS) (Figure 1).

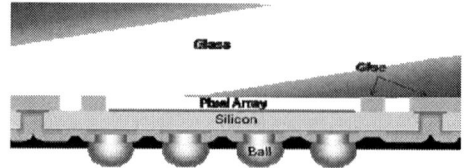

Figure 1. Schematic illustration of TSV technology

In the next part of this paper we will consider the connection area of our technology only. The main steps investigated in this study are presented in the Figure 2, the process flow is described below:

- RDL to silicon substrate insulation: SiN / SiON
- Metallization and backside metal rerouting: Barrier (Ti/TiN/Ti) and Cu RDL
  - o Barrier (Ti/TiN/Ti) / Seed (Cu) layers deposition
  - o Lithography for vias metallization and rerouting
  - o Cu electroplating
  - o Dry film stripping
  - o Barrier and Seed layers etching
- Metal protection: Passivation (PSV)
- Balling: 300µm SACN balls deposition by screen printing process
- Ball reflow (liquidus T°C = 217°C)

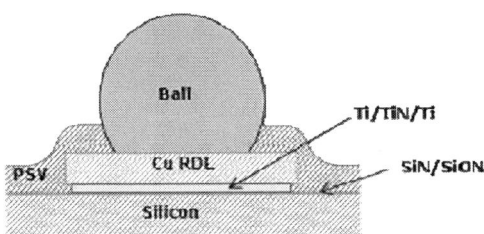

Figure 2. TSV technology - Cross Section – Focus on connection area

MECHANICAL TESTS & EVALUATION OF RELIABILITY:
Drop Test, Thermal Cycling Test and Total Ball Shear Test

Drop test and thermal cycling are common tests to evaluate the reliability of the solder joints. Those reliability tests have already been presented and described in various papers [3-4].

Various shear test methods are available in the industry: Single ball shear, High speed ball shear, Total ball shear, etc. Single ball shear is a well-known method to perform inline characterization of solder joint strength. Several studies have been performed to demonstrate correlation between high speed ball shear and mechanical drop test [5].

Total ball shear is usually performed at assembly plant on module or die. The goal of Total Ball Shear is to shear in a single run all the balls of one die using a specific shear tool. Then pads are observed to define failure mode, however shear value is not relevant for this test because whole pads are tested at once. Firstly, shear parameters of total ball shear need to be evaluated in order to select the expedient value depending on the test to implement. Increasing the test speed will increase the depth of tested interface whereas the height will change the way the load is applied to the joint.

For this study, the test has been adapted to wafer level in order to predict the pad lift during drop test. In fact, the implementation of this test at wafer level provides quick answers and thus prevents from some binding steps like sawing, etc. which are usually performed at backend plants.

In order to match failure rate between die and wafer level, shear parameters such as shear speed and shear height have been tuned and set-up to specific values.
Following parameter values have been defined:
- Shear Speed: 20mm/sec
- Shear Height: 40µm

For each wafer, five locations are tested (center, right, left, top & bottom of the wafer) and 5 dies are sheared for each location, so 25 dies are tested. As criteria, it has been decided to count a die as "failed" when at least one pad (of 30 bumps) of the die is lifting during shear test. Standard fracture mode occurring inside solder ball or at IMC interface.

PROBLEMATIC: PAD LIFT MECHANISM

A study was initiated to increase mechanical robustness of the solder joint. For this study, ball placement process is used to put balls directly on the copper surface of the RDL, without use of complex stack of UBM layers. One solution to improve the strength of the assembly is to increase opening diameter of the passivation layer. In such a way, the intermetallic compound (IMC) surface is maximized. As shown in Figure 3, the ball shear value is directly linked to IMC surface. Therefore, this new design improves ball shear value but early failure occurred during Thermal Cycling (TC) and Drop Test (DT) assessments.

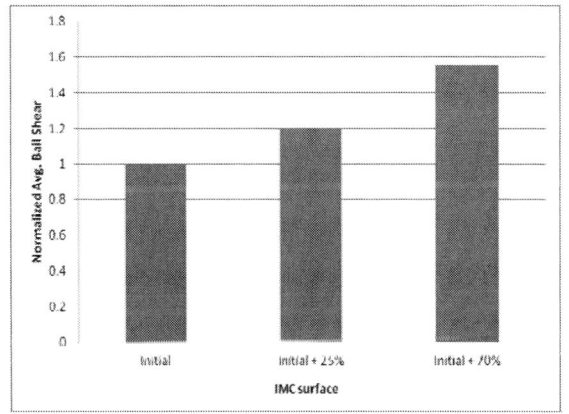

Figure 3. Increase of single ball shear value with IMC surface

The Figure 4 shows the Weibull plot observed during Thermal Cycling evaluation. Two different populations are obtained, the first one including early failure corresponds to pad lift and the second one corresponds to standard failure with cracks in solder near component or at component interface.

101

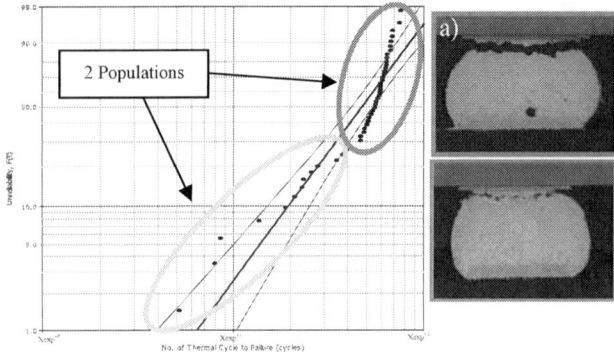

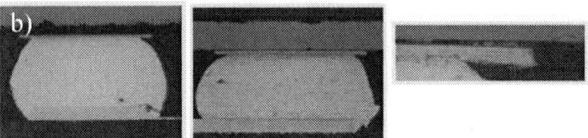

Figure 4. Weibull plot of TSV module after Thermal Cycling test showing two populations **a)** Standard failure: crack in solder near component or at component interface; **b)** Pad lift (crack between pad component and die)

Pad lift phenomenon corresponds to a brittle pullout of the whole RDL plate when the device is placed under mechanical stress condition (as shown in broad outline in Figure 5).

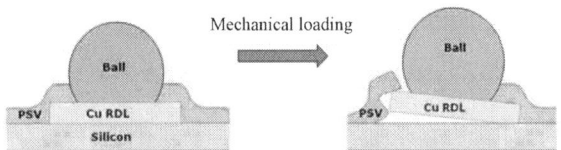

Figure 5. Schematic illustration of Pad lift mechanism under mechanical loading

The same failure mechanism is also observed during total ball shear. The main advantage of this test is that it gives a quick and overall answer using a basic and easy to put in place test which can be performed at die, module or wafer level. The figure 6 shows a top view inspection of a die tested with total ball shear test for enlarged PSV mask (310µm). The yellow layer revealed after a pad lift is the insulation layer SiN/SiON.

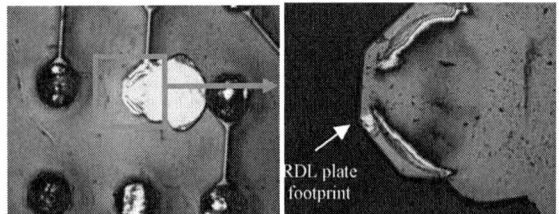

Figure 6. Optical Inspection - Top view of pad lift after total ball shear test

A FIB inspection of a pad lift confirms that the whole RDL plate of copper has been taken out, leaving the SiN/SiON insulation exposed (Figure 7). According to FIB-SEM inspection, the crack during the pad lift spreads between the SiN and SiON layers.

Figure 7. Pictures of pad lifting after total ball shear test **a)** SEM image; **b)** FIB-SEM inspection

METHODOLOGY

Numerous root causes may induce such failure. Un-optimized solder bumping process, pad structure weakness, high CTE mismatch between components, weak interfaces or inappropriate design are the main detractor/detrimental factors. Couplings between parameters and options (geometrics, materials, technological…) are complex, that induce tricky analysis. This paper aims to get a better understanding of the cracking mechanism and propose solutions to fix such fails. To do so, simulations and experimental investigations have been carried out. In this paper, focus is done on:
- Ti/TiN/Ti barrier undercut length
- Presence of the SiN insulation layer
- PSV enclosure length

Experimental Details
Ti/TiN/Ti barrier undercut length:
The barrier of the seed layer metallization is composed of a multi-layer of Ti (0.05µm), TiN (0.1µm) and Ti (0.05µm) which is deposited by sputtering (PVD), also named TiN in the next part of this paper. The Ti improves adhesion whereas TiN is a barrier layer to diffusion. During wet etching process of the barrier/seed layers, an undercut is created inside the Ti/TiN/Ti multi-layers (as shown in Figure 8).

The barrier ensures the contact between the copper and SiON layers. Moreover, interactions between the bump edge and the region nearby the undercut are suspected. The distance between those regions is assumed to be a key parameter. In order to confirm this hypothesis, a new etching condition has been defined to decrease undercut length. Main modifications have been done by removing some etching steps (HF chemistry) and decreasing step time. This condition has been compared thanks to total ball shear with wafers processed with standard etching sequence. For this experiment a passivation opening of 310µm has been used for a 335µm RDL pad size.

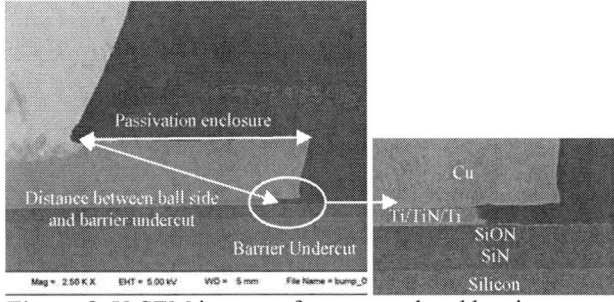

Figure 8. X-SEM images of copper pad and barrier undercut

Presence of the SiN insulation layer:
The insulation stack is composed of a multi-layer of thin SiN (50nm) and thick SiON (1500nm) which is deposited by a low temperature PECVD process. Silicon Nitride (SIN) is widely adopted in BEOL integration as a barrier layer against copper diffusion. Silicon Oxinitride (SION) provides the dielectric function to isolate the RDL and bump connection to the silicon substrate.

From previous observations, it has been demonstrated that during pad lift, the crack is initiated at the interface between the passivation and the RDL plate (near base of the ball), then the crack follows the side of the RDL plate and the propagation is finally located under RDL plate through the SiN/SiON layers. SiN material is more brittle than SiON, on top of that SiN layer is not mandatory for this low temperature technology (balling reflow around 240°C max). So the copper barrier function can be achieved by the single SiON layer. Thus, a study has been done to evaluate the impact of SiN on the pad lift phenomenon.

In order to evaluate SiN layer influence in pad lift mechanism, some wafers have been processed without this layer. Those wafers have been tested using total ball shear test (25 dies per wafer from 5 locations, 3 wafers per condition) and results have been compared to wafers processed from same lot with standard TSV flow, so with SiN/SiON insulation layers. For this experiment a passivation opening of 310μm has been used for a 335μm RDL pad size.

PSV enclosure length:
The passivation sequence is achieved by using a standard photo-sensitive dielectric polymer (negative tone) applied by spin coating. Then the pad is opened using a 1X lithography followed by a 190°C ex-situ cure to complete the cross-linking reaction.

The resist used is a fluoropolymer with low dielectric constant, low moisture uptake and good planarization properties. The goal of this layer is to protect copper routings electrically and chemically with offering openings on pads. The final layer thickness is 7 to 9 μm and 2 to 4μm depending on wafer topology.
However, regarding our investigation about pad lift mechanism, the passivation layer could also have a

mechanical function. Along intrinsic mechanical properties of the material itself like strain, elongation rate, etc., the enclosure is also a parameter to be considered. The enclosure is in fact (as shown in Figure 8) the PSV length overlapping the plate of copper RDL (CD Passivation & RDL).

In order to evaluate this hypothesis, a misalignment of 10μm along X-axis has been manually created during lithography exposition step, PSV opening diameter (CD: critical Dimension) remaining unchanged. As a result, different enclosures have been evaluated depending on total ball shear direction of test (as shown in Figure 9 on a pad before ball placement process). Using direction of test along the X-axis, 15μm and 35μm enclosures were assessed, while the standard PSV overlap of 25μm was tested along the Y-axis. For this experiment a passivation opening of 295μm has been used for a 335μm RDL pad size.

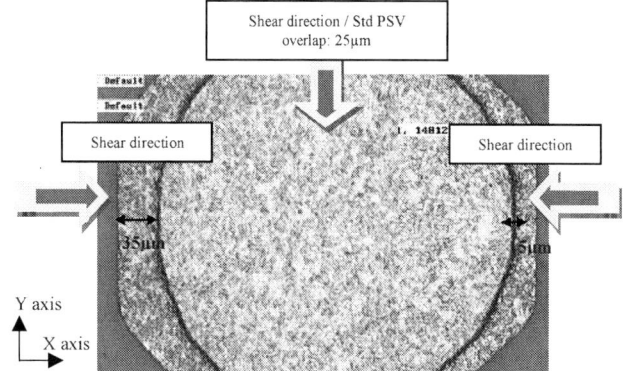

Figure 9. Top view inspection of pad before balling – PSV/RDL distance assessment as a function of shear direction set-up
For X-axis: Passivation enclosure: 35μm or 15μm
For Y-axis: Passivation enclosure: 25μm

In order to better understand the mechanical mechanism involved for each parameter, an numerical study is done thanks to the Ansys Finite Element (FE) software. The FE model and the corresponding assumptions are described in the next section.

Thermal and Mechanical Modeling and Simulation
Model description
The main features of the models are as follows: an 2D Plane strain model is chosen to reproduce shear and thermomechanical tests. Since fails are located nearby the connection area, attention is paid on this region (see Figure 10). Model geometry and materials are similar to those of the failed product, except the thin SiN layer, which is not modelled because of the very distinct scale compared to the others components.

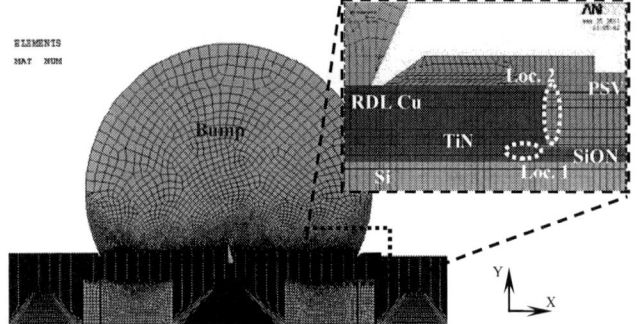

Figure 10. Finite element model of the connection area: bump and its landing pad. Two locations are more particularly studied (see circle, Locations 1 & 2). Pad is meshed with hexahedron elements (except enclosure of the PSV layer), bump is meshed freely.

In this work, a linear material properties approach is adopted. The main drawback of this linear approach is that no cumulative effects can be considered. However, the trends and the failure mechanisms are allowed to be found. Note that this approach has been previously validated on others topics [6-7].The residual stress of each material is taken into account by setting a stress free state at their deposit temperature.

The main geometrical dimensions and materials properties are resumed in Tables 1 and 2, respectively. To allow further industrial design optimizations and easier investigations, model is fully parameterized.

Table 1. Main dimensions used in the FE model

Standard values	Dimension (µm)
Bump Diameter	325
RDL Cu Thickness	7
RDL Cu Pad size	335
PSV opening	295
PSV Thickness	9

Table 2. Mechanical properties of the main materials used in the FE model

	Young Modulus (GPa)	Poisson ratio	CTE (ppm/°C)
Silicon	130	0.3	3
SiON	84	0.17	2.5
RDL Cu	117	0.35	17
Bump (SnAg)	22.7	0.3	15.1
PSV (Spin On Dielectric)	1.3	0.3	60

Loading

In order to be able to depict both tests (thermo cycling and shear tests), two distinct loadings have been applied to the model. Note that the drop test was not modelled since it involves a very different physical phenomena (i.e. dynamics versus static) and thus, a numerical scheme more adapted is required (i.e. explicit versus implicit). However, it has been experimentally found that the failure mechanism occuring

during drop test is similar to those of the shear and thermal cycling tests (§ "Problematic: Pad lifts mechanism"). It is thus expected that the critical parameters remain unchanged with this test. The simulated loadings are:

- Thermal cycling: To investigate the thermomechanical effects within the structure (i.e. thermal cycling test), a temperature loading from 240 ° C (reflow temperature of the ball) to 25°C (room temperature) is applied. The difference between deposit and room temperatures and the CTE mismatch between materials will induce thermo-mechanical stress.

- Shear test: In order to mimic the shear test, a constant displacement is applied to the bump from 40µm to the RDL Cu (see Figure 11). This loading is assumed to represent, roughly and at the first order, the shearing tool displacement. The aim is to reproduce qualitatively the shear test loading. Since no delamination or crack propagation is modelled, only the first steps of the test are investigated. The displacement value is chosen arbitrarily in order to obtain standard strain values in the frame of the elastic linear theory. In this paper, the interest concerns mainly the contribution of each parameter and the corresponding trends. By consequence, the stress values have to be considered with caution because they are depending on this displacement value. The reader may refer to [8] for an numerical analysis of the shear test compared with experimental data.

The table 3 resumes the experimental and numerical parameters investigated and the associated range.

Table 3. Parameters and values investigated

Unit: µm	Undercut	SiN layer	PSV enclosure
Experimental	2.83 - 6.3	w - w/o	15-25-35
Numerical	4-8-12	-	12.5-20

RESULTS & DISCUSSION
1\ Ti/TiN/Ti Barrier Undercut
Numerical results

Based on the model described above, stress and strain cartographies in the structure are obtained. On one hand, by applying thermal loading, due to the CTE mismatch and different deposit temperatures, residual stress appears within the structure at room temperature. The RDL layer tends to shrink and is then in a tensile state. As shown in Figure 11, the highest strain values appear near the geometric singularities: ball step and the undercut.

On the other hand, with a mechanical loading (i.e. shear test), the bump is submitted to the shear tool displacement. The stress is mainly transmitted by the RDL Cu layer to the underneath stack, as shown by the arrows figure 11.

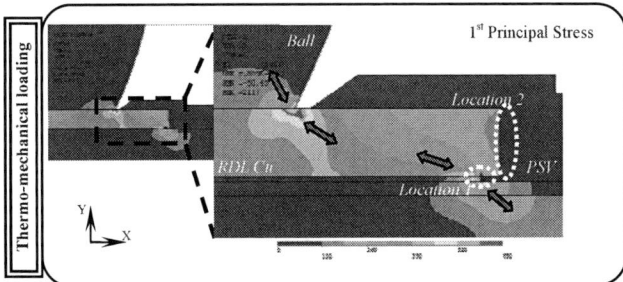

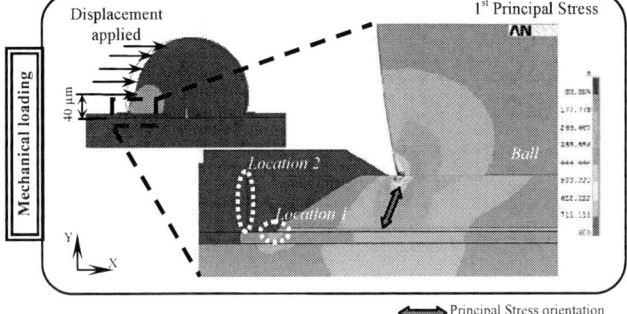

Figure 11. Stress field for the thermo-mechanical and mechanical loadings.

By modifying the undercut geometry, dissimilar trends are found according to the zone and loading observed. During the thermal loading, by increasing the undercut (i.e. 12 μm), the RDL Cu is more free to evolve (location 2 more stressed). Indeed, the PSV layer has poor mechanical properties and its behavior is mainly driven by the RDL Cu. By similar behavior, the location 1 experiences higher stress with lower undercut.

During the mechanical loading, the stress is transmitted mainly by the RDL Cu. By consequence, the RDL Cu/TiN contact has to be the highest to spread the stress over the interface. Numerical results of both the loadings and locations concerning the undercut parameter are resumed in table 4.

Table 4. Maximum stress difference (σ_1) for a variation of the undercut parameter from 4 to 12μm

	Location 1 (edge of TiN layer)	Location 2 (edge of the RDL Cu layer)
Thermal loading	- 7.4 %	+ 20 %
Mechanical loading	+ 54 %	+ 106 %

The reader has to keep in mind that the displacement has been chosen arbitrarily for the mechanical loading. As a consequence, the stress variation cannot be compared from one loading to another. Only the trends should be considered.

As described in table 4, the trends are inverted according to the loading applied. The guideline has then to be considered according to the most critical test. Moreover, these results may imply that others parameters and/or couplings are involved.

Experimental results

As described previously (§ "Methodology/Experimental Details"), by changing etching sequence and step duration, the undercut length has been decreased to 2.83μm compared to 6.3μm with the standard condition (see Figure 12).

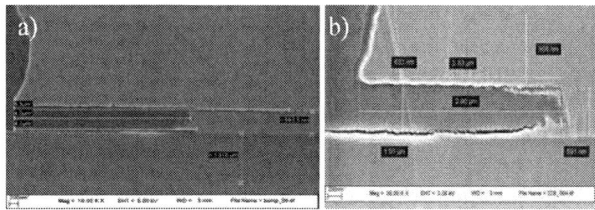

Figure 12. X-SEM images of undercut generated during barrier/seed layer etching
a) Standard etching condition, undercut: 6.3μm
b) Specific etching sequence to decrease undercut, undercut: 2.83μm

Total ball shear has been performed on both conditions. The rate of defective die (pad lift) is approximately the same between the new etching condition and our standard process (Figure 13).

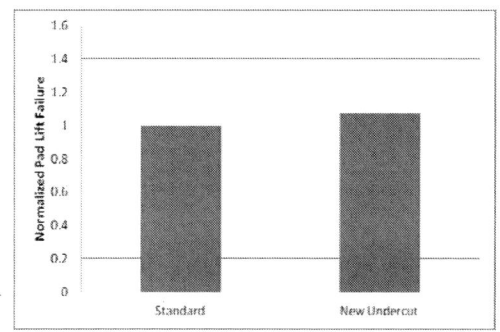

Figure 13. Normalized defective die after total ball shear test vs. undercut conditions

Experimental investigations have then shown that the undercut length is not a critical parameter in pad lift mechanism and has no impact on the reliability. This is in contradiction with numerical results. A possible explanation could be the limited range of the undercut values experimentally tested, which is not large enough to reproduce the behavior numerically seen. Moreover, this observation highlights the lack of threshold value or numerical failure criteria as things stand at present. More precisely, the stress variation obtained numerically cannot be interpreted in terms of fail/no fail response since no formal criteria are available and thus, no clear statement can be made on the impact of the modification. Physics governing nucleation and criteria are still a great challenge for fracture mechanics and fundamental work is still on-going, see for example [9].

2. Influence of SiN Insulation Layer
Numerical results

This configuration has not been studied numerically. Indeed, it is expected that the main factor is the adhesion properties of the interface rather that a different mechanical behavior of the structure induced by the SiN barrier removal. In fact the SiN thickness is too low with respect to the whole stack SiN + SiON to drastically change the global mechanical behavior. For such analysis, experimental characterization of the interface (such four point bending measurement [10]) would be more relevant. Note that these measurements were not available during this work.

However, numerical results have shown that high stress peaks appear nearby these layers (see Figure 11), suggesting that a particular attention must be paid to this location.

Experimental results
Figure 14 shows the normalized defective dies (pad lift) occurrence depending on barrier type: SiN/SiON or SiON. The rate of pad lift increased to more than 6 times when the SiN layer is removed.

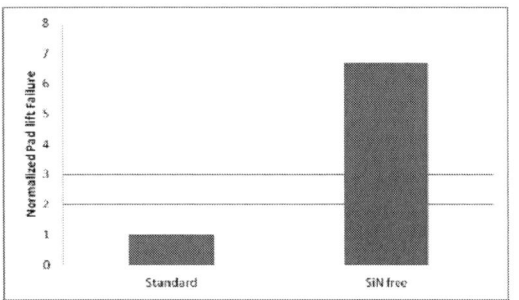

Figure 14. Normalized Pad lift failure after total ball shear test vs. insulation type

From this experiment, the barrier seems to be a key parameter to limit pad lift. Without SiN, pad lifts are more present. The difference of behavior between the two configurations can come from stress inside layers or adhesion issue between SiN and SiON layers.

A TEM analysis of a pad lift has been initiated to define precisely the crack path within SiN/SiON stack.

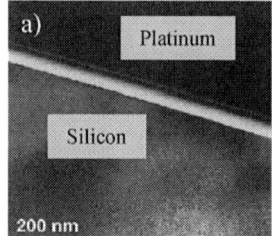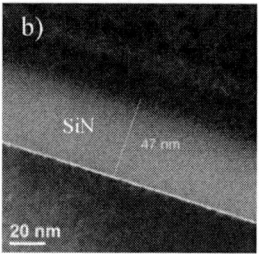

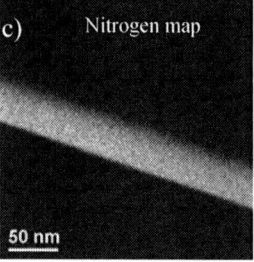

Figure 15: TEM observation of a pad lift confirms that crack occurs at SiN/SiON interface during failure **a)-b)** TEM observations; **c)** Nitrogen map

The TEM observation (Figure 15) confirms that crack occurs at SiN/SiON interface. Indeed, Figure 15 c) shows that the remaining layer at pad surface is a 47nm layer thickness including nitrogen component which corresponds to the SiN layer.

One way of improvement could be to do engineering study on this process in order to increase adhesion strength of those layers, for example by improving cleanliness or preconditioning during deposition.

3. Influence of PSV Enclosure
Numerical results
According to the different enclosure values, modelling shows that a short enclosure option induces the highest stress within SiON & TiN layers. Indeed, the PSV layer plays an encapsulation role, as shown picture 16. The behavior of the PSV layer is completely driven by the RDL Cu layer due to its poor mechanical properties.

Figure 16. 1st principal strain (deformed shape scaled 40 times) during the thermal loading. The PSV layer acts as an encapsulation layer and allows to limit the stress

By increasing the enclosure, this encapsulation effect is increased. It allows to spread the stress along the structure and then limits the stress nearby the SiON & TiN layers, see table 5. Similar results are found whatever the locations and loadings considered. Note that in case of the mechanical loading, only the presence of the PSV was simulated.

Table 5. Maximum stress (σ_1) difference for a variation of the PSV enclosure value from 12.5 to 20 μm and the effect of the PSV presence

Increasing of the PSV enclosure from 12.5 to 20 μm	Location 1 (edge of TiN layer)	Location 2 (edge of the RDL Cu layer)
Thermo-mechanical loading	- 15 %	-8.1 %
No PSV vs. PSV		
Mechanical loading	- 25 %	-

Experimental results

The figure below shows the rate of dies failing with pad lifting as a function of the enclosure length. 50 dies have been tested per enclosure dimension.

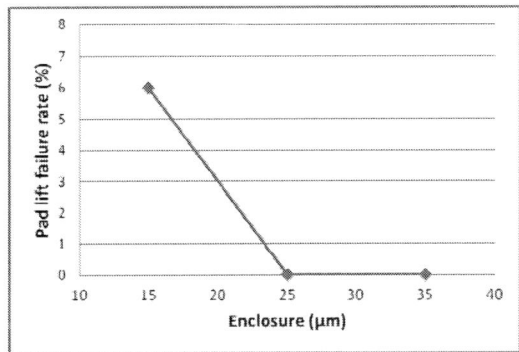

Figure 17. Defective die (%) after total ball shear vs. enclosure length

The PSV/RDL distance is critical; a minimum enclosure distance seems to be required to prevent pad lift failure. This length will depend on PSV material properties but also on technology design. For our configuration, this value is defined in the range of 20-25μm (i.e. 20μm drawn distance as the opening is then reduced due to the slope of the passivation at open areas). A fine screening should be needed to define accurately the minimum enclosure acceptable.

CONCLUSIONS

A study was started to improve the mechanical robustness of the solder joint. However, early failures corresponding to pad lift were observed during reliability assessment (drop test & thermal cycling). Pad lift phenomenon corresponds to a pullout of the whole RDL plate. In this paper, this mechanism has been more particularly studied.

Firstly, in order to be able to reproduce such fails, total ball shear test has been used at wafer level with special set of parameters. This test is a convenient method as it can be easily operated at wafer level. The benefit is to early assess reliability performances of the connection depending on process conditions applied. Cycle time for process optimization is shortened avoiding long and heavy product level reliability trials (sawing, assembly and cycling). Therefore, product qualification is getting smoother.

Dedicated experiments and simulations have been performed in order to give a first guideline set and recommendations to decrease event rate of this brittle and unacceptable failure.

Main parameters have been investigated:
- Barrier undercut (generated during barrier/seed layer etching): In summary, from experimental study, the barrier undercut does not play a key role in pad lift phenomenon. For this parameter, results from simulation are not in line with experimental data. This mismatching can come from simplification applied to the simulation model but also from the range of values studied. Improvement could be brought by including in the model adhesion properties between layers.

- Insulation layer: This parameter has to be considered in terms of material properties, adhesion and stress inside the layer. Experimental trials have shown the significant effect of insulation stack on pad lift. Indeed, the rate of pad lift increased strongly when the SiN layer is removed. Optimization of the SiN/SiON interface could mitigate the pad lift event. Since the current FE model was not able to take into account the adhesion properties of the insulation layer, no simulation was performed. However, high stress peaks appear nearby these layers, suggesting that a particular attention must be paid to this location.

- Passivation enclosure (overlap length of the passivation on the copper RDL plate): Finally, the simulation fits well with experiment results for passivation overlap. Numerical and experimental results underline the strong effects of the PSV enclosure which plays an encapsulation role. This distance is a key parameter, a minimum of 20-25μm seems to be required to prevent pad lift occurrence.

In conclusion, here-above process and design recommendations have been proven to be useful to prevent the pad lift mechanism. Same methodology would be efficiently applied to optimize the mechanical robustness of new connection solutions.

ACKNOWLEDGEMENTS:

The authors would like to thank people from Imaging team at ST Grenoble: Hervé Dubus, Olivier Le-Briz, Sylvie Vergez, Frederic Poilane for their cooperation during this study as well as our colleagues in ST Toa Payoh: David Gani and Daniel Yap for reliability tests.

We are grateful to Veronique Guyader for characterization, Vincent Fiori from simulation team and Patrice Loiodice from CEA Leti for wet etch experiment. A special thanks to Abdelouahab Behillil and Francois Mariani for their supports during the correction of this paper.

REFERENCES:

1. Xuejun Fan and Qiang Han, Design and Reliability in Wafer Level Packaging, 2008 EPTC Electronics Packaging Technology Conference

2. Henry D. *et al*, Through Silicon Vias Technology for CMOS Image Sensors Packaging, 2008 Electronic Components and Technology Conference

3. Pandher R. and Healey R., Reliability of Pb-free Solder Alloys in Demanding BGA and CSP Applications, 2008 Electronic Components and Technology Conference

4. Anderson R. *et al*, Advances in WLCSP technologies for growing market needs, 2009 SMTA's 6th Annual International Wafer Level Packaging Conference

5. Fubin Song *et al*, High-Speed Solder Ball Shear and Pull Tests vs. Board Level Mechanical Drop Tests: Correlation of Failure Mode and Loading Speed, 2007 Electronic Components and Technology Conference

6. Gallois-Garreignot S., Fiori V., Nelias D., Fracture phenomena induced by Front-End/Back-End interactions: Dedicated failure analysis and numerical developments, 2010, Microelectronics Reliability, vol.50, pp.75–85.

7. Fiori V., Gallois-Garreignot S., Tavernier C., Jaouen H., Juge A., Chip-package interactions: some combined package effects on copper/low-k interconnect delaminations, 2008, In: 2nd Of Electronics System integration Technology Conference, pp. 713-18

8. Xingjia Huang, S.-W. Ricky Lee, Chien Chun Yan, and Sam Hui, Characterization and Analysis on the Solder Ball Shear Testing Conditions, Proceedings of ECTC, 2001

9. Y. Mishin, M. Asta, Ju Li, Atomistic modeling of interfaces and their impact on microstructure and properties, Acta Materialia, vol.58, 2010, pp.1117–1151

10. Moore T.M., Hartfield C.D., Anthony J.M., Ahlburn B.T., Ho P.S., Miller R., Mechanical Characterization of Low-K Dielectric Materials, In: 15th Proc. Of Characterization and Metrology for ULSI Technology Intern. Conf., 2000, pp.431-438

CHARACTERIZATION OF EWLB POP STRUCTURES

Tom Strothmann
STATS ChipPAC, Inc
Tempe, AZ, USA
tom.strothmann@statschippac.com

ABSTRACT

eWLB fan-out package technology provides a flexible package platform for 3D package options. Vias placed into the fan-out area provide a convenient method to enable an electrical connection to a top package without the use of TSV's. This basic capability enables existing silicon designs to be incorporated into high performance PoP structures with a minimal package height. Since there is no laminate used in the bottom package, a total stacked package height of <1.0mm can be achieved. This supports the increasing demand for ever thinner mobile products. The small package size is also advantageous for improved performance. The very short interconnect routing length with thick plated Cu provides superior electrical performance and flexibility in the design. Recent work with traditional stripline features has enabled a 4 layer laminate FC design to be converted to a 2 layer design with eWLB. Critical design considerations for matched pair routing can be relaxed as the line lengths are decreased. Fine pitch lines and spaces used in the package can also reduce the number of interconnect layers required on the board, thereby reducing board cost. Two methods are examined to achieve these advanced PoP structures. The first method uses low density vias that are laser drilled from the topside of the package, enabling a top package connection using 0.4mm pitch and above. The second method uses pre-formed high density vias placed into the fan-out area, enabling a topside redistribution layer. The key attributes of form factor, reliability, warpage, electrical performance, and relative cost will be reviewed for both package types and compared with conventional laminate PoP packages.

eWLB PACKAGE BACKGROUND

Wafer Level Packaging is experiencing disproportionate growth that is driven by the reduced form factor required by advanced mobile products. The Wafer Level Package has the smallest possible package size and excellent performance due to the very short routing length in the fan-in redistribution. The eWLB fan-out process augments the WLP by retaining the inherent benefits of the package while extending the Si die area through the addition of mold compound surrounding the die. The fan-out technology is a powerful concept in its basic implementation because it enables the die size to be disconnected from the package size required for a standard WLP format. This allows the die size to shrink to the optimum die size for the technology node, thereby reducing the fab cost of the silicon. Low cost and minimum form factor are achieved since the substrate is eliminated from the package.

The mold compound surrounding the die also creates a flexible platform for novel package structures. The process allows for multiple die to be embedded into the same wafer level package and the close proximity of the die to each other enables the die to be interconnected with extremely short Cu RDL. The short routing length reduces the signal latency between embedded die, thereby improving performance. The flexibility of the platform allows for discrete components, such as decoupling caps to be placed in the Wafer Level Package adjacent to the active die to minimize the board footprint and improve performance. The fan-out area can also be used to create novel 3D packages without the use of TSV.

THE STRUCTURE OF 3D eWLB PACKAGES

There are two basic methods to achieve a 3D capability in the fan-out package. The first is to use a laser to ablate the mold compound from the top of the package, allowing a connection to be made to the Cu redistribution layer in the fan-out area. This type of 3D package can be referred to as a eWLB Molded Laser ablated PoP Package or eWLB-MLP package. The second method is to create metal vias in the fan-out area. There are several methods to achieve the through-mold-via including laser drilling and metal fill or the placement of preformed vias made of glass, silicon, or laminate materials as a multi-chip package. This type of package can be referred to as a 2S eWLB-PoP. The 2S eWLB-PoP package enables the connection of redistribution on both sides of the package, fully enabling the package capability as a 3D platform. The 2S eWLB-PoP package discussed here is enabled with the use of preformed laminate via blocks.

COMMON ATTRIBUTES OF eWLB PoP PACKAGES

There are common structural elements that are used in the formation of eWLB PoP packages and the manufacturing is done on the same production line. The common elements of the process are listed as follows:
1) The reconstitution step at the start of the process
2) The material sets supporting the Cu redistribution layers
3) The photolithography capability with respect to line and space.

Reconstitution
The die are initially placed face down onto an adhesive film and a compression molding process is used to lock the die into position and to create the basic package. The placement accuracy is critical on both a micro and a macro level since photolithography is required to form the redistribution layers connecting the pads on each die or via structure.

Material Sets
The material sets used for the construction of the layers use the same dielectrics and plated Cu runners. The dielectrics selected for use in the eWLB package have demonstrated excellent performance in component level tests, including thermal cycle, HTS, multiple reflow, and ball shear. The dielectrics are able to achieve good TC performance not only for the temperature range of -40 to 125C used for TCB but also to -55 to 125C at CLR tests. In all cases the components test are successfully passed with MSL1 pre-conditioning.

The dielectric and Cu thickness can be selected to meet the needs of a given design and the specific requirements of the customer product. Two layers of Cu redistribution have been qualified for the bottom side of the package whereas the top redistribution is typically just 1 RDL. The line /space capability of the process is currently 10um/10um with process capability expected to extend to 5um/5um in the future.

Signal Integrity
A benefit of the short RDL routing length and tight L/S routing is the ability to reduce the number of layers in the design. Comparing the thickness between a 4L-laminate approach and the 2L-eWLB approach, a 4L-laminate package usually has 45um to 60um (or ever thicker) for prepreg and core thicknesses, respectively. The 2L-eWLB only has 5um between RDL1 and RDL2, which is about 10 times thinner/smaller. Converting from 4L design (laminate) to 2L design (either laminate or eWLB) roughly reduces the power and ground plane area by half, if patterns are proportionally routed or treated. The overall effect makes the power-plane capacitance (for decoupling purpose) about 5 times higher than in laminate design, which helps to yield a better power integrity (PI) performance from 2L-eWLB designs. The maximum package size permitted by the current eWLB design rules is 14mm x 14mm and chip-to-chip and chip-to-ball signal traces seldom exceed 5mm in length. This is much shorter than the design operating length of common digital signal standards. The signal propagation time along these interconnections is very short compared with the Unit Interval (bit period) at rates up to many Gb/s. Signal reflections on such short lines do not create appreciable Inter-Symbol Interference (ISI) and controlled impedance lines are typically not needed in eWLB packages.

Thermal performance of eWLB
Thermal modeling of a eWLB PoP bottom package indicates the performance is comparable to fc PoP bottom package. As with other wafer level packages, heat is primarily transferred through the solder balls and dissipated into the board. The thermal modeling of a 14x14mm eWLB PoP bottom package as compared to a fcPoP bottom package is shown in Table 1 and Figure 1.

Table 1: Thermal modeling results

Package 14x14mm Pwr: 3W	T_A (°C)	T_J (°C)	Q_{JA} (°C/W)	Y_{JT} (°C/W)	Y_{JB} (°C/W)	Q_{JB} (°C/W)
eWLB-PoP	50.0	103.1	**17.7**	0.01	5.22	**5.61**
fcPoPb	50.0	103.6	**17.85**	0.05	5.89	**6.20**

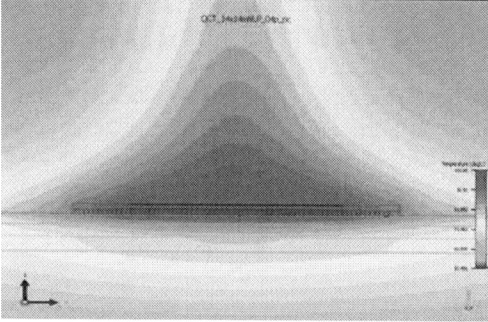

eWLB under natural convection and 3W

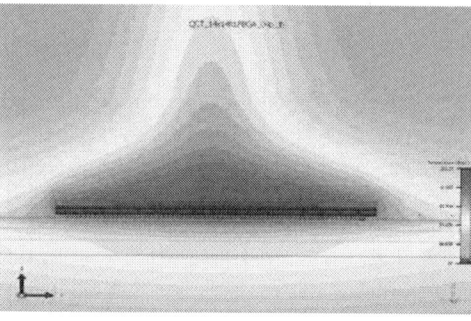

fcPoPb under natural convection and 3W

Figure 1: Thermal Models for eWLB and fcPoP

eWLB-MLP CHARACTERIZATION
The eWLB-MLP structure can achieve the thinnest package height since a very thin body thickness can be used and there are no top side metal and polymer layers. A total body thickness of 300um is typical with this package design for a body size up to 14x14mm. 250um solder balls placed on a 0.4mm pitch with a collapse height of 190um enable a total package height of 490um. The package is very cost effective since no substrate is required and RDL patterning is limited to one side of the package. The package has been demonstrated in 10x10mm, 12x12mm, and 14x14mm package sizes. Larger package sizes are possible but have not yet been demonstrated.

Figure 2: Schematic cross-section of eWLB-MLP

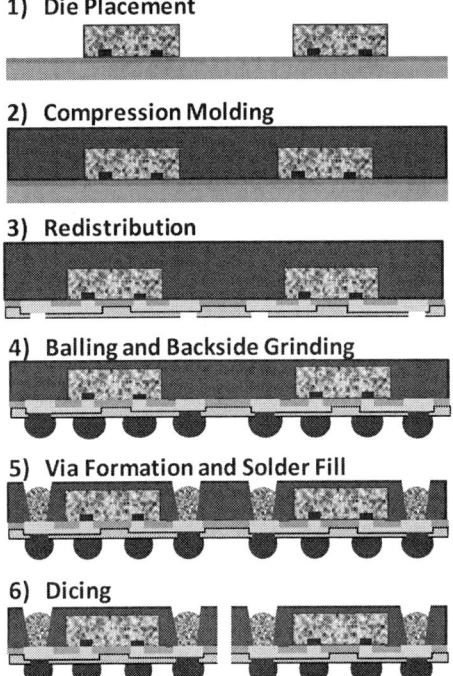

Figure 3: Basic process flow for eWLB-MLP

Structure

A schematic cross-section of the eWLB-MLP is shown in Figure 2 and the process flow is shown in Figure 3. The standard eWLB process is used to create a reconstituted wafer that can be processed with standard wafer level packaging equipment. Two Cu RDL layers are then plated in conjunction with 3 polymer layers on the bottom of the package to enable the redistribution. 10um line and 10um

space capability can be achieved with this process. The first Cu layer is plated directly onto the mold compound in the location of the desired via. Laser ablation is then used to remove mold compound above the Cu until the surface of the Cu is exposed. The laser via is subsequently filled with solder to within a few microns of the top surface of the package and the reconstituted wafer is then diced to provide the finished singulated packages. The design of the package limits via placement to the perimeter of the package and it is not possible to assemble a top package with a peripheral bump pattern size smaller than the bottom via placement.

The bottom package thickness eWLB-MLP presents one of the thinnest form factors possible in a PoP package. Total package thickness including bumps is <500um. Challenges with this package are warpage, BGA bump coplanarity, and via fill coplanarity. One of the challenges with a thin package of any construction is warpage due the mismatch of materials used in the assembly. Warpage observed with both the 12x12mm package and 14x14mm package are shown in the Figure 4. Even in a 14x14mm package the warpage is < +/- 80um.

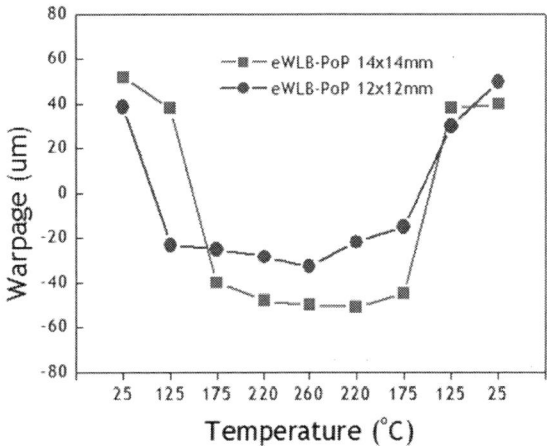

Figure 4: Warpage for thin eWLB-MLP

Reliability Performance

Reliability performance of the package is shown in Table 2. The board level data is from assembled PoP packages mounted onto the laminate without underfill.

Table 2: Component Level Testing 14x14mm eWLB-MLP

Test	Condition	Criteria	Status
MSL1	MSL1, 260C Reflow, 3x	-	Pass
Temperature Cycling after Precon	-55 C to 125C	1000x	Pass
HAST (w/o bias) after Precon	130C / 85% RH	96 hr	Pass
High Temperature Storage (HTS)	150C	1000 hr	Pass

Board Level Testing

Temperature cycling on board (TCoB)	-40C to 125C, 1 cyc/hr	500x	Pass
Drop Test	JEDEC	>30	Pass

2S eWLB-PoP CHARACTERIZATION

The 2S eWLB-PoP structure provides a very flexible platform by enabling a top side RDL layer to be added to the package. The structure evaluated here uses pre-formed PCB via blocks to connect the bottom of the package to the top. The via blocks can be placed into the eWLB using the same techniques as are used with a multichip eWLB package. The technique does not provide the thinnest body thickness in a PoP package, but it does allow the size of the top package or packages to decoupled from the size of the bottom package. Although the via blocks must still be placed outside of the area required for the bottom device, the top pads can be redistributed to any desired location. This package has been demonstrated in a package sizes up to 14x14mm. Larger packages are possible but have not yet been demonstrated.

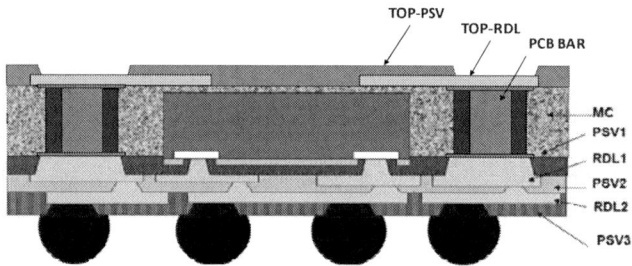

Figure 5: Schematic cross-section of 2S eWLB-PoP

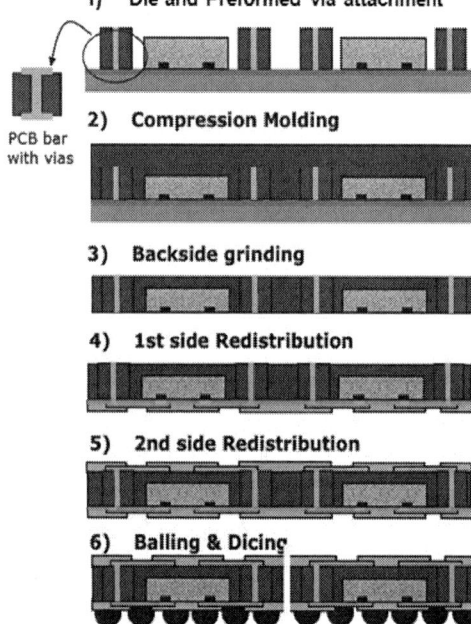

Figure 6: Schematic cross-section of 2S eWLB-PoP

Structure

A schematic cross-section of the 2S eWLB-PoP is shown in Figure 5 and the process flow is shown in Figure 6. The standard eWLB process is used to create a reconstituted wafer that can be processed with standard wafer level packaging equipment. The preformed via blocks are placed at the same time as the active die at the start of the process.

Two Cu RDL layers are then plated in conjunction with 3 polymer layers on the bottom of the package to enable the redistribution. 10um line and 10um space capability can be achieved with this process. The first Cu layer is plated directly onto the bottom of the preformed vias through openings in the first polymer layer. When the RDL patterning on the bottom of the package is complete, similar techniques are used to deposit polymer and plated Cu layers on the top of the package. No special carrier is required if the body thickness is > 450um. Once all RDL patterning is complete, solder balls are placed and reflowed onto the bottom of the package. The reconstituted wafer is then diced to provide the finished singulated packages.

The design of this package is very flexible as a result of the top redistribution layer. One or more packages can be assembled to the top of the package including additional active circuits or passive devices. This technique has been demonstrated with complex System in Package (SiP) structures including a stacked PoP package that included 5 die.

Warpage Performance

Warpage performance is very good in this thin package as compared with the fcPoP package in Figure 7.

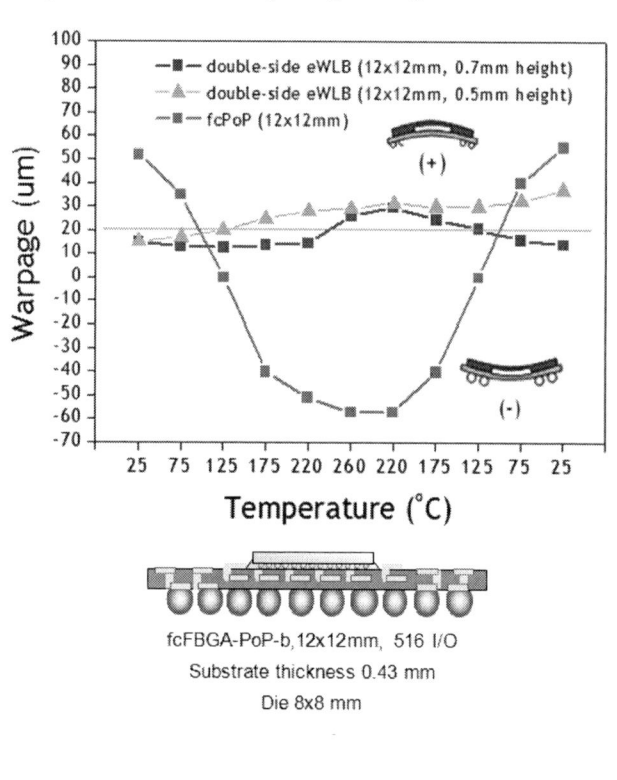

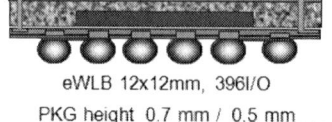

Figure 7: Warpage for eWLB and fcPoP

Table 3: Component Level Testing 12x12mm 2S eWLB-PoP

Test	Condition	Criteria	Status
MSL1	MSL1,260C Reflow (3x)	-	Pass
Temperature Cycling after Precon	-40 C to 125C	1000x	Pass
HAST (w/o bias) after Precon	130C / 85% RH	96 hr	Pass
High Temperature Storage (HTS)	150C	1000 hr	Pass
Ball shear test after multiple reflow	5x, 10x, 20x reflow	Value/Mode	Pass

Board Level Testing

Temperature cycling on board (TCoB)	-40C to 125C, 1 cy/hr	500x	Pass
Drop Test	JEDEC	>30	Pass

Reliability Performance

Reliability performance of the package is shown in Table 3. Board level data is from assembled PoP packages mounted onto the laminate without underfill

Summary of competing package types

eWLB PoP technology provides a packaging solution that is competitive with established MLP PoP packages. Thermal performance is similar to the fc PoP packages and the eWLB PoP packages have the benefit short signal routing

length which may improve electrical performance in high speed devices. The eWLB-MLP structure offers a cost effective package on package solution that is <1.0mm in total package height. The 2S eWLB-PoP package offers a solution that is similar in height to other packages, but has the flexibility of RDL routing on both top and bottom surfaces of the package. Both of the eWLB PoP package technologies offer unique advantages for mobile phone and tablet applications requiring very small form factor packages.

MARKED RELIABILITY INCREASE OF PLASTIC-CORED SOLDER BALL FOR LARGE SIZE WAFER-LEVEL CSP

Hiroya Ishida, Kiyoto Matsushita
Sekisui Chemical Co., LTD, Japan
Koka, Shiga, Japan
ishida030@sekisui.com

ABSTRACT

A unique plastic-cored solder ball (PCSB) developed by Sekisui Chemical Co., Ltd consists of plastic core, copper layer and solder layer. It enables high reliability as a solder bonding material in Wafer-level Chip Size Package (WLCSP) by dispersing the occurred stress via its flexible plastic core. Currently, PCSB has been used commercially in analog devices within mobile phones. However, the demand for higher reliability is increasing as manufacturing shifts to smart phones that utilize packages of larger body size, a finer ball pitch, and thinner chip. Particularly, the reliability performance of board-level temperature cycling tests (TCT) is a great concern to the use in a wide flexibility of package designs.

In this study, we focused on the composition of PCSB in order to maximize its reliability properties and obtain greater reliability in larger-sized WLCSP that use fine ball pitch. We evaluated various compositions of PCSB and compared the reliability performance with conventional solder ball technology. The advanced type of PCSB achieved over 1000 temperature cycles in WLCSP with a package size of 10.2×10.2 mm square, and a ball pitch of 0.3-mm in a temperature range of -40 deg. C. to 125 deg. C. According to the observation of cross sections after TCT, the major crack of the advanced PCSB occurred along the copper electrode, while one occurring using conventional PCSB occurred along its copper layer. SEM analysis revealed that the intermetallic compound (IMC) was formed in the interfaces of the copper layer and the electrode. The IMC of the advanced PCSB was $(Cu_xNi_y)_6Sn_5$ and fine in form, whereas the IMC of conventional PCSB was Cu_6Sn_5 and rough in form. This results in the advanced PCSB demonstrating better TC life.

This technique contributes to excellent reliability and underfill-resin-less in WLCSP, and proves that advanced PCSB is capable of large-size packaging design applications.

Key words: WLP, CSP, Reliability, Solder Ball

INTRODUCTION

Wafer-level chip-scale packaging (WLCSP) that is larger than 5×5 mm^2 has been mainly used for mobile phone applications such as wireless basebands, network processors, and RF connectivity. As wireless handsets shift to smartphones and tablet PCs, there is an increasing demand for advanced, multifunctional applications. To incorporate such applications in a single package, large WLCSP for body sizes of about 8×8 mm^2 is being adopted in smartphones. Future applications are expected to be integrated in single packages that employ WLCSP with body sizes of over 10×10 mm^2 and a fine ball pitch. The package configuration will be vertical such as the three-dimensional package (3D-PKG) in which several large WLCSP are stacked. However, there are several potential problems in adopting large WLCSP.

Board-level temperature cycling reliability is a major concern in smartphones and tablet PC applications since huge amounts of data are rapidly transferred between silicon chips. Due to the different coefficients of thermal expansion (CTE) between the silicon die and the printing circuit board (PCB), more thermal stress is generated at the solder joint in larger WLCSP. Although underfill can be used to improve reliability, device manufacturers are reluctant to use underfill because it makes it more difficult to rework packages. It is also desirable to omit the underfilling process. Moreover, underfill does not give a high reliability because it does not completely fill all solder bumps. There is thus a strong demand to realize a high board-level reliability without using underfill. On the other hand, solder bridges form in fine ball pitch WLCSP because solder balls (SBs) are depressed by the PKG weight during multiple reflow. In addition, it is more important to control the stand-off controllability for 3D-WLP to control the PKG height. Copper pillars and copper-core SBs can be used to realize fine-pitch WLP and 3D-WLP, but they do not provide a sufficiently high board-level reliability. There is thus a need to realize both a high reliability and high stand-off controllability since these are key factors when selecting package materials and designs.

Plastic-cored solder ball (PCSBs) are promising for achieving a sufficiently high reliability without using underfill in large WLCSP with fine pitch since their flexible plastic cores absorb generated stress and disperse the stress over the solder joint interface. Moreover, they allow the stand-off height to be easily controlled by varying the plastic core diameter. Several studies have investigated their use [1–4]. They found that PCSBs give both a high reliability and a high stand-off controllability for packages. However, several problems with PCSBs still need to be overcome. Specific packages mounted on PCSBs are required to be assembled without using solder paste to permit rework processing of packages. In the rework process, faulty packages are removed from a PCB and packages are reassembled without solder paste as it is difficult to reprint them. In this rework process, a thick solder containing PCSBs is adopted for packages that are required to be assembled without solder paste. However, thick solder containing PCSBs is less reliable than conventional solders since it has a lower flexibility. Thus, there is a demand for a thick solder containing PCSB that is also reliable.

In this study, various PCSBs were prepared and their reliabilities were assessed to achieve large WLCSP with a fine ball pitch. The stand-off controllability and PKG assembly performance without solder paste (i.e., the assemblability) were also evaluated.

TYPES OF PLASTIC-CORE SOLDER BALLS

Two types of PCSBs were evaluated (Fig. 1): conventional PCSBs and advanced PCSBs (NN-PCSBs) that incorporates two new technologies, namely nickel barrier technology and nickel doping technology. The nickel barrier layer is expected to prevent copper diffusion and improve the reliability. Mottled nickel doping on the solder surface is used to realize a finer intermetallic compound (IMC) $(Cu_xNi_y)_6Sn_5$ in the electrode interface. A fine IMC helps anchor the solder connection, which is also expected to improve the reliability [5]. In addition, the nickel-doped PCSBs exhibited improved reliability [6].

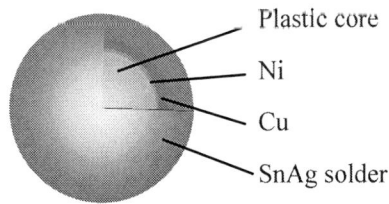

(a) Conventional PCSB

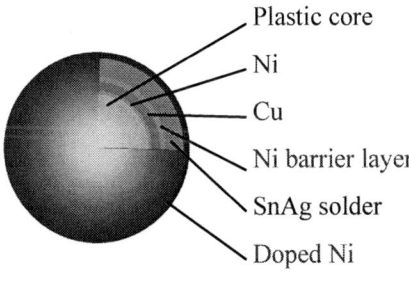

(b) Advanced PCSB (NN-PCSB)

Figure 1: Schematic diagrams of PCSBs

EXPERIMENTAL PROCEDURES
Test Element Group Design

Daisy chain devices with a 10.2×10.2 mm^2 PKG body size and a 0.3-mm ball pitch were used for PCSBs characterization. Table 1 lists the design dimensions of the daisy chain. Figure 2 depicts the structure of the cross-sectional test element group (TEG) that was used. Solder bumps were formed in a two-row array to evaluate the temperature cycling (TC) reliability under severe conditions. Bumps were also formed at the center as dummy balls to prevent PCB warpage occurring during reflow in PKG mounting. There is thus no electrical connection between the central solder bumps. Figure 3 shows a bottom view of a TEG-mounted PCSB. Ball mounting was conducted using a water-soluble flux, which was printed by pin transfer. Reflow was performed in a nitrogen atmosphere with a peak reflow temperature of 250°C. The PCB dimensions are 132 mm × 77 mm × 1 mm. The PCB is made of FR-4. The PCB pad design is solder-mask defined (SMD). The SB pad and solder mask opening have dimensions of 250 and 200 μm, respectively. An organic solderability preservative (OSP) was used as the PCB surface finish. Table 2 lists the PCB design. PKG assembly was conducted using a SAC305 solder paste, which was printed with an 80-μm-thick stencil. Reflow was performed in air using the same temperature profile as for ball mounting. Figure 4 shows a cross-sectional SEM image after balling and PKG assembly.

Table 1: WLCSP TEG parameters

Parameter	Description
Die Size (PKG Size)	10.2 × 10.2 mm
Die Thickness	300 μm
Bump Pitch	0.3 mm
Passivation Opening	100 μm
Passivation Thickness (PBO)	8 um
UBM Size / Thickness	150 μm / 10 μm
UBM Composition	Copper
DNP	7.00 μm
Solder Ball Diameter	190-210 μm
Solder Ball Count	292 bumps

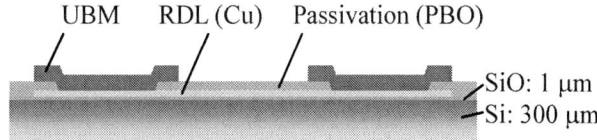

Figure 2: Schematic of WLCSP TEG cross section

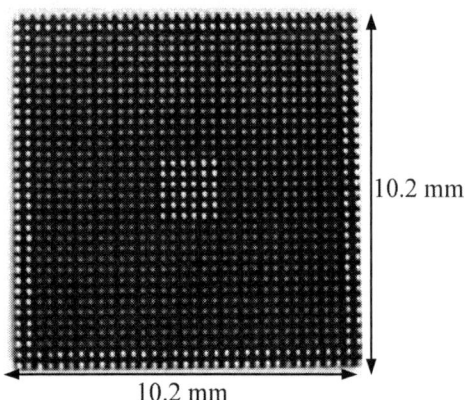

10.2 mm

10.2 mm

Figure 3: Bottom view of a PCSB-mounted TEG.

Table 2: PCB parameters

Parameter	Description
Board Size	132 × 77 mm
Thickness	1.0 mm
Material	FR-4
Land Size	250 μm
Land Type	SMD
Solder Resist Opening Size	200 μm
Solder Resist Thickness	20 μm
Surface Finish	Copper-OSP

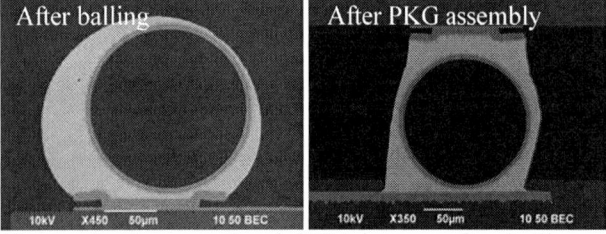

Figure 4: Cross-sectional SEM images after balling and PKG assembly.

Board Level Reliability Test

Board-level TC tests (TCTs) were conducted in accordance with JEDEC specifications JESD22-A104D (condition G). The temperature was cycled in the range –40°C to 125°C with a 10 min dwell on each end. The transition time on each side was approximately 2 min. The resistance of each side was monitored in real time. A failure was considered to have occurred when the resistance increased by 10% from the initial resistance of 1–1.1 Ω. Distributions were evaluated in terms of the shape parameter m and the characteristic life η (the minimum duration for 63.2% accumulative failure). Optical microscopy was used to analyze the failure mode. In addition, a board level drop test was conducted in accordance with JEDEC specifications JESD 22-B111 (group F) with a peak acceleration of 1500 G and a half-sine shock pulse duration of 0.5 ms.

RESULTS AND DISCUSSION
Comparison of Reliabilities of PCSBs

We compared the characteristics of NN-PCSBs with those of conventional PCSBs. Two types of PCSBs were

evaluated. Table 3 shows the PCSB configurations used in this study. Each PCSB was 200 μm in diameter. The solder composition is tin–3.5 wt.% silver.

Table 3: PSCB configurations

Type	Diameter	Core	Copper layer	Solder layer
PCSB 200-20	200 μm	150 μm	5 μm	20 μm
NN-PCSB 200-20	200 μm	150 μm	5 μm	20 μm

Scanning electron microscopy (SEM) was used to observe the solder joint interfaces. Figure 5 shows the interfacial IMC microstructure of each solder joint after balling and PKG assembly, respectively. The IMC in NN-PCSBs has different constituent elements and morphology from those of PCSB. EDX analysis revealed that the IMCs of NN-PCSBs and PCSBs were $(Cu_xNi_y)_6Sn_5$ and Cu_6Sn_5, respectively. This implies that the doped nickel was incorporated in the solder and reached the copper electrode during solder convection in reflow. In addition, the NN-PCSBs had finer IMCs than PCSB. The finer IMC is expected to better anchor the electrode and to improve the reliability.

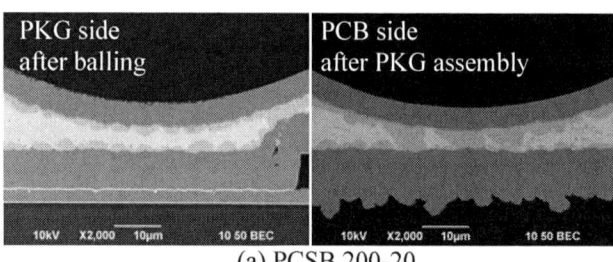

(a) PCSB 200-20

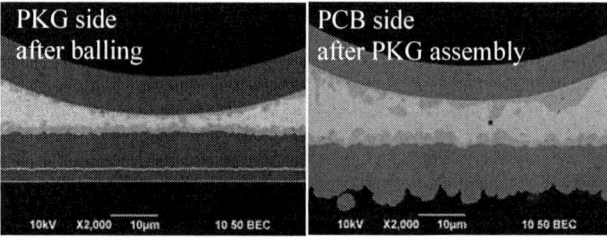

(b) NN-PCSB 200-20

Figure 5: SEM images of IMC morphology.

TCT was performed to verify the effect of the finer IMC. In addition to the PCSBs, three solid SBs were evaluated: SAC305, SAC405, and nickel-containing SAC125 (SACN). The results reveal that NN-PCSB has a much higher reliability than PCSB and the other SBs, as shown in Fig. 6. Its first failure occurred at 1001 cycles with η of 1167 cycles, as shown in Table 4. The TC reliability of NN-PCSB is over 350% higher than that of the conventional SB SAC305. The crack generated during TCT propagates laterally from the weakest interface due to the lateral stress due to the CTE mismatch between the silicon die and the PCB. Therefore, the failure mode occurs at the weakest solder connection in the interface. Figure 7 shows cross-sectional optical microscopy images of the failure mode. These images reveal that NN-PCSB has a different failure mode from conventional PCSB. In PCSB, the crack occurred along the IMC interface near the copper layer on the PKG side (Fig. 7(a)). The IMC interface near the curved copper layer is assumed to be more stressed than that near the parallel copper under bump metallization (UBM) by the lateral stress generated in TCT. In contrast, the crack occurred along the IMC interface near copper UBM in the PKG side in NN-PCSB (Fig. 7(b)). This implies that the nickel barrier layer strengthens the IMC of the NN-PCSBs near the copper layer.

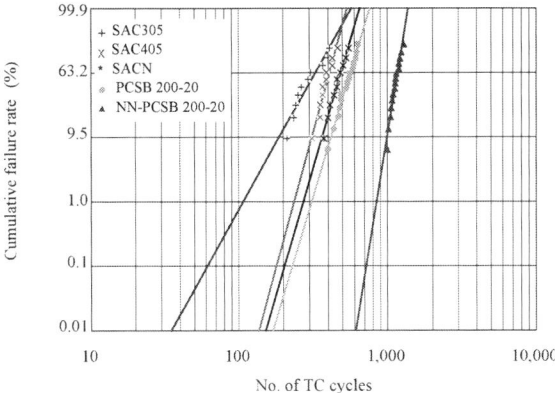

Figure 6: Weibull plot of TCT of PCSBs.

Table 4: Comparison of TCT results for PCSBs and SBs

Symbol	Type	First failure	m	η	vs. SAC305	vs. PCSB
+	SAC305	212	4.1	328	1.0	-
X	SAC405	313	8.4	407	1.2	-
*	SACN	374	7.9	488	1.5	-
◉	PCSB 200-20	399	7.7	565	1.7	1.0
▲	NN-PCSB 200-20	1001	14.3	1167	3.6	2.1

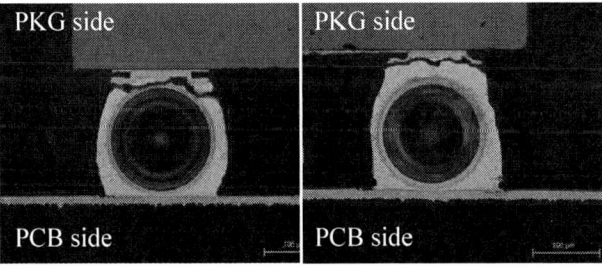

(a) PCSB 200-20   (b) NN-PCSB 200-20

Figure 7: Cross-sectional optical microscopy images of TCT failure mode

We also performed drop tests, which revealed that NN-PCSB has a much higher reliability than PCSB and SACN (Fig. 8 and Table 5).

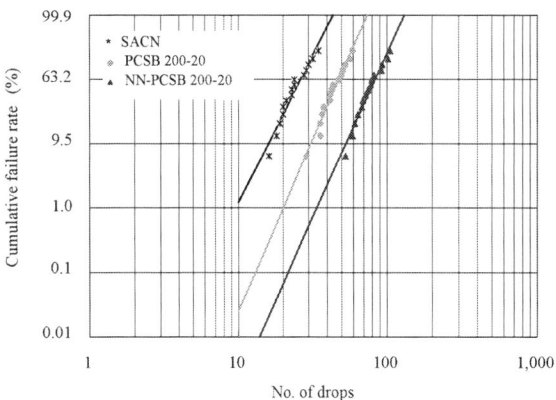

Figure 8: Weibull plot of drop tests for PCSBs.

Table 5: Comparison of drop test results for PCSBs and SBs

Symbol	Type	First failure	m	η	vs. SACN	vs. PCSB
⋆	SACN	16	4.5	26	1.0	-
◉	PCSB 200-20	29	5.3	48	1.8	1.0
▲	NN-PCSB 200-20	53	5.1	83	3.2	1.7

Stand-off Controllability by Plastic Core

To investigate whether the stand-off could be controlled by the plastic core diameter, 200-μm-diameter NN-PCSBs with various configurations were prepared. Table 6 shows the types of NN-PCSB used in this study.

Table 6: Configurations of 200-μm-diameter NN-PCSBs

Type	Diameter	Core	Copper layer	Solder layer
NN-PCSB 200-20	200 μm	150 μm	5 μm	20 μm
NN-PCSB 200-25	200 μm	140 μm	5 μm	25 μm
NN-PCSB 200-30	200 μm	130 μm	5 μm	30 μm

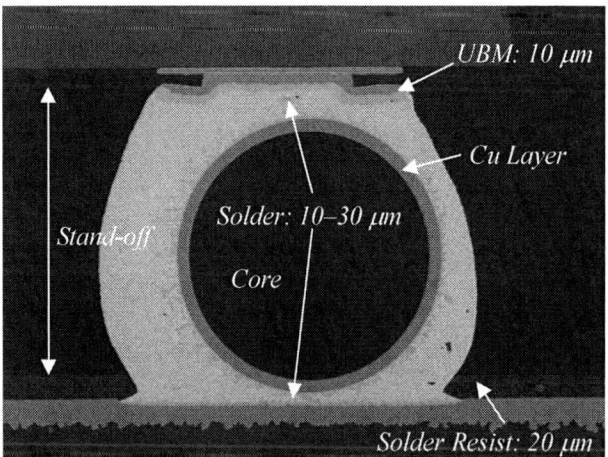

Figure 9: Estimated stand-off height

In PCSBs, the stand-off height was estimated from the plastic core diameter, the copper layer thickness, the amount of solder, UBM thickness, and the PCB solder resist thickness (Fig. 9). The estimated stand-off height (ESH) is defined as:

$$ESH = core + 2 \times Cu + solder + UBM + resist \quad (1)$$

The solder amount is estimated to be between 10 and 30 μm based on our experience. Therefore, ESH is defined as:

$$core + 2 \times Cu < ESH < core + 2 \times Cu + 20 \quad (2)$$

Figure 10 shows cross-sectional images after balling and of the PKG assembly of NN-PCSB interconnections.

As the core diameter was smaller and the solder layer was thicker, the bump fillet changed from cylindrical to elliptical and the stand-off decreased. Table 7 compares the estimated and measured stand-off heights. It demonstrates that the stand-off height can be easily controlled by the core diameter. Moreover, the stand-off height was almost the same as the ball bump height in NN-PCSB 200-20. In contrast, the stand-off height was much lower than the ball bump height in SACN since PCSB was not pressed down by the PKG weight, unlike for the SBs. Thus, PCSBs are expected to contribute to the 3D-PKG height, especially when larger or heavier PKG are used to maintain a specific stand-off between the die and PCB. In addition, the stand-off height varies little despite the PKG weight varying somewhat. This implies that PCSBs give an adequate margin for PKG height design.

After balling

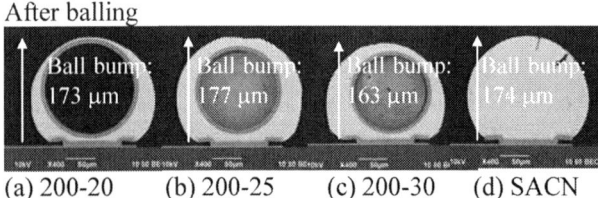

(a) 200-20   (b) 200-25   (c) 200-30   (d) SACN

After PKG assembly

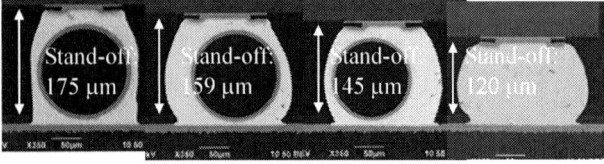

(a) 200-20   (b) 200-25   (c) 200-30   (d) SACN

Figure 10: Comparison of ball bump and stand-off heights for various NN-PCSBs.

Table 7: Comparison of estimated and measured stand-off heights.

Type	Estimated stand-off height	Measured stand-off height
NN-PCSB 200-20	160-180 μm	175 μm
NN-PCSB 200-25	150-170 μm	159 μm
NN-PCSB 200-30	140-160 μm	145 μm

TCTs were performed to investigate the influence of the NN-PCSB configuration on reliability. The results revealed that NN-PCSBs with smaller cores had a lower reliability due to the lower flexibility of the solder joint; however, they showed much higher reliabilities than the other SBs (Fig. 11 and Table 8).

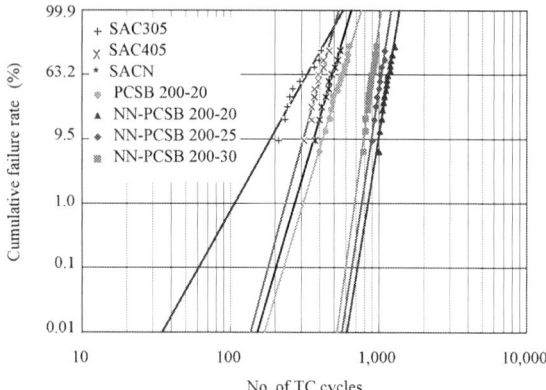

Figure 11: Weibull plot of TCT results for 200-μm-diameter NN-PCSBs.

Table 8: Comparison of TCT results for various 200-μm-diameter NN-PCSBs

Symbol	Type	First failure	m	η	vs. SAC305	vs. PCSB
+	SAC305	212	4.1	328	1.0	-
×	SAC405	313	8.4	407	1.2	-
*	SACN	374	7.9	488	1.5	-
●	PCSB 200-20	399	7.7	565	1.7	1.0
▲	NN-PCSB 200-20	1001	14.3	1167	3.6	2.1
◆	NN-PCSB 200-25	903	15.2	1038	3.2	1.8
▦	NN-PCSB 200-30	786	17.0	903	2.8	1.6

Drop reliability exhibits a similar trend as the TCT results, (Fig. 12 and Table 9).

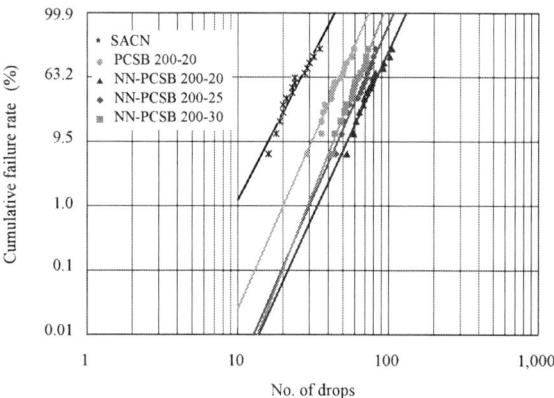

Figure 12: Weibull plot for drop test of 200-μm-diameter NN-PCSBs

Table 9: Comparison of drop test for various 200-μm-diameter NN-PCSBs

Symbol	Type	First failure	m	η	vs. SACN	vs. PCSB
★	SACN	16	4.5	26	1.0	-
◉	PCSB 200-20	29	5.3	48	1.8	1.0
▲	NN-PCSB 200-20	53	5.1	83	3.2	1.7
◆	NN-PCSB 200-25	45	5.4	71	2.7	1.5
▦	NN-PCSB 200-30	42	6.0	63	2.4	1.3

PKG Assembly Performance without Solder Paste and Board Level Reliability

To investigate the PKG assembly performance without solder paste, various solder thicknesses of NN-PCSBs were evaluated (Table 6). In this experiment, PKG was assembled with flux instead of solder paste, as shown in Fig. 13. The PKG mounted NN-PCSB 200-20 was not assembled on PCB without solder paste, whereas NN-PCSB 200-25 and NN-PCSB 200-30 were assembled. Figures 14(a) and (b) show cross-sectional SEM images of interconnections of NN-PCSB 200-20 and NN-PCSB 200-30 after balling and PKG assembly, respectively. In NN-SOL 200-30, since sufficient solder remained on the top of bump after balling, PKG could be assembled even without solder paste and a perfectly cylindrical bump fillet was formed.

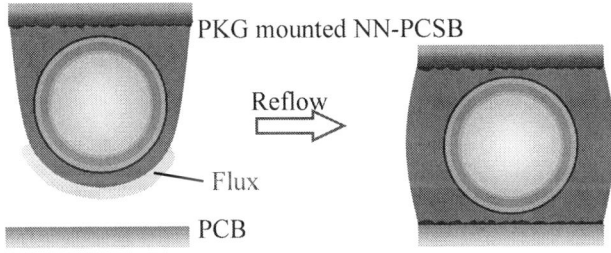

Figure 13: PKG assembly without solder paste.

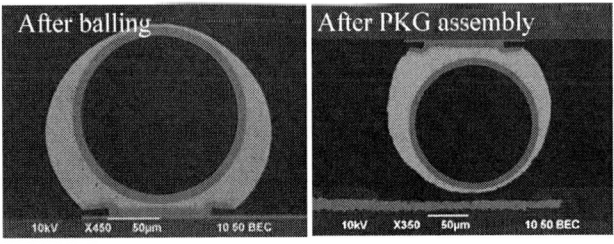

(a) NN-PCSB 200-20

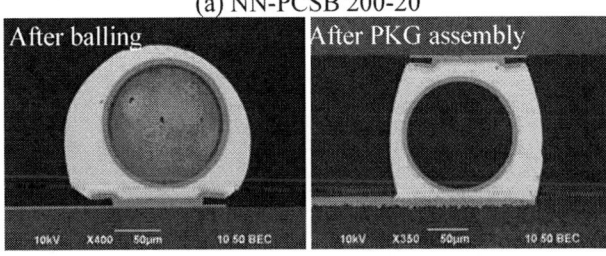

(b) NN-PCSB 200-30

Figure 14: Cross-sectional SEM images after balling and PKG assembly w/o paste.

TCTs and drop tests were performed to investigate the influence of the solder paste on reliability. For SACN, the

TC reliability without paste was 15% lower than that with paste, and the drop reliability without paste was 24% lower than that with paste. For NN-PCSB, there was no significant difference in TCT between the reliabilities with and without paste for NN-PCSB, and the drop reliability without paste showed the similar trend as SACN (Fig. 15, 16 and Table 10, 11).

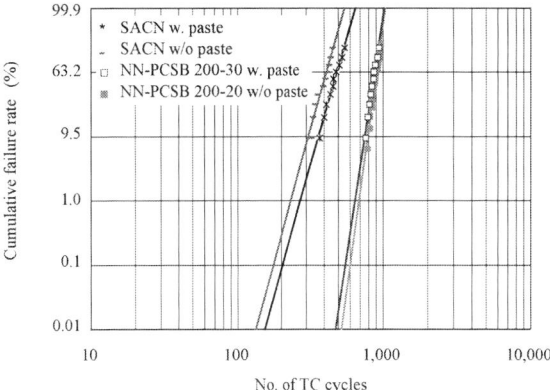

Figure 15: Weibull plot of TCT results for NN-PCSB 200-30 without paste.

Table 10: Comparison of TCT between with and without paste

Symbol	Type	First failure	m	η	vs. w. paste
★	SACN w. paste	374	7.9	488	1.00
-	SACN w/o paste	325	8.1	414	0.85
□	NN-PCSB 200-30 w. paste	786	17.0	903	1.00
▦	NN-PCSB 200-30 w/o paste	763	15.2	875	0.97

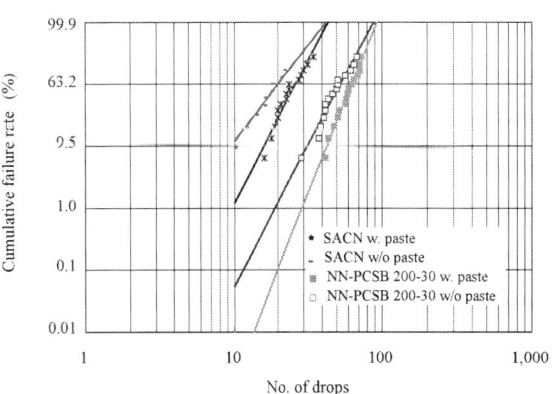

Figure 16: Weibull plot of drop test results for NN-PCSB 200-30 without paste.

Table 11: Comparison of drop test results with and without paste.

Symbol	Type	First failure	m	η	vs. w. paste
★	SACN w. paste	16	4.5	26	1.00
-	SACN w/o paste	10	3.0	20	0.76
□	NN-PCSB 200-30 w. paste	42	6.0	63	1.00
▦	NN-PCSB 200-30 w/o paste	29	4.5	53	0.84

To evaluate these results, we compared the stand-off heights obtained with and without paste (Fig. 17). The stand-off height of SACN decreased from 120 μm with solder paste to 111 μm without solder paste, whereas the stand-off heights of NN-PCSB 200-30 were the same with and without paste. Thus, for NN-PCSB, it's considered that the solder paste didn't affect the reliability critically.

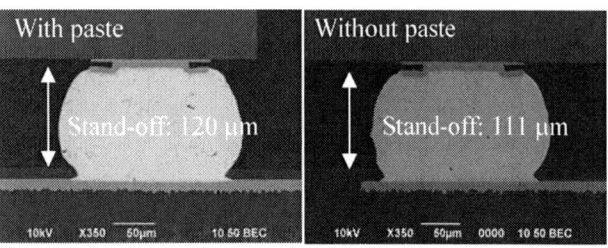

(a) SACN 200 μm

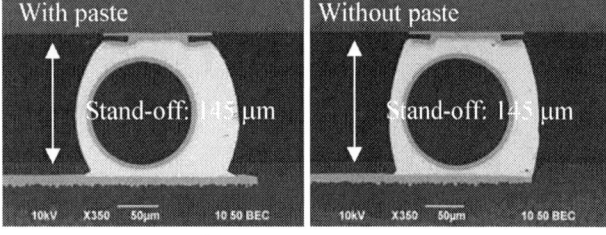

(b) NN-PCSB 200-30

Figure 17: Comparison of stand-off heights with and without paste.

Dependence of Board Level Reliability on Various NN-PCSB Configurations

Various configurations of NN-PCSB were evaluated to maximize the NN-PCSB quality and to search for a configuration that optimizes the PKG quality and process for 0.3 mm pitch WLCSP. Table 12 shows the types of NN-PCSB used in this study. The PCSBs have diameters in the range 190–210 μm.

Table 12: Configurations of NN-PCSB

Type	Diameter	Core	Copper layer	Solder layer
NN-PCSB 190-20	190 μm	140 μm	5 μm	20 μm
NN-PCSB 190-25	190 μm	130 μm	5 μm	25 μm
NN-PCSB 190-30	190 μm	120 μm	5 μm	30 μm
NN-PCSB 200-20	200 μm	150 μm	5 μm	20 μm
NN-PCSB 200-25	200 μm	140 μm	5 μm	25 μm
NN-PCSB 200-30	200 μm	130 μm	5 μm	30 μm
NN-PCSB 210-20	210 μm	160 μm	5 μm	20 μm
NN-PCSB 210-25	210 μm	150 μm	5 μm	25 μm
NN-PCSB 210-30	210 μm	140 μm	5 μm	30 μm

Table 13 compares TCT, Drop and PKG assembly performance without paste. NN-PCSB 210-20 exhibited the highest reliability. NN-PCSB 210-30 exceeded over 1000 TC cycles and realized PKG assembly performance without paste.

Figure 18 shows the relationship between the core–volume ratio and the reliability. The reliability increased with increasing core–volume ratio and increasing ball size due to the larger cores having greater flexibility and higher stand-off heights to the solder joint. A solder thickness of over 25 μm and a ball size of over 200 μm are required to conduct PKG assembly performance without paste and achieve a high TC reliability (Fig. 19).

Table 13: PKG assembly performance without paste and TC for various NN-PCSB configurations

Type	Assemblability w/o paste	TCT (η)		Drop (η)	
		w. paste	w/o paste	w. paste	w/o paste
SAC305	pass	328	NA	NA	NA
SAC405	pass	407	NA	NA	NA
SACN	pass	488	414	26	20
PCSB 200-20	fail	565	-	48	-
NN-PCSB 190-20	fail	959	-	NA	-
NN-PCSB 190-25	pass	862	NA	NA	NA
NN-PCSB 190-30	NA	NA	NA	NA	NA
NN-PCSB 200-20	fail	1167	-	83	-
NN-PCSB 200-25	pass	1038	1038	71	NA
NN-PCSB 200-30	pass	903	875	63	53
NN-PCSB 210-20	fail	1209	-	NA	-
NN-PCSB 210-25	pass	1142	1098	NA	NA
NN-PCSB 210-30	pass	1051	1038	NA	NA

Core-volume ratio vs. reliability

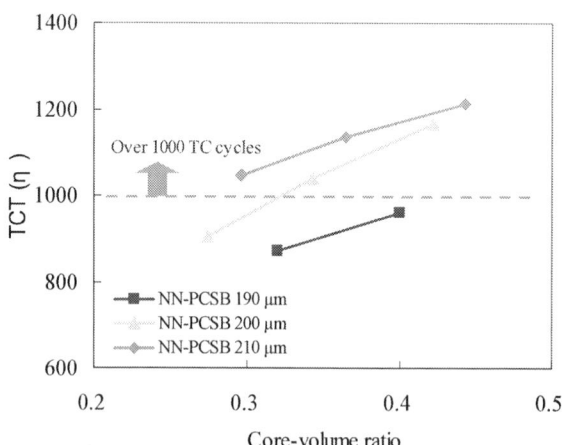

Figure 18: Relation between core–volume ratio and TCT (η).

Solder thickness vs. reliability

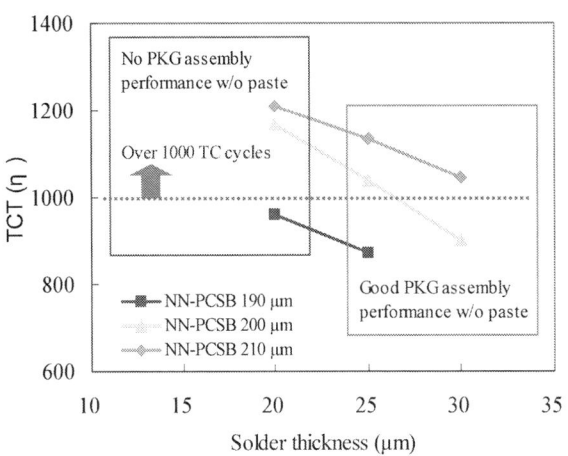

Figure 19: Relation between solder thickness and TC (η).

PKG Yield for Various NN-PCSB Configurations

In WLCSP with a fine ball pitch, solder bridges significantly reduce the PKG manufacturing yield. To consider the effect of solder bridges, the bump gap after balling and the fillet width after PKG assembly were evaluated. Since 210-30 and 190-20 have the widest and narrowest bumps respectively, we compared NN-PCSB 210-30 and 190-20 with SACN and considered whether they affect the PKG yield.

In the balling process, a solder bridge forms when balls are moved by air or nitrogen convection during reflow; this reduces the PKG yield of a wafer singulation. Figure 20 shows the bump gaps after balling. Measurements were performed for several tens of gaps and the average values were compared. The gap width in SACN was 97.5 µm, whereas the gap widths in NN-PCSB 210-30 and 190-20 were 95.8 and 103.8 µm, respectively. A dicing width of over 50 µm is generally required for package singulation. Therefore, the measured gap widths are acceptable.

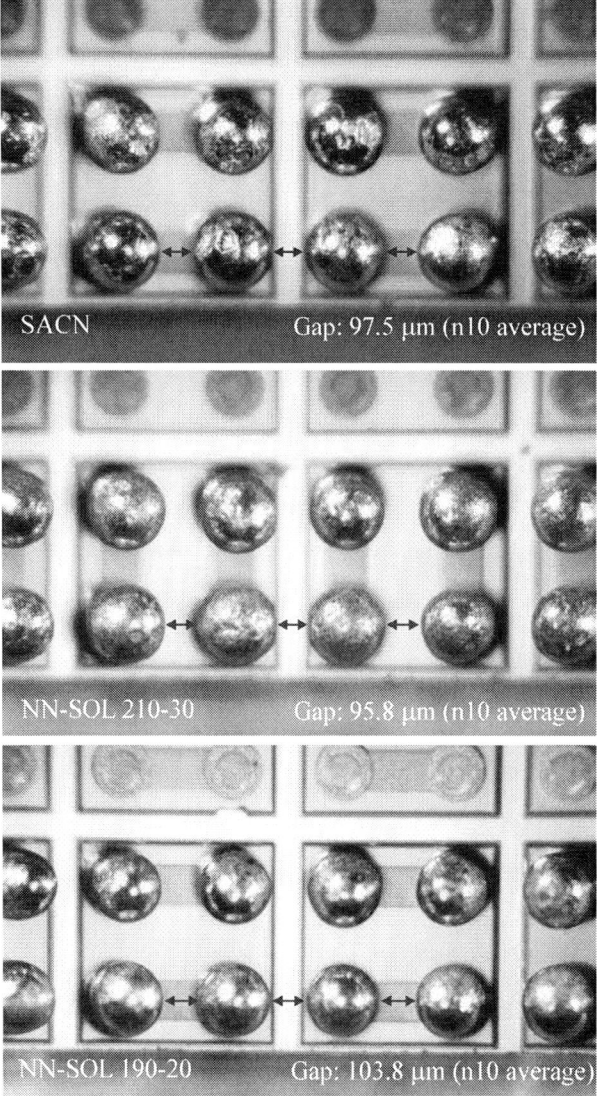

Figure 20: Bump gap after balling

In the PKG assembly process, solder bridges also formed through bridging between balls pressed together by the PKG weight during reflow, which also reduces the PKG yield. In addition, misalignment of PKG-PCB causes solder bridges to form. Therefore, a fillet width after PKG assembly is a key factor for obtaining a high PKG yield. Figure 21 compares the bump width after balling and the fillet width after PKG assembly with solder paste for different balls. In SACN, the width increased by 7% (209 µm to 224 µm) after PKG assembly, whereas the widths were considerably narrower and the values were smaller than SACN for NN-PCSBs. This implies that the risk of solder bridges forming in NN-PCSB is much lower than SACN, despite NN-PCSB 210-30 having larger balls than SACN. Moreover, both NN-PCSBs are expected to have much higher reliabilities than SACN. Both a high reliability and a high PKG manufacturing yield are required.

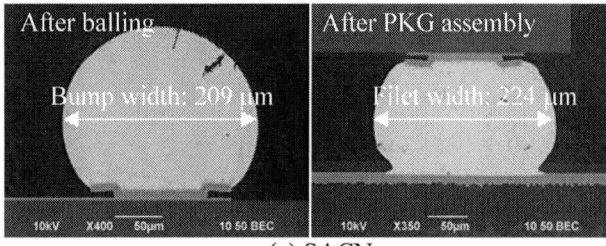

(a) SACN

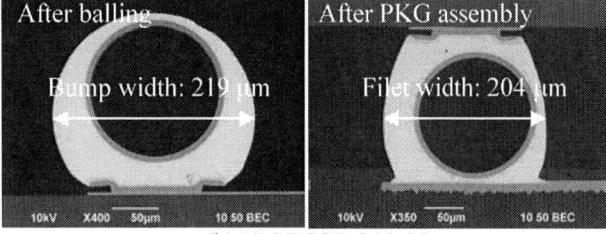

(b) NN-PCSB 210-30

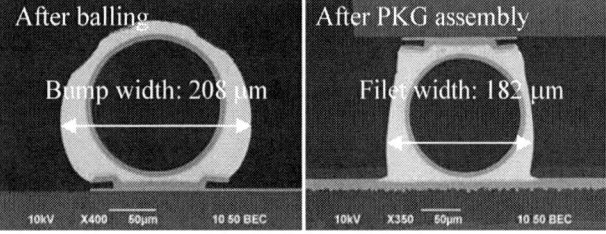

(c) NN-PCSB 190-20

Figure 21: Comparison of bump and fillet widths

CONCLUSION

NN-PCSB techniques allowed over 1000 temperature cycles to be performed without using underfill, even for large WLCSP with a fine pitch. This suggests that NN-PCSB can enable several functions to be integrated into a single PKG and that further functional applications can be achieved in a single PKG.

In addition, it should permit the specific stand-off height to be easily controlled by the core diameter in the 3D-PKG design of even larger and/or heavier PKG since NN-PCSB was not depressed by the PKG weight, unlike SB. Moreover, NN-PCSB exhibited an excellent reliability with TEG and solder bumps formed in just two rows in a peripheral array. Thus, NN-PCSB is promising for 3D-WLP designed for CMOS sensors whose solder bumps are fabricated in a peripheral array consisting of one or two rows. This is because plastic cores are very effective in maintaining a high stand-off height between the die and the PCB.

On the other hand, the thicker NN-PCSB solder achieved package assembly without solder paste and also exhibited an excellent reliability. It is thus not necessary to use solder paste in a board assembler for PKG assembly or PKG reworking. Thus, PKG mounted NN-PCSB are feasible for assembly in a narrow area where it is difficult to print solder paste.

Depending on the PKG quality, various configurations could be chosen. It gave excellent reliability, stand-off controllability, and PKG assembly capability without paste

for any NN-PCSB. Moreover, NN-PCSB is expected to give a higher PKG yield than SAC series.

NN-PCSB technologies can realize highly versatile packaging designs such as multi-die FOWLP, 3D-WLP, and simple structures without UBM. Moreover, they can be utilized in applications requiring high reliabilities such as automotive applications.

REFERENCES

[1] Yeng-Ping Wang, "High Drop Performance Interconnection: Polymer Cored Solder Baa" in Proc. ECTC 2008, P1208-1211

[2] Jani Miettinen, "Stacked 3-D MCP with Plastic Ball Vertical Interconnections" in ECTC 2003, P1101-1105

[3] Masato Sumikawa, "Reliability of a Wafer Level Packaging Method with Plastic-core Solder bumps: -- Utilizing Sn-Ag solder at 0.3 mm diameter --" in IMAPS,

[4] Suminoe Shinji, "OPTIMIZATION DESIGN FOR WAFER-LEVEL CSP" in ISTC 2002, P800-811

[5] Shinichi Terashima, "Improvement on Thermal Fatigu Properties of Sn-1.2Ag-0.5Cu Flip Chip Interconnects by Nickel Addition": Materials Transactions, Vol. 45, No.3 (2004), P673-680

[6] Ren-De Sun, "Study on Improving the Drop Impact Reliability of Plastic Core Solder Balls" in ICEP 2009 Proceedings, P869-874

3D TSV MICRO CU PILLAR CHIP-TO-SUBSTRATE/CHIP ASSEMBLY/PACKAGING TECHNOLOGY

Seung Wook Yoon, *K. T. Kang, W. K. Choi, * H. T. Lee,
Andy C. B. Yong and Pandi C. Marimuthu
STATS ChipPAC LTD, *STATS ChipPAC Korea Ltd.
Singapore, Singapore, *Bubal-eub, Korea
seungwook.yoon@statschippac.com

ABSTRACT

Increasing demand for new and more advanced electronic products with a smaller form factor, superior functionality and performance with a lower overall cost has driven semiconductor industry to develop more innovative and emerging advanced packaging technologies.

Memory bandwidth has become a bottleneck to processor performance and lower power consumption for high performance computing needs. To reduce obstacles, a revolution in device architecture and package technologies is require. 3D TSV (Through Silicon Via) stacking is believed to be one of the technologies that can meet those requirements. In advanced 3D stacking technologies, one of the important steps is to develop and assemble fine pitch, high density microbumps. This type of microbump in flip chip interconnection provides a high wiring density in silicon die with a high-performance signal and power connection.

This paper addresses Cu pillar technology for 3D TSV assembly/packaging process. Latest developments in the key elements of thin TSV wafer dicing and thermo-compression bonding with Cu pillar bumps are presented. 3D TSV interconnects as an interim play prior to full production of the active Si TSV approach is reviewed with specific example of configurations approaching volume production in real products. TSV backside via reveal process including thin wafer handling will be discussed. TSV Packaging challenges, thermocompression bonding and its reliability experimental results will be presented for CTC (Chip-to-Chip), CTS (Chip-to-Substrate) bonding with ultra fine 40um pitch microbump interconnections in this paper.

Key words : Cu pillar, micro bump, 3D TSV packaging, wider IO memory, Chip-to-chip bonding, Chip-to-substrate bonding

INTRODUCTION

The synergies and intersections among three parallel developing areas of packaging technology (i.e. traditional die and package stacking on substrates, fan-in and fan-out wafer level packaging and 3D Si integration), and the resulting path for advanced packaging technology are quite critical for future microelectronics system packaging.

One of the hottest topics in the semiconductor industry today is 3D Packaging using Through Silicon Via (TSV) technology. Driven by the need for improved performance and the reduction of timing delays, methods to use short vertical interconnects have been developed to replace the long interconnects found in 2D packaging. The industry is moving fast the feasibility phase for TSV technology into the commercialization phase, where economic realities will determine the technologies that can be adopted. Low-cost, high aspect ratio, reliable via formation and via filling technologies are the need of the hour [1]. Choosing the right process equipment and materials with innovative design solutions addressing thermal and electrical issues will be the key winner. As functional integration requirements increase, assembly and wafer fabrication companies are looking to 3D TSV technology, which allows stacking of LSIs thereby enabling products to be made smaller with more functionality. 3D technology realizes miniaturization by 300-400% compared to conventional packaging.

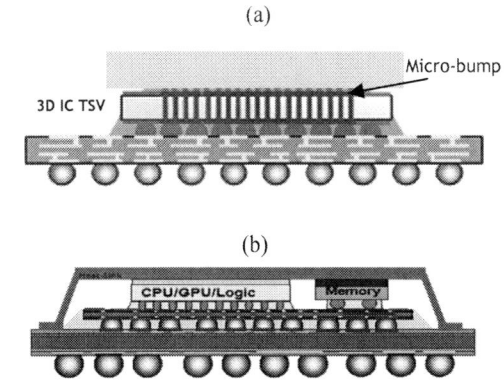

Figure 1. 3D integration packaging with microbump and TSV, (a) 3D-TSV IC and (b) 2.5D TSV interposer

3D integration is progressing on three fronts starting with package-level (die, package stacking), wafer level (die-to-wafer bonding, fan out WLP) and more recently at the Si level (TSV). The demand for high density and multifunctional microelectronics leads to the development of 3D and wafer level packaging, which provides an optimal solution for the shortened interconnects, increased performance and functionality, miniaturization in size and weight, integration of heterogeneous technologies and complex multi-chip systems as well as reduced power consumption. Such packaging technology normally requires the use of ultrathin devices (less than 100μm in thickness).

The key benefits from thinned wafers include improved heat dissipation and reduced electrical resistance which offers better flexibility for 3D stacking. However, it brings up a challenge for assembly and packaging; thinning and handling ultrathin semiconductor devices in both front-end and back-end processes due to its fragility and tendency to warp.

Fig.2 shows schematic process flow of MEOL (Mid-End of Line) and BEOL (Back-End of Line) assembly process of 3D IC TSV after TSV formation in fab processes.

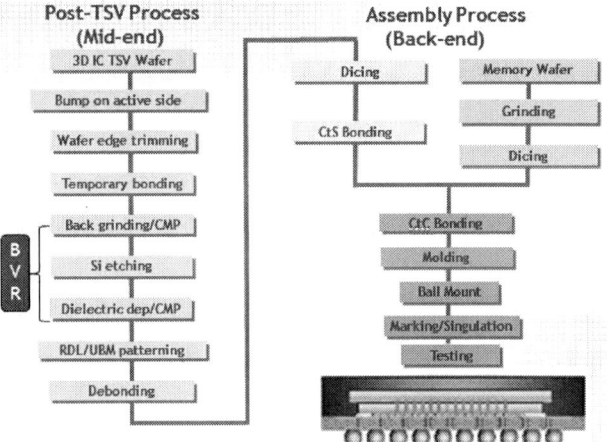

Figure 2. Schematics of process flow of 3D TSV MEOL and BEOL process

3D TSV MEOL PROCESS

TSV MEOL process flow that occurs between the wafer fabrication and back-end assembly process. MEOL processes support the advanced manufacturing requirements of 2.5D and 3D TSV as well as wafer level packaging, flip chip and embedded die technology.

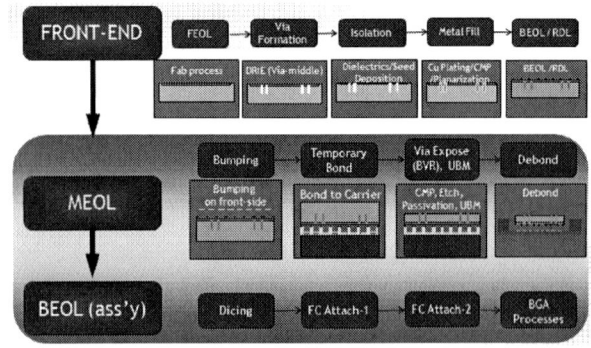

Figure 3. TSV-MEOL & BEOL assembly steps in overall 3D TSV process flow.

Flip chip and wafer level packaging are important drivers of mid-end processing in addition to the anticipated growth in 3D solutions utilizing TSV technology, particularly with the integration of memory and logic devices at advanced technology nodes. The initial markets that are expected to embrace 3D TSV technology are mobile applications, memory stacking and high performance processors for the computing segment.

Cu Pillar Microbump Process

Micro bumping technology where bump pitches are less than 50 micrometers using solder is explored extensively in industry for realization of miniaturized 3D IC integration. For Wide IO microbump in JEDEC 42.6 standards, it has 50/40um bump pitch in x/y direction, respectively. Cu pillar with solder cap micro-bumps have been studied with the objective to develop reliable fine pitch solder micro joints at low cost. Microbump fabrication is based on photolithography and electroplating processes, which is compatible with conventional IC fabrication. The fabrication process of wafer level process starts with bare Si wafer using Ultra Violet (UV) - light lithography of spin on dielectric material. Secondly, Redistribution line (RDL) layer plating to re-route the Al/Cu bond pads to microbump locations. Thirdly, passivation of RDL layer using spin on dielectric coating and UV lithography to open the RDL metal pads at the bump pads. Fourthly, deposition of Ti/Cu seed layer and patterning of thick photoresist film using lithography to copper pillar plating and then Ni/solder plating. SEM micrographs of microbump are shown in Fig.4 for 20μm diameter and 40μm height microbumps for 80/40μm staggered bump pitch.

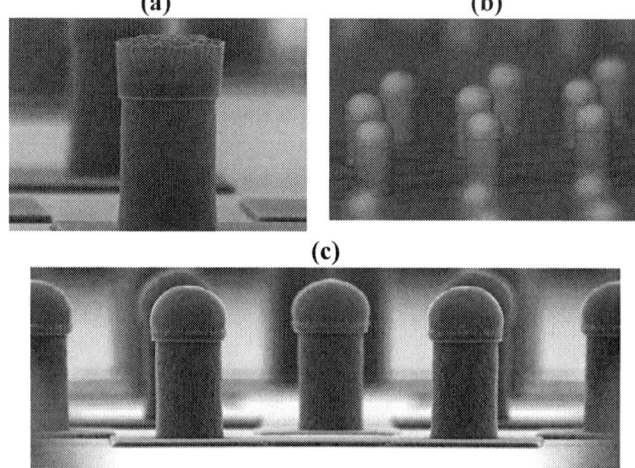

Figure 4. Micrographs of micro bump (a) after micro-bump fabrication and (b & C) after reflow

Backside Via Reveal (BVR) Process

As shown in Fig.5 (a), TSV is revealed to backside for 3D vertical interconnection after front-end TSV formation. With temporary bonding/debonding system, TSV wafer from fab is to be back grinded and Si etched to expose Cu via with fab process. There was TOF SIMS (Time-of-Flight

124

Secondary Ion Mass Spectroscopy) analysis for Cu contamination on Si wafer during CMP (Chemical Mechanical Polishing) process and verified non-detectable Cu content after chemical composition analysis along whole 12" TSV wafer. Fig. 5(b) shows SEM cross-section view of solid Cu filled TSV of 10um diameter and 50um depth after backside via reveal process.

(a)

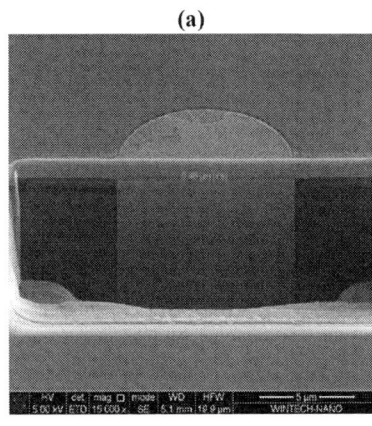

(b)

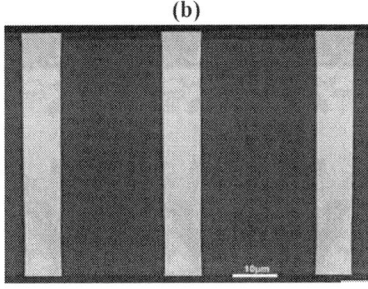

Figure 5. SEM micrographs of backside TSV via revealed after CPM and Si etching process in 12" 3D TSV wafer. (a) FIB titled view of TSV via revealed and (b) cross-sectional view of 10um diameter and 50um depth TSV

3D TSV ASSEMBLY AND PACKAGING

Compared to conventional flipchip process, TSV assembly process is more complex due to TSV wafer as well as microbump. As shown in Fig. 6, there are additional materials, like as additional encapsulation in between bump and flipchip die or bump and TSV die. There are quite critical challenges for assembly view point both in materials and assembly process.

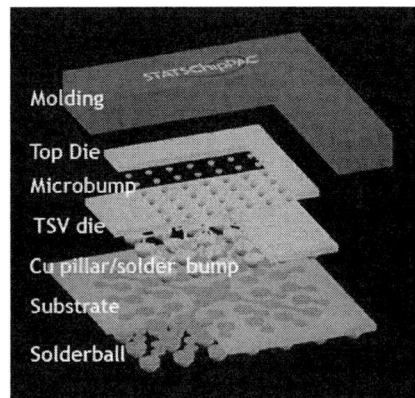

Figure 6. Schematics of 3D TSV assembly and packaging.

In advanced 3D TSV stacking technologies, one of the important steps is to develop and assembly fine pitch and high density solder microbumps. Solder microbumps for flip-chip interconnections allow high wiring density in the Si-carrier, as compared to organic or ceramic substrates, and enable high-performance signal and power connections [2].

Flip chip assembly was carried out to establish bonding process and investigate the reliability with Cu pillar microbump. After microbump test vehicle fabrication with bump, the flip chip attachment was carried out. Several DOE (design of experiments)s were carried out to find optimized flip chip attach process conditions as functions of time, temperature and pressure. Assessments by checking fractural surface and mechanical shear strength were conducted to evaluate DOEs of bonding parameters.

(a)

(b)

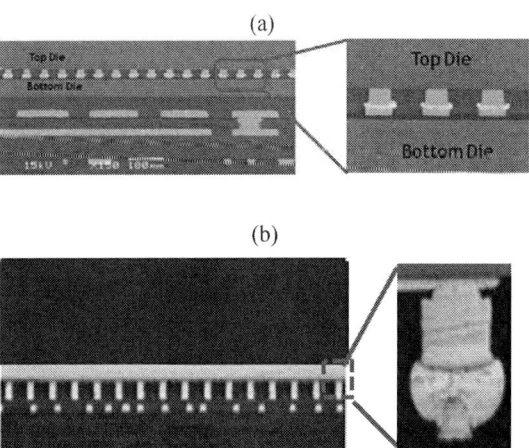

Figure 7. Micrographs of (a) 40μm pitch of chip-to-chip bonding and (b) 40/80μm pitch of chip-to-substrate bonding.

Fig. 7 shows the micrographs of cross-section of the chips joined for 40μm pitch microbumps. A misalignment of about <2μm was observed between the Si chip and 3D TSV chip after assembly. This misalignment was a result of accuracy limitation of the bonder equipment. After assembly, x-ray image was observed and found successful 3D TSV flipchip bonding without voids in between Chip-to-Chip, Chip-to-Substrate, respectively.

Reliability Test Results
After BEOL assembly process, assembled 3D TSV stacked (top die and bottom die) test vehicles were sent to JEDEC standard reliability tests. The electrical test (open/short) was performed to check the daisy chain interconnection. Details of reliability results for 3D TSV packages are described in Table 1.

Reliability samples passed both MSL (Moisture Sensitivity Level)-2aa and MSL-3 with 3x reflow process at Pb-free 260°C peak temperature. All samples passed TC (Temperature Cycling), unbiased HAST (Highly Accelerated Stress Test) and HTS (High Temperature Storage) reliability tests. There was no failure found after 1000 T/C.

SAT (scanning acoustic tomography) also was performed to check interfacial delamination or defects for all samples before and after reliability tests. With final test (open/short) and SAT results both show robust packaging reliability of 3D TSV packaging with thermal compression of 40um pitch Cu pillar bumps.

Table1. Summary of component level reliability test of 3D TSV flipchip packaging with micro Cu pillar bump.

Reliability Test Type		Performance Data	
		FT(O/S)	SAT
Reliability	MSL2aa	Pass	Pass
	MSL3	Pass	Pass
	HAST 192hrs after MSL3	Pass	Pass
	TC 1000x after MSL3	Pass	Pass
	HTST 1000hrs	Pass	Pass

CONCLUSION
In this paper, recent development of Cu pillar and TSV packaging/assembly were discussed. For successful implementation of TSV technology to microsystem products, TSV MEOL process and TSV packaging/assembly were both successfully developed and established with close collaboration of seamless integration and clear understanding from process, design, materials and reliability perspective.

TSV technology enables the integration of semiconductor device fabricated in different technology nodes with diverse testing requirements. The short vertical TSV interconnections through the silicon wafer achieve greater space efficiencies for a smaller form factor and higher electrical performance. 3D TSV must help to reduce overall semiconductor cost, improve performance or reduce form factor and reduce the time-to market.

ACKNOWLEDGEMENT
Authors appreciate TSV MEOL & BEOL team in STATS ChipPAC for their support of test vehicles fabrication and characterization.

REFERENCE
[1]. Seung Wook Yoon, "3D integration with TSV technology," Invited talk in 3D TSV technology session, SEMICON Singapore, May 5-7, 2008 (2008)
[2]. Knickerbocker, J. U. et al, "Development of next generation system-on-package (SOP) technology based on silicon carriers with fine-pitch interconnection," IBM J. Res. Dev. Vol. 49, No. 4/5, pp. 725-754 (2005)

VERIFICATION OF BACK-TO-FRONT SIDE ALIGNMENT FOR ADVANCED PACKAGING

Warren W. Flack, Manish Ranjan, Gareth Kenyon, Robert Hsieh
Ultratech, Inc.
San Jose, CA
mranjan@ultratech.com

John Slabbekoorn, Andy Miller
IMEC
Leuven, Belgium
millera@imec.be

ABSTRACT

Leading edge consumer electronic products relentlessly drive demand for enhanced performance and small form factors. This in turn defines manufacturing requirements for all aspects of semiconductor device fabrication. As the cost of front end manufacturing continues to escalate rapidly with each new technology node, semiconductor manufacturing companies are increasing their focus on packaging technology such as silicon interposers with through silicon vias (TSV) to deliver improved performance and reduced form factor.

Lithography is one of the critical process steps that affect the final device performance and associated yield for TSV manufacturing. One of the unique lithography requirements is the need for back-to-front side alignment. Obtaining precise metrology for measuring back-to-front side overlay performance is an industry challenge. Unlike front end manufacturing where automated metrology tools are widespread, metrology options are limited for back-to-front side overlay. This paper will discuss a metrology package which has been developed to evaluate and qualify back-to-front overlay performance using the lithography tool itself. The package is unique in its capability to measure any location of the wafer and model the acquired data to provide detailed insight in back-to-front side overlay performance.

Silicon test wafers were fabricated over a range of thicknesses to evaluate the stepper self-metrology for back-to-front side overlay. The reference layer is defined in a standard damascene copper process and protected with a passivation layer. Next the wafers are flipped, bonded, and thinned to various thicknesses. Wafers were produced with coarse and fine grinds to compare with chemical mechanical polish (CMP) to assess the impact of surface interference on the alignment system of the lithography tool. Experimental back-to-front alignment metrology data is shown as a function of silicon thickness and surface finish using the lithography tool self-metrology. The accuracy of the tool self-metrology is verified independently using external infrared (IR) microscopy.

Key words: Metrology, Overlay, 3D Packaging, TSV, back-to-front side alignment

INTRODUCTION

Over the last four decades semiconductor device manufacturers and wafer foundries have utilized shrinking gate dimensions and decreasing operating voltage to enhance device performance. This approach has driven broad industry growth. However as customers transition to sub-28 nm manufacturing technology, traditional front end scaling is becoming increasingly complex with significant cost impacts. Semiconductor manufacturing companies are focusing on various packaging technologies to play an important role in delivering improved system level performance in a cost effective manner. The manufacturing approach utilizing three dimensional (3D) TSV technology alleviates interconnect delay considerations by reducing global interconnect wiring length. In addition, TSV delivers superior bandwidth performance, power management and addresses some device latency issues. Many companies have research activities in 3D TSV technology and numerous demonstration vehicles have been developed [1,2]. Some of these TSV processes sequences require the need for back-to-front alignment solutions during the lithography process step.

Interposers for packaging also require TSV technology. Devices under consideration for use with interposers include graphics processors, high-end ASICs, and FPGAs. The drivers are mainly partitioning large die, integrating single chips into a module, reducing die size where substrate density is the constraint, and the use of the interposer to minimize the stress on large die that is fabricated with extra-low-k dielectrics (ELK).

While image sensors have utilized TSV technology in mass production over the last few years, it is projected that DRAM manufacturers will begin utilizing the TSV packaging technology within several years [1,3]. Adoption of this technology by memory companies can play a significant role in driving technology adoption. However the requirements for memory manufacturers are much more stringent and require development of improved back-to-front alignment capability.

TSV technology is the primary driver to improve IR back-to-front capability. However, other technologies can benefit from this capability. The design possibilities for MEMS, microfluidics and bio-technology devices can be greatly enhanced with the improved alignment capability and metrology examined within this study [4,5].

ALIGNMENT SYSTEM
The naming convention used in this study is that the wafer device side is the front side and the silicon side is the back side. The side facing up on the lithography tool is the back side of the TSV wafer as shown in figure 1.

Various methods have been investigated for viewing embedded lithography alignment targets [6]. The method of top IR illumination, shown in figure 1, provides practical advantages for integration with stepper lithography. Since the illumination and imaging are directed from the top, this method does not interfere with the design of the wafer chuck, and does not constrain where alignment targets can be located on the wafer.

The top IR alignment method illuminates the alignment target from back side using an IR wavelength that can transmit through silicon (shown as light green in figure 1) and the process films (shown in blue). For this configuration the target (shown in orange) needs to be made from an IR reflective material such as metal for best contrast. The alignment sequence requires that the wafer move in Z in order to shift alignment focus from the wafer surface to the embedded target.

The lithography system used in this investigation has a top IR alignment system that supports back-to-front side alignment applications. The back-to-front side alignment or dual side alignment (DSA) application was originally developed for CMOS image sensor applications [3]. This application required alignment to embedded metal targets below silicon, and the system was designed to achieve back-to-front side overlay of less than 2.0 μm (|mean| + 3sigma) over a full 300mm wafer. However, a large portion of this budget is consumed by the error uncertainty of the back-to-front metrology.

Unlike many front side alignment systems which view the wafer through the projection lens (TTL), the DSA alignment system views the wafer in an off-axis configuration. Therefore the calibrated offset between the exposure and alignment systems is maintained using a common stage fiducial that is measured by both the DSA and TTL alignment systems. Note that the DSA alignment camera is in a fixed position. The wafer stage on the stepper moves in X and Y to enable alignment at any location on the wafer, and the wafer stage also provides the range of Z travel to allow focus on both top surface targets and embedded targets across the practical range of silicon thicknesses.

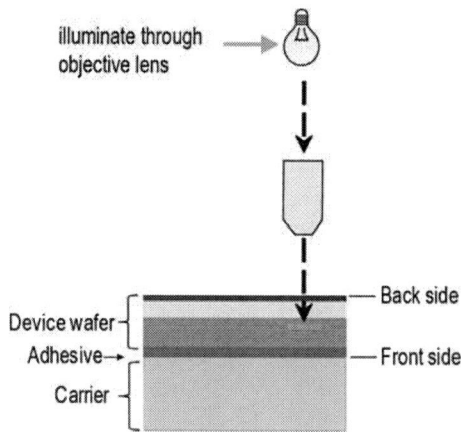

Figure 1: Off axis alignment configuration with infrared illumination and imaging from above the wafer. This configuration is the most flexible, providing access to the entire wafer for target alignment.

METROLOGY
The wafer processing sequence consists of off-axis alignment of the embedded targets on the lithography tool followed by an exposure of the reticle pattern to form resist patterns on the back side. However finding suitable methods for evaluating back-to-front side overlay performance with available metrology is a challenge. Typical registration metrology systems measure topside planar structures, and for initial DSA investigations various overlay tests were constructed to accommodate this metrology restriction [7]. The Single Pass overlay test, which uses monitor wafers with an etched pattern at the surface, allows detailed characterization of the off-axis alignment system. The Single Pass overlay test is very useful for monitoring performance, but this test does not view targets through silicon and does not require a Z move between alignment and exposure [6,7]. The Double Pass embedded overlay test addresses this shortcoming by aligning to embedded targets and forming resist patterns at the top surface using two separate passes at 0 and 180 degree orientation respectively. The overlay between the two passes can then be measured using a conventional topside metrology system [6,7] The Double Pass test provides useful information about mean performance and stability. However this test cannot be used for detailed overlay analysis because linear modeling terms cancel out in the double pass operation. The limitations of the Single Pass and Double Pass overlay tests can only be address by direct measurement of back-to-front alignment. Direct metrology provides the overlay data required for effective process monitoring and overlay optimization.

To eliminate the inherent restrictions in topside planar metrology, a direct back-to-front side registration metrology package was developed to run on the lithography system, utilizing the off-axis IR alignment system. This metrology method incorporates separate focus offsets and separate pattern recognition models to achieve optimum pattern capture and localization for each pattern. The implementation includes utilities for recipe management;

and is capable of measuring many sites across the wafer and on multiple wafers. The new metrology package enables realistic testing of back-to-front alignment.

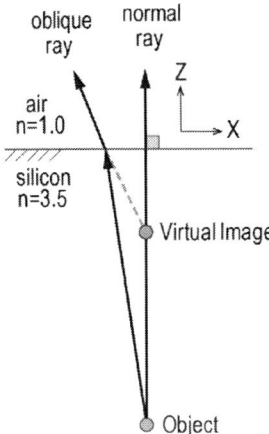

Figure 2: Ray trace of the principal ray shows that the apparent target position viewed from the air side appears at different Z position, but has the same XY position as the actual target position.

To understand the potential issues affecting back-to-front metrology, a simple model for imaging an embedded target under the silicon can be constructed from ray tracing analysis, as shown in figure 2. Viewed from the air side, light refraction at the air/silicon interface makes the virtual image (green) appear higher in Z than the actual object position (orange). This estimation problem does not exist in the lateral direction: the virtual image and object positions are the same in X and Y. Therefore the virtual alignment target provides a stable reference for overlay metrology.

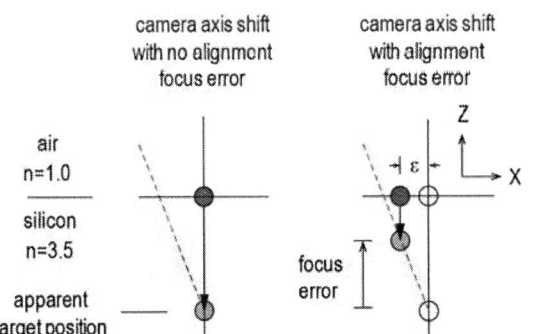

Figure 3: Tilt of the alignment system can create a lateral error (ε) in captured position if there is an alignment focus error. Focus error will cause the perceived image position to shift along the axis of the alignment system shown as a dotted line. Arrow shows the Z shift translation from the surface position to the corresponding alignment target position.

There are several potential error sources that need to be considered in the design of the metrology system. If the principal ray of the alignment system is tilted from the wafer normal, then this pointing error in the alignment system can produce a lateral measurement error (ε) if the alignment target is not viewed at best focus as shown in figure 3. Therefore maintaining best focus is important in minimizing the effects of an alignment system pointing error. Focus is generally well controlled in a lithography system since errors in focus will also affect the image appearance and localization by the pattern recognition algorithm.

Another potential source of registration error is tilt of the Z-axis. An angular error of the Z-axis results in a lateral translation (ε) proportional to the size of the Z move. This can have a large effect when there is a large Z shift from the alignment focus position to the exposure focus position, as shown in figure 4. Note that this error cancels out in the new back-to-front side measurement method since the same sequence is used for both wafer alignment and metrology. Therefore this error needs to be independently characterized to verify that it is not a significant effect. This error can be effectively measured with the double pass test [6].

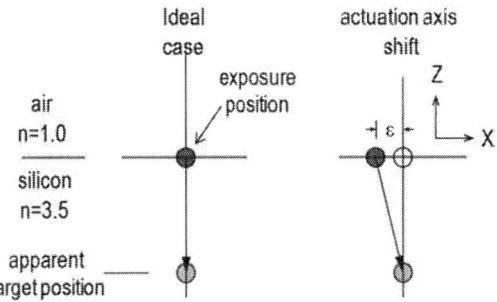

Figure 4: Diagram showing the effect on Z-axis tilt on overlay. A Z-axis tilt shifts the target during the Z-axis move from the surface position to the corresponding alignment target position.

EXPERIMENTAL METHODS
The lithography system used for back-to-front side alignment and metrology is an Ultratech AP300 DSA stepper. The system has a 0.16 NA, 1X Wynne-Dyson projection lens design with broadband ghi-line Hg illumination [8,9]. The stepper is used in high volume advanced packaging applications and provides a stable platform for DSA operation. In this investigation a new back-to-front side metrology capability is evaluated.

The previous stepper self metrology package was designed for all metrology targets to be at the same Z height. This results in metrology targets from one level being out of focus compared to the other level when measuring DSA structures as shown in Figure 5. The new back-to-front side metrology package used in this study provides flexibility to measure targets at different Z heights with different designs at each level as shown in figure 6. This requires sequential measurement of two patterns at each measurement site and the ability to specify focus offset and pattern recognition

129

model for each pattern. A precision mechanism in the wafer stage provides Z travel normal to the wafer, which is required in order to minimize registration error as discussed in the Metrology section of this paper. Since the metrology package uses the same target capture mechanism as the alignment system, the error analysis has similar terms. For example, since the Z travel angle is common to alignment and metrology operations, the effect of this angle cancels and cannot be characterized by this method.

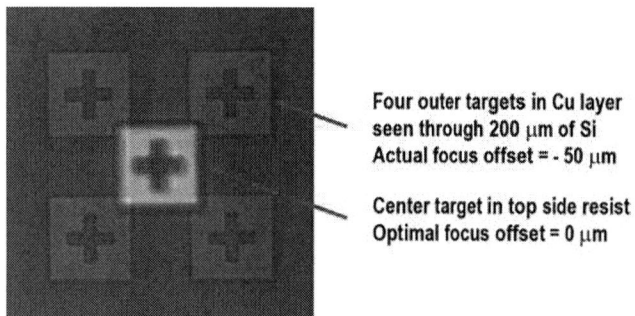

Figure 5: The orginal SSM measurement clover leaf target expects all five sub targets in the same focal plane. This approach does not work for DSA applications: top side and embedded targets are in different focal planes.

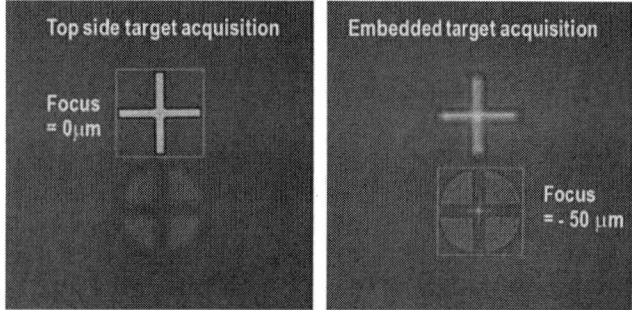

Figure 6: New Stepper Self Metrology for DSA changes focus between acquisition of top side and embedded targets. The embedded target is 200 μm into the Si. Note that light refraction at the air/silicon interface makes the virtual image appear higher in Z (-50 μm) than the actual object position (-200 μm).

Monitor test wafers for topside overlay baseline testing were prepared. The first pattern level was created using an ASML PAS5500/750 deep UV scanner with a specially designed mix-and-match test reticle containing alignment targets and various metrology structures. These patterns were etched 500 nm into the silicon surface to make artifact wafers, Baseline monitoring for the single pass test can be performed using metrology for topside planar structures. These wafers were used to calibrate the off-axis alignment to a calibrated reference [10].

The embedded target test wafers were prepared using a copper damascene process. A dielectric layer consisting of 250 nm of SiO_2 was deposited followed by an etch-stop layer and 600 nm of SiO_2. Then an ASML PAS5500/750 deep UV scanner was used to image a first pattern level with a specially designed mix-and-match reticle. After etching the 600 nm SiO_2 layer, 1000 nm of copper was deposited to fill the trenches and Chemical Mechanical Polish (CMP) was used to expose the underlying oxide, leaving a flat surface with copper filled trenches. This was covered by a SiO_2 passivation layer.

The copper damascene wafers were inverted and glued to a silicon carrier. The wafers were thinned using a grinding system to four Si thicknesses (50, 100, 200 and 300 μm). The last step was polishing the surface by CMP to remove surface damage leaving an optically smooth surface. The TTV (Total Thickness Variance) of the wafers was measured to be below 3 μm. A drawing of the wafer cross section is shown in figure 7. Some wafers were also prepared and left un-polished to evaluate the impact of the coarse and fine grinding scratches on the embedded target alignment performance.

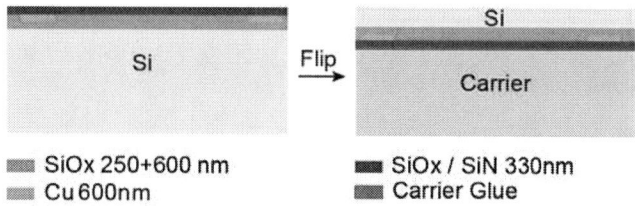

Figure 7: Cross section of test wafers with embedded damascene Cu targets. Wafers are inverted, glued to a carrier, and then the silicon is thinned and polished.

A mix-and-match test reticle designed to match the PAS5500/750 patterns was used on the AP300 for printing a second level matching pattern for back-to-front registration testing. The metrology structure compares the embedded copper damascene pattern to the topside resist pattern.

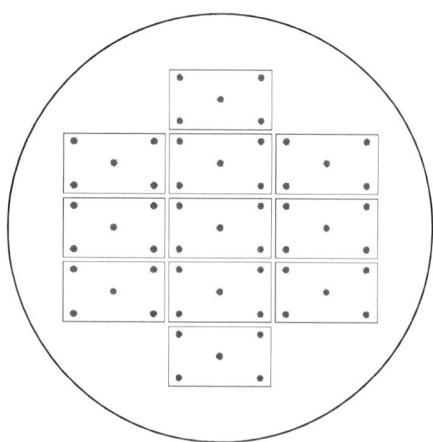

Figure 8: Metrology sampling plan on 200 mm wafer. Dots in the wafer map denote measurement locations.

The stepper self metrology sampling plan is shown in figure 8. It consists of five sites per field in eleven fields for a total of 55 sites across the wafer. The metrology measurements can be collected quickly which allow a large number of measurements for inter and intrafield characterization.

The polishing sequence consists of a grind operation followed by CMP. To study the effect of polishing quality, wafers were produced with coarse grind and fine grind to compare with the CMP wafers. Photos of coarse and CMP images are shown in figure 9. Severe scratches are visible on the coarse grid wafers while the CMP provides virtually perfect images. Since the visible surface scratches can obscure targets depending on where they fall, target enhancement and backup strategies can be used to eliminate alignment risk. However, the target image capture and overlay performance was good for all three finishes.

 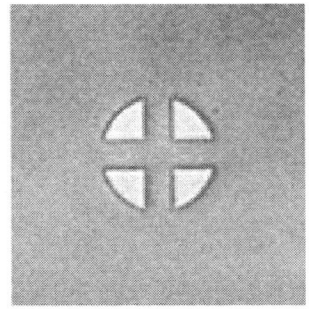

(a) Coarse grind   (b) CMP polished

Figure 9: DSA camera view of (a) wafer after coarse grinding, (b) wafer after grinding and CMP. The wafer is 100 μm thick silicon.

The effectiveness of the IR DSA camera for viewing embedded metal targets is shown in figure 10 for 100, 200, and 300 μm thick silicon. Image quality for alignment can be maintained across this practical range of silicon thickness. Even at the thickest 300 μm thick film the target image quality was good and no problems were observed for alignment capture or overlay performance.

(a) 100 micron   (b) 200 micron   (c) 300 micron

Figure 10: Images of DSA targets through three thicknesses of Si.

To provide an independent check of the accuracy of the embedded wafer alignment, an Olympus LEXT OLS4000 laser scanning infrared confocal microscope with 300 mm motorized stage was used. This microscope uses a Z move to capture surface and embedded images, and manual analysis of 0 and 180 degree orientation measurements was performed on the composite images. The two orientations were used to calibrate a tool bias for each individual wafer. The automatic mode was not used since the intensity variation of the two image positions prevented reliable

measurement. Manual data collection and analysis greatly limit the amount of data that can be collected. Measurement sample for each wafer consists of five fields with five sites per field for a total of 25 sites.

For embedded target measurement, a particular challenge for accurate TIS measurement is to perform a pure 180 degree rotation about the wafer normal. This is a challenge for most microscope systems because rotation of the wafer chuck by 180 degrees typically changes the wafer tilt relative to the optical axis. To calculate a correction for tilt on the microscope, the relative tilt of a flat wafer at zero and 180 degree chuck rotation was measured by mapping Z-height readings at best focus at several positions across the wafer. This tilt correction method assumes that the measured tilt versus chuck orientation repeats for all wafers. Obtaining a consistent wafer tilt is not a problem for the stepper measurement since the global tilt compensation sets the wafer tilt to the same reference during each wafer cycle.

RESULTS AND DISCUSSION

Preliminary testing was done to verify that metrology data collected on the off-axis IR camera matches data measured with the previous method using the on-axis TTL alignment system. The two methods closely match and slight differences were attributed to the specific offsets in the different pattern recognition models used for each test. An attempt to measure tool induced shift (TIS) was unsuccessful due to by the asymmetry in the target design shown in figure 11 [10]. Future plans call for evaluating TIS using a reticle set containing an appropriate structure.

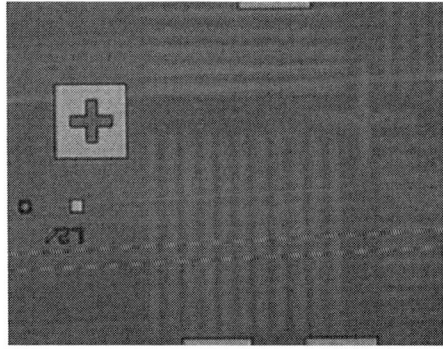

Figure 11: View from off-axis camera of resist feature (left cross in box) in focus and embedded metal mark under 200 μm thick silicon which is out of focus. Focus offset is needed to get the embedded mark in proper focus for measurement. Note that the large offset between marks and the non-symmetric nature of the embedded mark prevented reliable TIS measurement from this structure.

For back-to-front side alignment, mean versus silicon thickness data from double pass testing are shown in figure 12, and data from the LEXT microscope for the same wafers are shown in figure 13. Although the measured mean variation is not significant for the current design overlay requirements, this is a potential area for future improvement. The two measurement methods suggest a small slope in X mean versus silicon thickness; however there is significant uncertainty in these estimates based on the data sample size. Since the silicon thickness correlates to the run time parameter of alignment focus offset, the measured slopes can be incorporated into the run time lithography system calibration to null out the effect of silicon thickness on means. Additional data collection is required to establish a reliable calibration for the system.

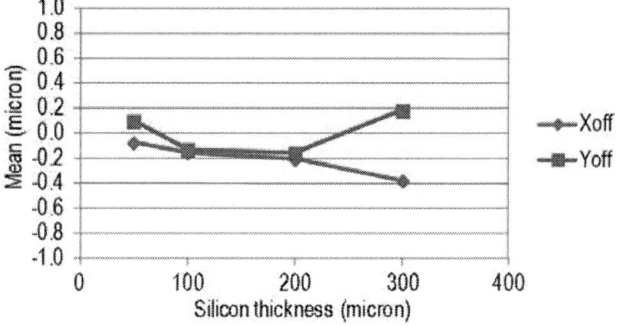

Figure 12: Plot of mean versus silicon thickness for double pass test.

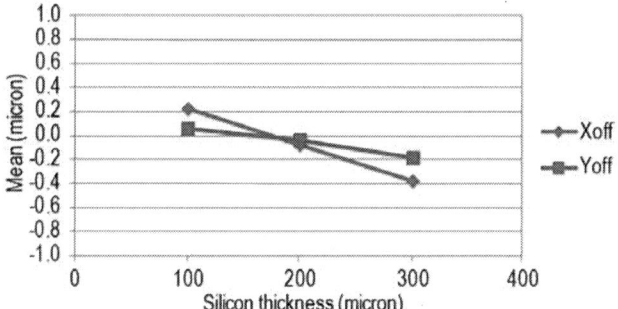

Figure 13: Plot of mean versus silicon thickness for LEXT microscope measurement.

DSA wafers with silicon thicknesses of 50, 100, 200 and 300 μm were aligned and exposed. The same off-axis alignment system was used for metrology. A plot of mean versus silicon thickness is shown in figure 14. Back-to-front overlay means are not significantly affected by silicon thickness. The results were calibrated using a fixed offset to best match the double pass result for 100 and 200 μm thick silicon. This calibration accounts for errors such as pattern training offset.

A plot of three sigma versus silicon thickness is shown in figure 15. Back-to-front three sigma is not significantly affected by silicon thickness 100 μm or greater. However the 50 μm thickness exhibited a large three sigma.

The linear terms in the overlay analysis reveals that the 50 μm thick silicon wafers have a large isotropic (17 ppm) intra-field scaling compared to thicker wafers. This scaling is suspected to come from distortion of the thin silicon film during the bonding process. Further study with more wafers having less than 100 μm silicon thickness is necessary.

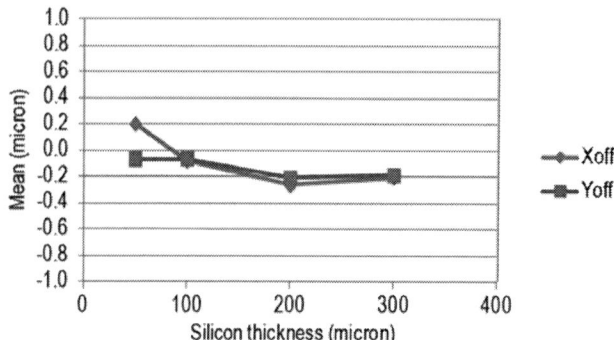

Figure 14: Plot of mean versus silicon thickness for back-to-front side measurement.

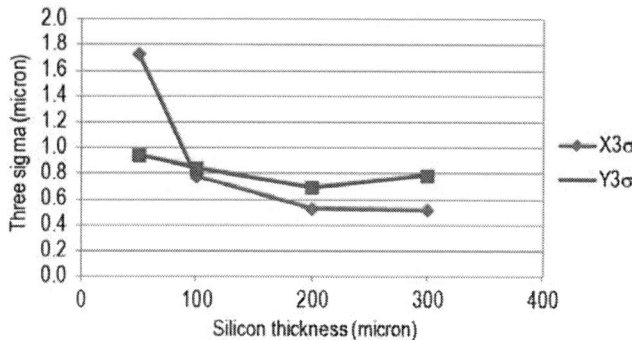

Figure 15: Plot of three sigma versus silicon thickness for back-to-front side measurement. Increased variation in X for the 50 μm thick silicon is due to a larger scaling of the wafer pattern, suspected to be caused by distortion of the thin silicon film during the bonding process.

Three different grades of surface polishing were investigated. The three polishing grades, in order of increasing surface quality, are coarse grind, fine grind and CMP. In extreme cases the surface scratches can interfere with the capture of particular targets; however miscapture can be avoided through proper setting of pattern capture criteria and the use of backup targets [7]. Figure 16 shows a plot of three sigma versus silicon thickness for front-to-back side measurement for 200 μm silicon and various polishing techniques. The three grades of polish give similar results for three sigma.

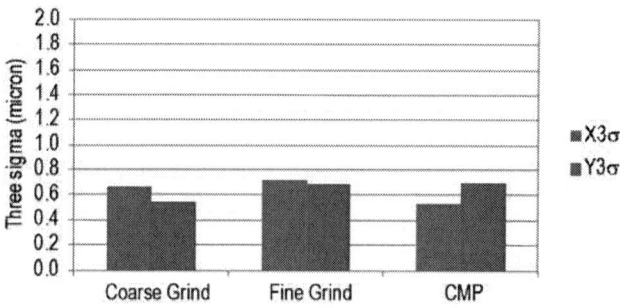

Figure 16: Plot of 3-sigma versus polishing technique for 200 μm thick silicon.

CONCLUSIONS

A new back-to-front metrology method was developed to provide direct feedback for overlay optimization and monitoring. It provides flexibility to measure targets at different Z heights with different target designs.

Silicon test wafers have been fabricated with a range of silicon thicknesses and surface polish to evaluate the stepper self-metrology for back-to-front side overlay. Mean versus silicon thickness data using the self metrology double pass test was compared with the LEXT microscope metrology. The level of variation is not significant for current design overlay requirements.

Using the new metrology method, back-to-front overlay means are not significantly affected by silicon thickness. Back-to-front three sigma is not significantly affected by silicon thickness 100 μm or greater. However the 50 μm thickness exhibited a large three sigma due to isotropic scaling. This scaling is suspected to come from distortion of the thin silicon film during the bonding process, and further study with more wafers having less than 100 μm silicon thickness is necessary.

Three different grades of surface polishing were investigated. The quality of surface polish did not affect the overlay performance. However severe surface scratches can potentially obscure targets requiring that backup target strategies be used for wafer alignment mapping.

Ongoing work will look at collecting more DSA overlay data to improve the lithography system calibration to null out the effect of silicon thickness on means. A revised reticle set with symmetric targets will be evaluated for TIS measurement. Additional work is required to address the observed TIS error with the infrared microscope due to variation of wafer tilt relative to the optical axis.

REFERENCES

[1] Vardaman, Jan et. al., *TechSearch International: 3D TSV Markets: Applications, Issues and Alternatives*, August 2012.

[2] Visser, J. "3D Interconnect Integration Success and Challenges", *2011 Eighth International Wafer-Level Packaging Conference*, Santa Clara, CA, November 2011.

[3] Denda, S., "Through Silicon Via for MEMS Packaging and Photo Sensors", *ICEP 2008 International Conference on Electronics Packaging*, Tokyo, Japan, June 2008.

[4] J. Zhu et. al., "Through–silicon via technologies for interconnects in RF MEMS", *Microsyst. Technol.* **16** (2010).

[5] Lietaer, N. et. al., "Integration Technologies For Miniaturized Tire Pressure Monitor System (TPMS)", *2009 IMAPS Device Packaging*, Scottsdale, Arizona, 2009.

[6] Flack, W. et. al, " Development and Characterization of a 300mm Dual-Side Alignment Stepper", *Optical Lithography XX Proceedings*, SPIE **6520** (2007).

[7] Flack, W. et. al., "Lithography Challenges for Leading Edge 3D Packaging Applications", *2011 Eighth International Wafer-Level Packaging Conference*, Santa Clara, CA, November 2011.

[8] Flores, G. et. al, "Lithographic Performance of a New Generation i-line Optical System," *Optical/Laser Lithography VI Proceedings*, SPIE **1927** (1993).

[9] Flack, W. et. al, "Process Characterization of One Hundred Micron Thick Photoresist Films", *Advances in Resist Technology and Processing XVI Proceedings*, SPIE **3678** (1999).

[10] Coleman, D. et. al, "On the Accuracy of Overlay Measurements: Tools and Mark Asymmetry Effects", *Integrated Circuit Metrology, Inspection, and Process Control IV Proceedings*, SPIE **1261** (1990).

A STUDY OF A DEVELOPMENT LITHOGRAPHY PROCESSES FOR 3DI PLATING APPLICATIONS

Patrick Kearney[1], Kirsten Ruck[1], Kathleen Nafus [2], Tetsushi Miyamoto[2], Patrick Jaenen[3], Andrew Miller[3]

[1] Tokyo Electron Europe Limited, Dresden, Germany patrick.kearney@europe.tel.com; kirsten.ruck@europe.tel.com

[2] Tokyo Electron Kyushu Limited, 1-1 Fukuhara, Koshi-shi, Kumamoto, Japan testsushi.miyamoto@tel.com

[3] Inter-University Micro-Electronics Centre, Leuven, Belgium millera@imec.be; jaenenp@imec.be

ABSTRACT

The emergence of FBEOL (Far Back End of Line) processes such as MEMS and 3Di has shifted the importance of processing thick films to the forefront in semiconductor production. Finding more efficient processes that satisfy the industry's stringent requirements for low chemical consumptions, fast processing times, and CDU is an ongoing task. FBEOL processes often have large topographies, (10 to 100µm deep) that are difficult to develop and often have substrates that are prone to undercut or peeling. Long process times and high chemical consumption need to be addressed. For both environmental and economic reasons it is important to strive to attain the most efficient systems available.

This paper aims to show a comprehensive developing study into 60µm photoresist films applied to a 3D plating process on 300mm wafers. The end result being the determination of control knobs that facilitate a more efficient and controlled.

Keywords: 3Di, Films for Plating. Developing thick photoresist films

INTRODUCTION

The work in this paper focuses on the development process pertaining to the FBEOL Cu pillar interconnects. The wafers were processed using a TEL CLEAN TRACK ACT12™ 300mm coater/developer specially adapted for processing thick photoresist. Exposure was completed using an Ultratech AP300 1x stepper [1]. The resist was coated at 60um and was supplied by AZ Electronic materials, and coated in a single coat step. The material supplied was a chemically amplified ghi line sensitive resist specifically developed for large topography Cu plating processes. The use of chemically amplified resist allowed for lower exposure doses and shorter developer times. Additionally the resist allowed the use of standard normality TMAH developer. All of the work was conducted at the IMEC facility in Belgium. The ultimate goal was to improve the developer process time and developer consumption without sacrificing process performance. In order to achieve such a broad based objective, we needed to make hardware changes as well as systematically adjust different recipe parameters to attain the most sensitivity for our process.

DEVELOPER PROCESS OPTIMIZATION

The experimental approach was separated into two distinct segments:

The first segment covered a rigorous review of the POR (Process of Record) developer recipe for the IMEC's 3D lithography process. This process utilized the H-nozzle, common to TEL developer processes. Here we systematically looked at various elements of the developer recipe. The key goals of this section of the work were to understand the impact of the various parameters within the recipe with respect to resist residue, pattern fidelity and CD. Once these were optimized for the target process, the impact of the control knob adjustments was assessed for material consumption and wafer throughput.

The second segment compared this optimized process recipe with an alternative developer nozzle specifically designed to deliver further improvements in terms of material consumption and wafer throughput.

SECTION 1 – POR PROCESS OPTIMIZATION

After setting up a baseline recipe, (std) with a fixed exposure dose, a series of tests were run on different recipe parameters to find the process knobs for CD control.

The first parameter investigated was developer contact time. **Figure 1,** shows a schematic of the developer recipe with a puddle segment. During the puddling, the wafer is spun at very low spin speed and the developer is allowed to react with the exposed resist.

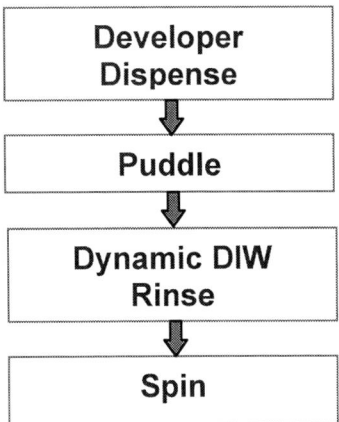

Figure 1. Schematic of A Developer Recipe

The objective of this experiment is to find the point when the developer solution stops being effective. **Figures 2 & 3**, shows the results of three FEM (Focus Exposure Matrix) wafers with different contact times. The data shows that even with an extra 57.5% added to the standard developer contact time, the developer solution is still reacting with the resist. This would indicate that the contact time can be increased significantly and the amount of developer solution applied can be decreased in order to reach the target CD, with only a slight decrease in wafer throughput. The tradeoff between wafer throughput and developer consumption can be optimized based on customer requirements.

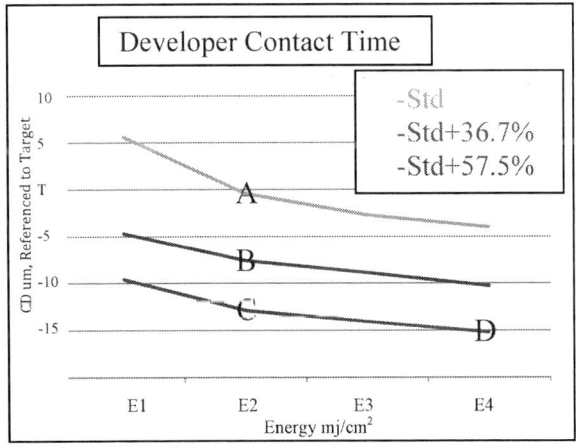

Figure 2. The Effects of Developer Contact Time

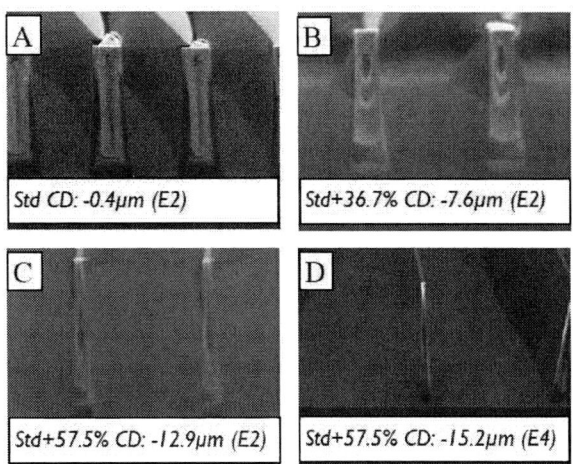

Figure 3. SEM micrograph for the Effects of Developer Contact Time

The effects of developer flow rate and changing the spin speed while dispensing were also investigated. **Figures 4, & 5,** show the results from the varying flow rate and spin speed during dispense.

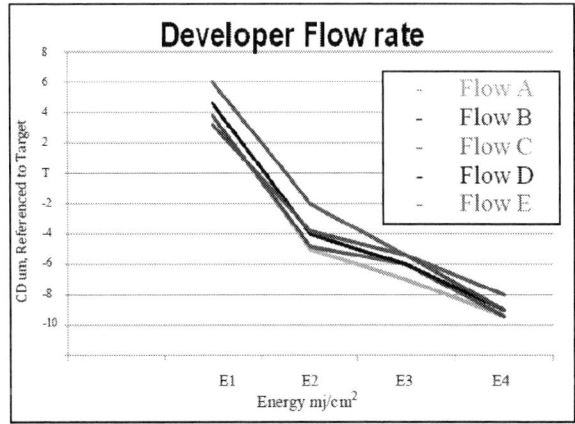

Figure 4. The Effects of Developer Flow Rate

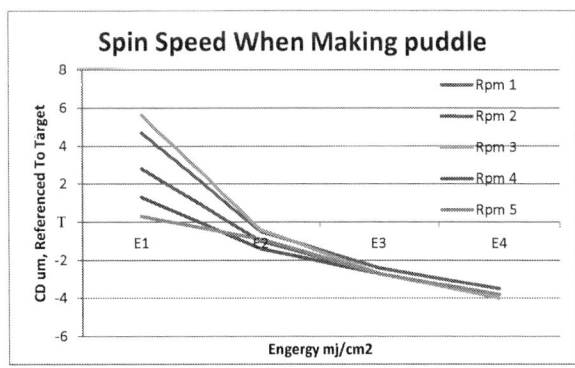

Figure 5. The Effects of Spin Speed during dispense

In both cases the process window was evaluated with respect to the baseline recipe. However, despite relatively large changes in these parameters, very little sensitivity to CD was observed.

Therefore, it can be concluded that these parameters have a small impact on CD and do not make good control knobs within the developer recipe.

The next item that we investigated was a comparison of 57% longer contact time, 50% less developer solution against the standard baseline recipe. The data in **Figure 6**, was collected via x-SEM. This data was limited by the number of data points that could be practically taken as measuring with X-SEM is time consuming. In this case we measured at 5 points, (North, South, East , West and Center). The results show that even though there is a difference in average CD, the uniformity between the base line recipe and the long contact time recipe are comparable.

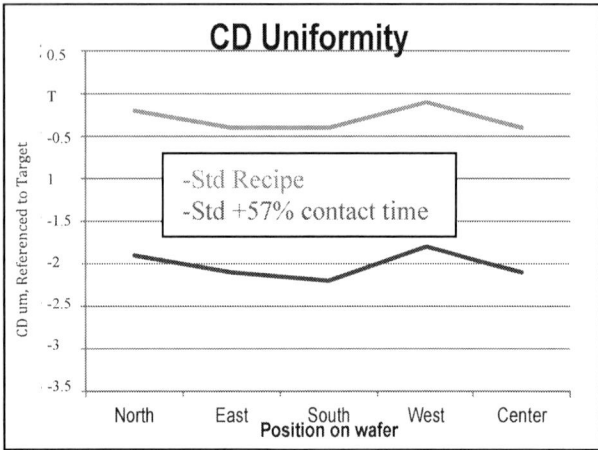

Figure 6. CD Uniformity: *Comparison of Std recipe to a recipe with Longer Contact Time & Less Developer solution.*

It is also important to understand what is happening on the unexposed areas of the wafer. For this reason a comparison of the standard recipe and the longer contact recipe was carried out, (**Figure 7.**). The thicknesses of both wafers were measured before and after development with no exposure. The data shows that with the standard recipe we have an average thickness loss of 3.7% . With the longer contact we have and average thickness loss of 11.1%. Although this is a significant difference, the Cu plating process has very large tolerances with regards to thickness, and there would still be enough resist on the unexposed areas of the wafer for this long puddle recipe to be acceptable.

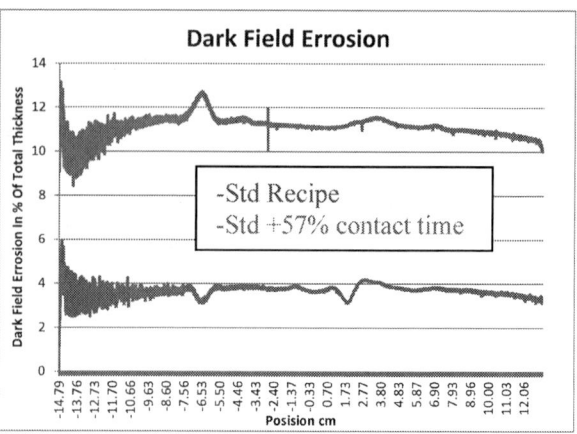

Figure 7. Dark Field Erosion

GP vs. H Developer Nozzle

A comparison was made between the two different types of developer nozzle. The nozzle that is used in our baseline test is a H Nozzle, and this uses a puddle type recipe. The GP. Nozzle however, uses a dynamic dispense recipe (*see* **Figure 8**). The GP nozzle is a nozzle that has been implemented with great effectiveness to the FEOL [2 & 3]. lithography development, but has not yet been applied to thick film processing. Our goal here was to match the structures created by the H-nozzle, and then compare the processing times and developer consumption. With both nozzles, the same coating and exposure conditions were applied, and TMAH developer solution was used for both samples. The testing was applied to two reticles, one on a test reticle with lines & spaces along with large open areas. The other reticle was mostly dark field with contact holes.

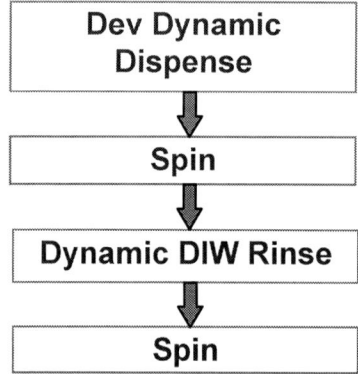

Figure 8. Schematic of a developer recipe with a dynamic dispense

In order to find a baseline recipe for the GP Nozzle, a series of tests were carried optimizing the contact time and dispense amount until the CD results of the H Nozzle were matched. **Figure 9,** shows the sem micrograph cross section results of the GP recipe compared to the H nozzle baseline results.

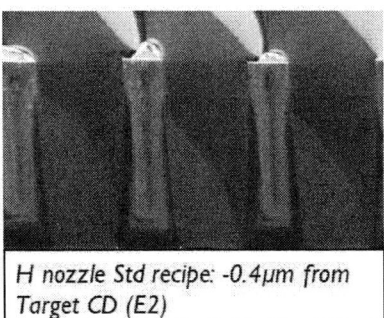

H nozzle Std recipe: -0.4μm from Target CD (E2)

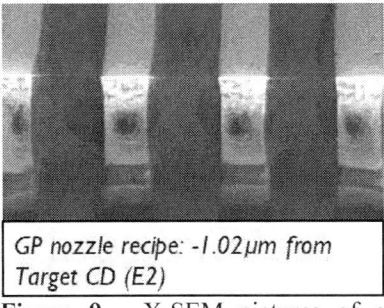

GP nozzle recipe: -1.02μm from Target CD (E2)

Figure 9. X-SEM pictures of comparison between H nozzle and GP nozzle

In both cases the image are resolved with normal profiles and no scumming, or pattern collapse.

Also a visual comparison was made using an optical microscope. **Figure 10**, shows images from the center die of the H nozzle and GP nozzle wafers. Each die was checked to make sure that there was no residue in the target lines and spaces. In addition to this an inspection of the large open areas was made were it was checked that no residue was remaining. With both the lines and spaces and the open areas, the images resolved clearly, showing similar results between the H nozzle and the GP nozzle.

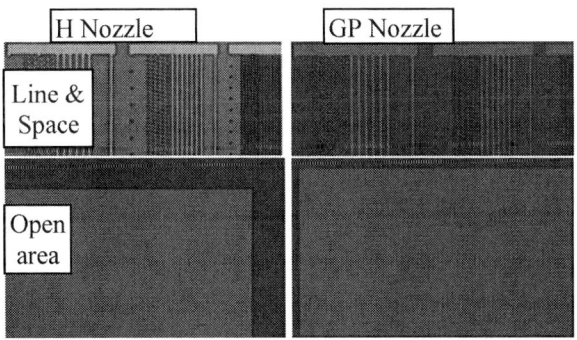

Figure 10. Optical microscope comparison of H nozzle and GP nozzle

In **Figure 11**, a comparison of CD uniformity between the H nozzle and the GP nozzle was made. With both wafers similar range values of 0.30um for the H nozzle versus 0.32um for the GP nozzle can be found.

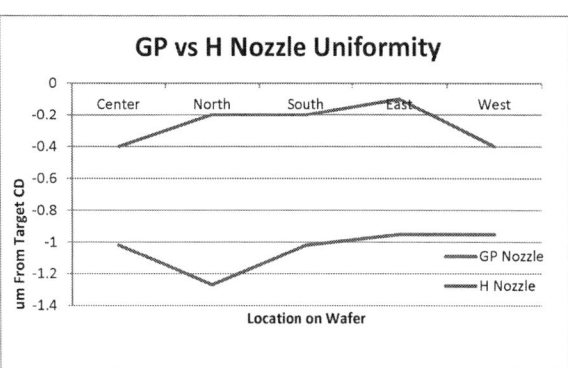

Figure 11. Comparison of CD uniformity between GP nozzle and H nozzle

One of the structures that is used in the 3Di plating process is a large contact hole formed using a dark field reticle. A comparison with this structure was also made and the results can be seen in **Figure 12**. The GP nozzle wafer was made with a modified recipe, where the contact time was decreased, but the dispense amount was increased with regards to the recipe used for the line/space pattern. As can be seen in the images, the GP nozzle shows a similar profile to the H nozzle with no residue or strange features on the structure.

With the line and space recipe, the GP nozzle showed a 48.8% improvement in developing time over the H nozzle, (the developing time is the time spent dispensing and developing the wafer and excludes the rinse time). This was done with a 10.4% increase in developer solution. With the contact holes there was an even greater improvement in developer time. Here we saw the GP nozzle was 83.3% faster than the H Nozzle. Also we used 20.7% less developer solution.

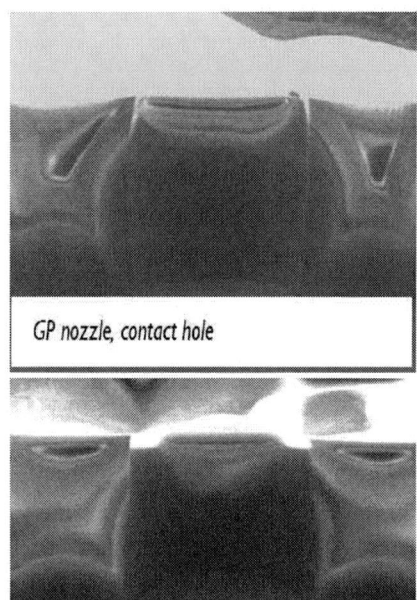

Figure 12. Comparison of GP & H nozzles with contact holes

In comparing the recipe optimization between the line space pattern and dark field contact holes, we can see the influence of the mask open area on the improvement level for wafer throughput and material consumption..

Future Work
Currently testing of the GP contact hole recipe on production wafers, and with a Cu substrate is underway. Here batch processing with wafer to wafer uniformity will be investigated. Also more cross sectional pictures will need to be taken along with repeatability tests.

Conclusions

In this paper it was demonstrated the different parameters of the H nozzle recipe showed the sensitivity of contact time and the possibilities of reducing the dispense time in order to reach the target CD. Also the comparison between H nozzle and GP nozzle, showed a significant improvement in process time and developer consumption.

Acknowledgements
The authors would like to thank Shane O'Neil from Tokyo Electron Europe Ltd, and Koichi Matsunaga from Tokyo Electron Kyushu and John Slabbekoorn of IMEC, and Richard Collett from AZ Electronic Materials, for their support during the setup and testing of these experiments.

References
[1] Warren W. Flack, Ha-Ai Nguyen, Elliott Capsuto, *Process Improvements for Ultra-Thick Photoresist Using a Broadband stepper*, SPIE Vol. 4346, 97 (2001).

[2] Tomoyuki Ando, Sho Abe, Ryoichi Takasu, Jun Iwashita, Shogo Matsumaru, Ryoji Watababe, Komei Hirahara, Yujiro Suzuki, Miki Tsukano, and Takeshi Iwai, *Topcoat-free ArF Negative Tone Resist,* SPIE Vol. 6923 69233W-4 (2009)

[3] Eitan Shalom and Shaike Zeid, *60 Seconds Puddle time- a tradition to overcome in CA resists: Process optimization and defect elimination.* SPIE Vol. 6923 69233W-4 (2008)

[4] www.microchemicals.eu/technical_information, *Thick Resist Processing,* Version: 2009-11-11

[5] www.microchemicals.eu/technical_information, Lithography Trouble Shooter, (2012)

INNOVATIVE 2.5D SOLUTION: EXTENDED / FLIP CHIP EWLB (EMBEDDED WAFER LEVEL BALL GRID ARRAY) TECHNOLOGY

Seung Wook Yoon, Yaojian Lin and Pandi C. Marimuthu
STATS ChipPAC LTD
5 Yishun Street 23, Singapore 768442
seungwook.yoon@statschippac.com

ABSTRACT

The market for portable and mobile data devices connected to a virtual cloud access point is exploding and driving increased functional convergence as well as increased packaging complexity and sophistication. This is causing an unprecedented demand for a variety of wafer level packages, thin Package-on-Package (POP) and 3D System-in-Package (SiP) solutions. We expect to see more exciting interconnect technologies in wafer level packaging such as 3D Through Silicon Via (TSV), 2.5D interposers, eWLB (embedded Wafer Level Ball Grid Array) / FOWLP (Fan-out Wafer Level Packaging) and innovative 2.5D/3D eWLB technology to meet these needs.

eWLB is a fan-out wafer level packaging (FOWLP) technology that enables a higher ball count by extending the package size beyond the area of the chip. By utilizing eWLB's fan-out packaging approach, the next level of multi-chip and thin packaging capability can be achieved. eWLB provides a robust packaging platform supporting very dense interconnection and routing of multiple die in very reliable, low-profile, low-warpage 2.5D and 3D solutions. The use of advanced eWLB designs such as a side-by-side configuration can replace a stacked package configuration or be utilized as the base for a 2.5D/3D TSV configuration in order to achieve a more cost effective packaging solution. Combining an analog or memory device with a digital device can provide an optimum solution for achieving the best performance in situations where thin, multi-die and heterogeneous integration is required for very high performance applications.

This paper will highlight the rapid trend towards extended eWLB and flip chip eWLB in high performance packaging technology. A study will be presented on flip chip substrate design optimization in combination with eWLB RDL technology. Mechanical simulation was carried out to investigate the stress on bumps and device stack layers with different bumping approaches. Extended eWLB / flip chip eWLB offers a cost-effective solution for advanced node technology with a smaller pitch required for high performance applications. Extended eWLB and flip chip eWLB technology provide the benefit of integration as well as a lower cost solution without sacrificing electrical performance.

INTRODUCTION

Wafer level packaging (WLP) applications are expanding into new areas and are segmenting based on I/O count and type of device. The traditional design layout of passive, discrete, RF and memory device is expanding to logic ICs and MEMS. The WLP segment has matured over the past decade with numerous companies delivering high-volume applications across multiple wafer diameters and expanding into various end-market products. With the WLP infrastructure and high volume production already in place, a major focus area now is cost reduction.

One of the most well known examples of a FOWLP structure is eWLB technology [1]. This technology uses a combination of front- and back-end manufacturing techniques with parallel processing of all the chips on a wafer which can greatly reduce manufacturing costs. The benefits of eWLB include a smaller package footprint compared to conventional leadframe or laminate packages, medium to high I/O count and maximum connection density as well as desirable electrical and thermal performance. eWLB also offers a high-performance, power-efficient solution for the wireless market [2]. Furthermore, 2.5D/3D eWLB technology enables 2.5D interposer packaging, 3D IC, and 3D SiP (System-in-Package) with vertical interconnection. 2.5D/3D eWLB can be implemented with heterogeneous integration of IPD, MLCC or discrete component embedding.

eWLB TECHNOLOGY

eWLB technology is addressing a wide range of factors. At one end of the spectrum is the packaging cost along with testing costs. On the other end there are physical constraints such as footprint and height. Other parameters that were considered during the development phase included I/O density which is a particular challenge for small chips with a high pin count, the need to accommodate SiP approaches, thermal issues related to power consumption and the device's electrical performance (including electrical parasitic and operating frequency) [3].

eWLB is an interconnection system processed directly on the wafer and compatible with motherboard technology pitch requirements. It combines conventional front- and back-end manufacturing techniques with parallel processing of all chips.

There are three stages in the eWLB process; i) reconstitution, ii) RDL and iii) backend & test. Additional fab steps create an interconnection system on each die, with a footprint smaller than the die itself. Solder balls are then applied and parallel testing is performed on the wafer. Finally, wafers are sawn into individual units, which are used directly on the motherboard without the need for interposers or underfill. Figure 1 shows an example of an eWLB wafer and package structures for 2D to 3D applications. A SEM photo of an eWLB package is shown in Figure 2.

(a)

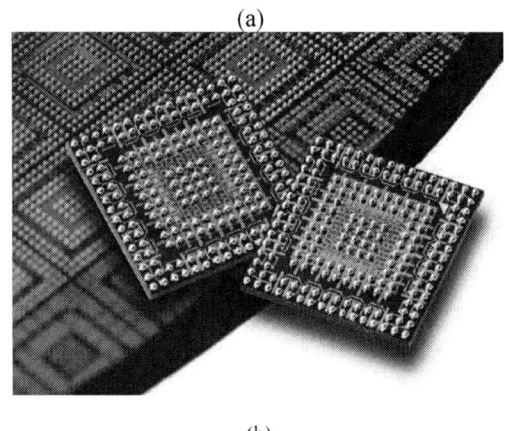

(b)

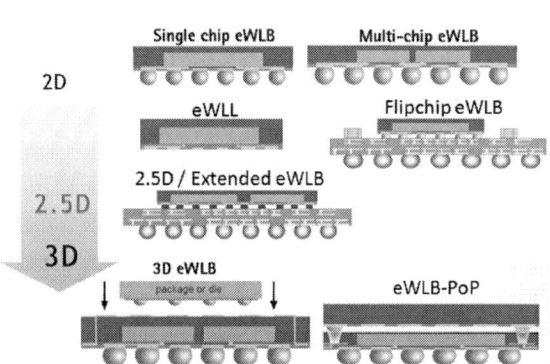

Figure 1. (a) eWLB wafer after packaging with reconstitution, RDL and backend processes and (b) schematics of innovative eWLB structures.

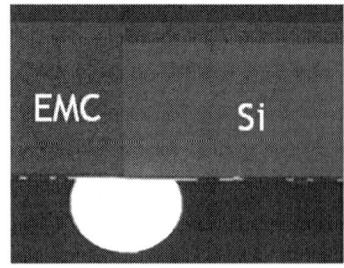

Figure 2. SEM micrographs of cross-section of eWLB. (total package body thickness ~500um)

Advantages of eWLB Technology

Today BGA package technology is limited by the capability of the organic substrate. eWLB helps to overcome such limitations and also simplifies the supply chain. Building the substrate on the package itself allows for higher integration and routing density in less metal layers. eWLB is an innovative packaging platform that will support future integration, particularly for mobile and high performance devices and this packaging technology has a number of important features. Transition to eWLB packaging technology enables a significant reduction in recurring costs by eliminating the need for expensive substrates. The advantage of eWLB packaging can be summarized in Table 1. BGA packaging also faces a challenge with technology nodes beyond 28nm as the device performance density drives the need for flip chip interconnect. However, extreme low-k (ELK) dielectric structures used in conjunction with advanced Si nodes are more fragile and are therefore more susceptible to cracking or delamination during flip chip assembly, resulting in flip chip becoming narrower in terms of packaging process margin. In addition, there is a growing trend in environmentally friendly packaging with lead free, halogen free, or green material sets. With ELK and interconnect pitches becoming smaller and smaller and the shift to lead free materials, the technical limitations faced by the packaging industry are becoming more challenging. eWLB technology provides a window for packaging next generation devices in a generic, lead-free/halogen free and green packaging scheme.

Table 1. Advantage of eWLB packaging.

• MCP configurations (down to 0.5mm)
• The thinnest 3D solution (stacked thickness down to 0.8mm)
• Scalable heterogeneous integration platform
• Leading cost/performance solutions (co-design optimized)
• Ultra fine ball pitch and maximum I/O density
• Excellent electrical and thermal performance
• Enhanced reliability with advanced dielectric materials
• PoP configurations - both single and double sided

EXTENDED eWLB / FLIP CHIP eWLB
Extended eWLB

The use of eWLB packages in a side-by-side configuration to replace a stacked package configuration or to function as the base for a 2.5D TSV interposer configuration is critical to enable a more cost effective packaging solution. Combining LSI with the wide I/O memory interfaces or high bandwidth memory (HBM) with the TSV packaging capability can provide an optimum solution for achieving the best performance in thin multi-die stacks for very high volume manufacturing. A comparison of a TSV interposer solution to an eWLB based solution is shown in Figure 3.

2.5D interposer technology offers several advantages including:

- IP block partition: De-coupling functional blocks in SoC (analogue, memory, I/O, RF)
- Heterogeneous package integration
- Lower power dissipation

- Lower total device cost

However, there are still many challenges to apply interposer technology to market applications. One of the key challenges is the supply of TSV interposers with cost effective solutions. Silicon wafers are one of the primary materials used and there are active research and development activities for glass, poly-Si or other materials.

eWLB has the capability of multi-die and multi-RDL structures as well as less than 10um/10um LW/LS capability. In addition, there have been advancements in eWLB integration with different components such as Si device, IPD, discrete, MEMS or glass based devices in a single package.

In terms of the assembly process, eWLB is well established with proven manufacturing yields. There are still some challenges in the TSV interposer assembly process flow as well as reliability/yield issues in multi-chip assembly with thin TSV interposers. From a test view point, extended eWLB has a number of advantages in final testing such as handling, logistics and compatibility of current test environments.

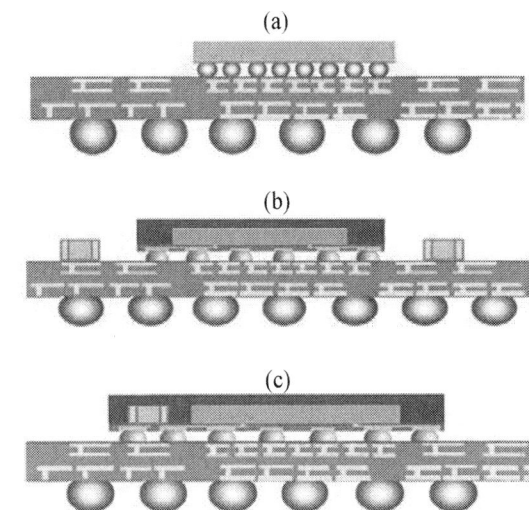

Figure 4. Schematics of (a) flip chip packaging and flip chip eWLB with decoupling capacitor for a high performance application; (b) capacitor on substrate and (c) capacitor embedded in flip chip eWLB.

TSV Interposer based solution

MLCC   IPD   TSV interposer

Organic substrate

eWLB based Solution

MLCC   IPD   Extended eWLB

Organic substrate

Figure 3. 2.5D interposer approach with eWLB technology. eWLB provides a 2.5D integration platform superior to conventional TSV Interposer based solutions in overall cost and process simplicity

Flip chip eWLB

As advanced technology nodes move beyond 28nm, there are more challenges in flip chip assembly with smaller bump pitches and ELK (extreme low-k) materials in ILD (interlayer dielectric materials) structures in a device. Standard flip chip assembly needs to address bump structure/materials, UBM designs, underfill processes/materials as well as substrate materials and package design to secure good solder joint reliability with flip chip bump interconnects.

Flip chip eWLB provides a fan-out area that has a larger pad pitch and RDL layer, providing an I/O reconfiguration that minimizes substrate layer numbers while optimizes electrical performance such as combined power and ground. As shown in Figure 4 (c) and Figure 5, flip chip eWLB has the option to integrate a decoupling capacitor and place it closer to the device for better electrical performance.

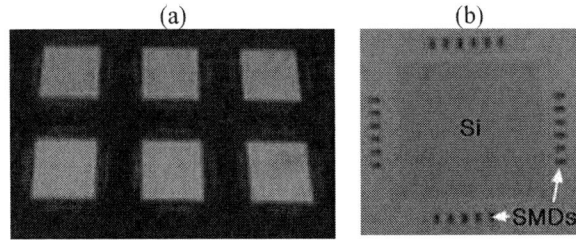

Figure 5. (a) Embedded decoupling capacitors and discrete SMDs with Si device in eWLB carrier and (b) X-ray image of SMD embedded eWLB from (a).; the black area in the peripheral represents for decoupling caps and SMDs, dark gray in center is for Si device.

SUBSTRATE DESIGN OPTIMIZATION WITH COMBINING eWLB TECHNOLOGY

eWLB technology has fine line width and spacing capability of 10um/10um as well as lower intrinsic electrical parameters, providing flexibility in routing designs as compared to substrate technology. Figure 6 shows the package routing design with eWLB (2-L RDL) for 4-layer (1-2-1) substrate. This was verified by a signal integrity study for high speed digital applications which included simulation and a functional test that proved a 2-L RDL eWLB design is comparable to at least 4-layer substrate designs.

(a)   (b)

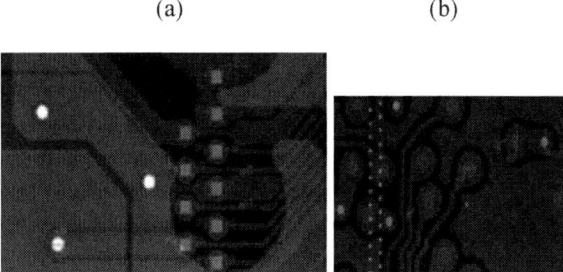

Figure 6. Design optimization with (b) 2-layer RDL in eWLB for (a) 4-layer organic substrate.

With the superior electrical performance of eWLB, it is possible to reduce the number of layers in organic substrates. As shown in Table 2, a 14-layer flip chip substrate would be replaced by an 8-layer substrate with flip chip eWLB technology. There are several variables in a flip chip substrate design such as Cu trace line width/spacing, substrate thickness, via pad and via hole size as well as flip chip/solder ball pitch. eWLB converted designs should be approached with more actual data to meet electrical performance and signal integrity. eWLB RDL designs provide an optimized and efficient signal integrity in interconnection routing with a coarse bumping pitch. With this coarse bump pitch in flip chip eWLB, we expect a lower cost organic substrate having a large pad pitch and reduced number of metal layers.

A previous study [4] shows a comparison of parasitic values of RLC for fcBGA and eWLB at 1GHz. For resistance, eWLB has 68% less value than fcBGA. Moreover, eWLB has a 66% less inductance value and 39% less capacitance value as compared to fcBGA. This is mainly due to the shorter interconnection in an eWLB package. For fcBGA, there are flip chip solder bumps and substrate interconnections that all contribute to signal delay. eWLB has shorter interconnections with the RDL process, thus it has improved electrical performance over fcBGA. Even in unit parasitics, eWLB has lower resistance and inductance values than fcBGA at all frequencies. eWLB shows less reflection noise and better transmission performance than fcBGA over all frequency ranges. fcBGA has a resonance near 7.5~8GHz due to the mutual factor of inductance and capacitance elements. This resonance affects crosstalk of neighbor signal, signal distortion/reflection, power integrity, signal integrity as well as EMI/EMC. However, eWLB shows no resonance and better electrical performance, therefore, eWLB can be applicable for higher frequency applications.

Table 2. Flip chip eWLB approach with less number of layers of organic substrate.

Flip chip PKG	Flip chip eWLB
flipchip	eWLB (2-layer RDL)
Organic substrate (14-layer)	Organic substrate (8-layer)

ENHANCED SOLDER JOINT RELIABILITY WITH RDL IN EXTENDED EWLB/ FLIP CHIP EWLB [5]

In this study, two different interconnect schemes of direct bump and RDL approaches were studied with mechanical simulation as shown Figure 7. Shearing force was simulated for each bump case. Figure 7 and Table 3 show the results from FEM (Finite Element Method) mechanical simulation. According to the results, the overall stress level was much higher in the direct bumped model. When it comes to a Cu / ELK ILD stacked area, an RDL approach produces a smaller stress value that is approximately 30% of the direct bump. The RDL approach was more stable and safer than direct bump with respect to the bump shearing condition. In this case, the additional dielectrics layers in an eWLB RDL provide stress relaxation and pad rerouting, resulting in lower stress on the ELK ILD area. As a result, better interconnection reliability can be accomplished with the RDL approach in eWLB technology.

Table 3. FEM results for maximum stress at solder bump and Cu/ELK ILD stacked area for direct bump and RDL model.

Simulation Models	Max. Shear Stress (MPa)
Direct C4-bump	114.2
C4-bump on RDL	25.9
Cu column on RDL	27.12
ELK ILD @ Direct C4-bump	151.2
ELK ILD @ RDL C4-bump	8.97
ELK ILD @ RDL-Cu column	3.529

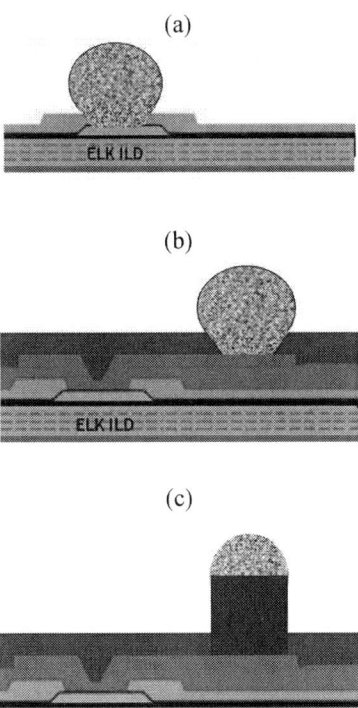

Figure 7. Schematics of bump models for mechanical simulation. (a) C4 bump on ELK ILD stack, (b) C4 bump on RDL and (c) Cu column on RDL

As shown in Figure 8 and Table 3, there is a significant reduction in Von-Mises stress and shear stress in the RDL approach as compared to direct bump on die pad. Therefore, the RDL bump in eWLB packages are also a promising approach to improving package reliability of Cu/ELK interconnects in flip chip packaging. Figure 9 shows the FEM simulation results of Figure 7(a) and Figure 7(c). With RDL layers, bump stress does not directly affect ELK stack layers, therefore, there would be less mechanical damage as compared to standard flip chip bump.

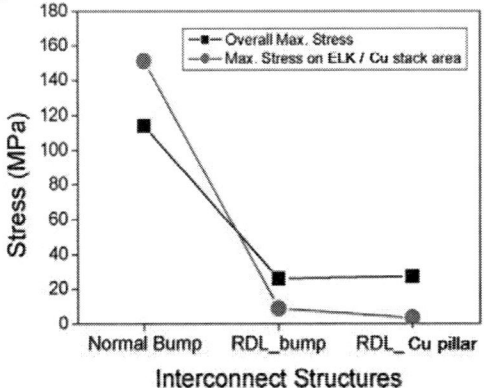

Figure 8. Plot of comparison of stress on bump and ELK ILD with different interconnect schemes of Figure 6.

(a)

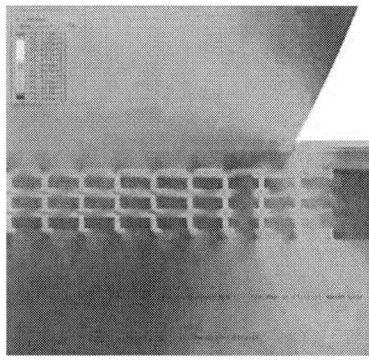

(b)

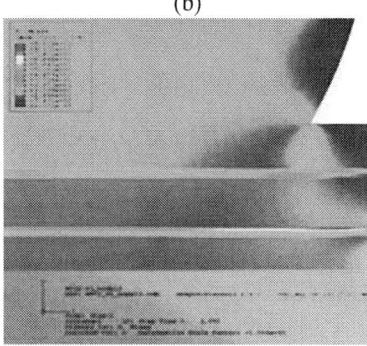

Figure 9. FEM mechanical simulation results for (a) directed C4 bumped (Figure 7(a)) and (b) bump on RDL approach (Figure 7(b)).

CONCLUSION

eWLB is a proven cost-effective manufacturing technology and is a versatile platform that is well-suited for a multitude of complex and highly integrated solutions that portable and mobile applications require. The advantages of eWLB include a multi-die package design, thin 3D solution, higher performance from reduced interconnect lengths, ultra fine pitch capability and superior warpage control.

Extended eWLB and flip chip eWLB technology provide a proven integration platform as well as a lower cost solution without sacrificing electrical performance. Extended eWLB can be an alternative solution for 2.5D TSV interposer technology for mid/low-end applications. Flip chip eWLB provides a cost-effective solution with a reduced number of layers and a more relaxed design rule in flip chip substrates. In addition, eWLB provides enhanced bump solder joint reliability with an RDL approach for advanced node Cu/ELK flip chip devices.

eWLB will provide more exciting developments in the future as the scalability of its carrier size will drive greater cost effectiveness. Extended eWLB and flip chip eWLB technology are successfully enabling semiconductor manufacturers to provide the smallest, highest-performing semiconductors to meet the ever increasing demands of the converging products in the market today and in the future.

ACKNOWLEDGEMENT

Authors appreciate Jang Tae Hoan, Kim Kyung Eun, Song Hyun Jin, Park Soo Han and Dr. Liu Kai in design and characterization team, STATS ChipPAC for design optimization study, electrical simulations and characterization study of eWLB technology.

REFERENCES

[1] M. Brunnbauer, et al., "Embedded Wafer Level Ball Grid Array (eWLB)," Proceedings of 8th Electronic Packaging Technology Conference, 10-12 Dec 2006, Singapore **(2006)**.

[2] Graham pitcher, "Good things in small packages," Newelectronics, 23 June 2009, p18-19 **(2009)**.

[3] S.W. YOON, Meenakshi PADMANATHAN, Andreas BAHR, Xavier BARATON and Flynn CARSON, *"3D eWLB (embedded wafer level BGA) Technology: Next Generation 3D Packaging solutions,"* San Francisco, IWLPC 2009 **(2009)**.

[4] S. W. Yoon, Roger Emigh , Kai Liu, Sin Jae Lee, Ray Coronado and Flynn Carson, "Thermal and Electrical Characterization of eWLB (embedded Wafer Level BGA)", ECTC2010, Las Vegas **(2010)**.

[5] 150-um Pitch Cu/Low-k Flip Chip Packaging With Polymer Encapsulated Dicing Line (PEDL) and Cu Pillar Interconnects, IEEE TRANSACTIONS ON ADVANCED PACKAGING, VOL. 31, NO. 1, pp58-65 FEBRUARY 2008 **(2008)**.

DEVELOPMENTS OF FAN-OUT WAFER LEVEL PACKAGING TECHNOLOGY FOR SYSTEM-IN-PACKAGE ON WAFER-LEVEL (WLSIP)

J. Campos; E. O'Toole; V. Henriques; A. Martins; A. Leão; A. Cardoso; A. Janeiro
NANIUM, S.A.
Vila do Conde, Portugal
jose.campos@nanium.com

ABSTRACT

The fan-out WLP Technology eWLB (Embedded Wafer Level Ball Grid Array) is running in high volume production since more than three years and NANIUM is one of the main Packaging companies providing this technology for its customers. Up to now this advanced packaging technology has been used for single die products and addressing mainly consumer mobile applications.

Leveraged by its advantages in performance; design flexibility; form factor and cost, this technology is being developed to better address the needs for System-in-package on wafer level (WLSiP).

This paper will reveal the last developments occurred at NANIUM that enable system-in-package solutions using eWLB fan-out WLP technology (FO-WLP).

One main enabler for SiP applications using eWLB is the possibility to design using multi-layer re-distribution layers (RDL) and using the finest possible line widths and spaces. During this paper we will present the results of the development of two RDL layers and on the development of finer RDL lines.

When considering SiP products, where several types of active and passive devices are designed and built-in inside one unique package every single space will count for the final package form factor. We will demonstrate the results of the developments that enable to achieve minimum distances between every single device and between device edges and the edge of the package.

Finally, in this paper we will show details about development work done for the integration of more than one active die on one package, as well as last results of development of 3D solutions using eWLB, that enable heterogeneous integration and PoP solutions, using 3D-PLUS wafer-stacking technology (W2W) called Wire-free Die-on-Die (WDoD™).

Key words: Wafer Level Package (WLP); Fan-Out Wafer Level Package (FO-WLP); Embedded Wafer Level Ball Grid Array (eWLB); System in Package (SiP); Wafer Level system in Package (WLSiP).

INTRODUCTION OF EWLB

Embedded Wafer Level Ball Grid Array (eWLB) is a leading FO-WLP technology initially developed by Infineon Technologies AG [1] and now licensed and used by NANIUM, SA in high volume since more than three years using 300mm reconstituted wafers.

This technology is based on a Backend Wafer on a temporary carrier (reconstituted wafer) with active know-good-dies (KGD) and/or passive elements, picked from original front-end silicon wafer or other packing media, prior to over-mold (**Figure 1** and **Figure 2**), and apply re-distribution layers (RDL) using thin film technology processes over its active side. Final interconnects are applied using wafer level preformed bumps drop processes and component singulation using mechanical blade dicing (**Figure 3**).

Figure 1 - Reconstituted 300mm wafer with different types of active/passive elements with dispensed Liquid Mold Compound before Molding

Figure 2 – Reconstituted 300mm wafer with different types of active/passive elements after Molding and Debonding from the temporary carrier

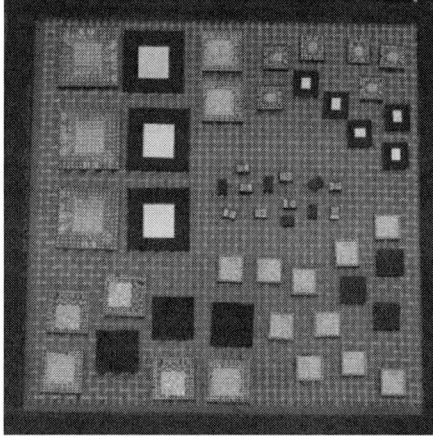

Figure 3 - Different single-die and multi-die FO-WLP (eWLB) packages in different sizes and configurations

Up to now this technology has been used mainly for single die products and addressing mainly consumer mobile applications. But supported on its last developments and leveraged by its advantages in electrical and thermal performance; design flexibility; form factor and cost, this technology is now being enabled for System-in-package on wafer level (WLSiP):

- by using a reconstruction scheme, this technology allows using only KGD and is not restricted to the original silicon wafer diameter and original die size, while covering the gap between the die pad size/pitch and motherboard/PCB pitch.
- by using a temporary carrier for reconstruction of a 300mm wafer, this wafer may be designed to include a include a wide range of elements like active KGDs; discrete passives (SMD capacitors and resistors); interconnect elements; sensors and optoelectronics [2]
- by using most advanced pick & place capabilities and thin film processes, the final form factor of a SiP based on WLSiP is reduced to the minimum thanks to the reduced space between all devices and package edges, and to the reduced dimensions of RLD lines and spaces and to the enabling of multi-layer metallization

ENABLING OF WLSIP

As part of the development of following generations of eWLB, NANIUM has been actively working, together with its customers, on the main enablers for WLSiP solutions using this FO-WLP technology:

- multi-layer RDL that will increase the design flexibility for our customers decoupling the chip pad design from the package ball matrix and will allow a better definition of different electrical planes;
- fine RDL line widths and spaces and fine ball pitch that will allow to increase the routing density and to match pad size/pitch of newer foundry technology nodes (40nm and below);
- reduced space between devices and between devices edges and the edge of the package that will contribute significantly to reduce the overall form factor of the package system;
- integration of multi-dies on one package, and integration of passive devices inside the package system;
- development of 3D solutions using eWLB, that enable heterogeneous integration and PoP solutions, using 3D-PLUS' wafer-stacking technology (W2W) called Wire-free Die-on-Die (WDoD™) [3].

MULTI-LAYER RDL

Higher level of routing, shield and power consumption required by most of SiP applications are being addressed in eWLB packaging technology at NANIUM, by developing its own capability to route multi re-distribution layers (RDL). Using RDL intrinsic characteristic for fine pitch traces NANIUM is also looking to provide smaller eWLB packages to its customers using its developed dual re-distribution layers technology.

For its 2 (two) RDL layers technology development, NANIUM considered to keep RDL layer dimensional characteristics similar to the ones used on its main volume products with single layer RDL (**Figure 4**).

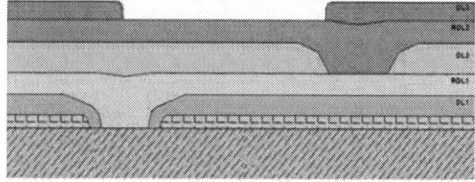

Figure 4 - Two Layer RDL Stack-up

During design of its daisy chain test vehicle, NANIUM main focus was directed to the vias between layers. Regarding contact vias requirements for manufacturability and its reliability, NANIUM design considered the following dimensions (**Table 1**):

Table 1 – *Via Dimensions for 2L RDL Development*

RDL Layer		Via Opening [μm]	Metal Overlay [μm]	Via On Via Offset (X/Y) [μm]
Die pad – to – RDL1	Via1	20	10	0/0; 75/75
		30	10	0/0; 75/75
RDL1 – to – RDL2	Via2	20	50	0/0; 75/75
		70	110	0/0; 75/75

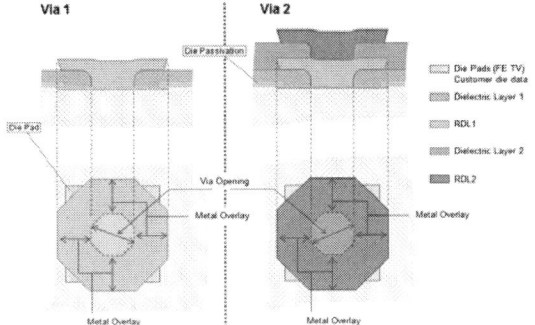

Figure 5 - Via-on-Via Stack-up

Although it is not represented in figures via stacking, via2 on via1 and via2 with offset were designed to evaluate process requirements due to dimple effect on via and reliability of both variants.

For electrical characterization typical structures for Leakage and Contact Resistance were integrated in the design. **Figures 6 - 9** show examples of typical contact resistance structures implemented using the 2 RDL layers.

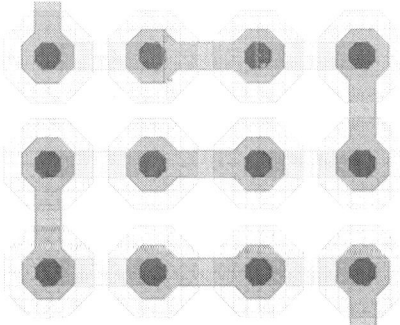

Figure 6 - Part of a RDL1 to Die chain contact resist
Blue: die pads;
Dark blue: via opening over die pad;
Red: Connection of 2 die pads in RDL1 layer

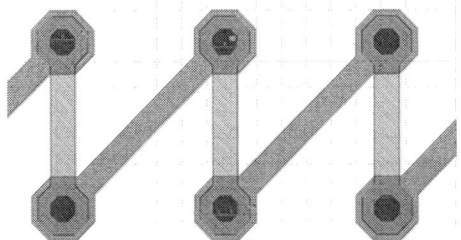

Figure 7 - Part of a RDL2 to RDL1 chain contact resist
Red: RDL1 structures (traces)
Pink: 20μm via opening over a 30μm land
Green: connection at RDL2 layer

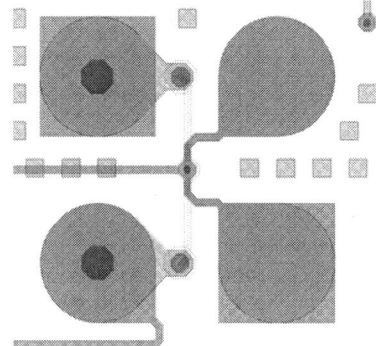

Figure 8 - Contact resistance measurement structure
Blue: die pads
Dark Blue: Via opening from RDL1 to die pad
Red: Connections in RDL1 layer
Pink: Via opening from
Green: Connections in RDL2 layer

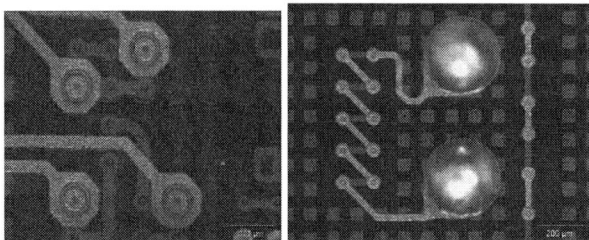

Figure 9 – Optical view of different Contact Resistance chains

These Contact Resistance chains were probed to confirm if 2L RDL technology could meet required specification for this critical electrical parameter. Either via contact resistance at 1st layer and at 2nd layer met NANIUM's requirements (**Figure 10** and **Figure 11**).

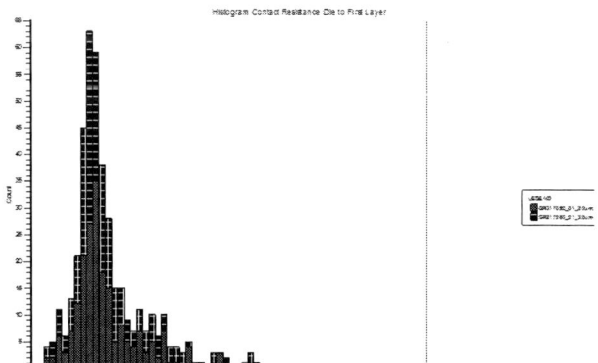

Figure 10 – Contact Resistance of Silicon Die to First RDL Layer for 20um and 30um via diameters

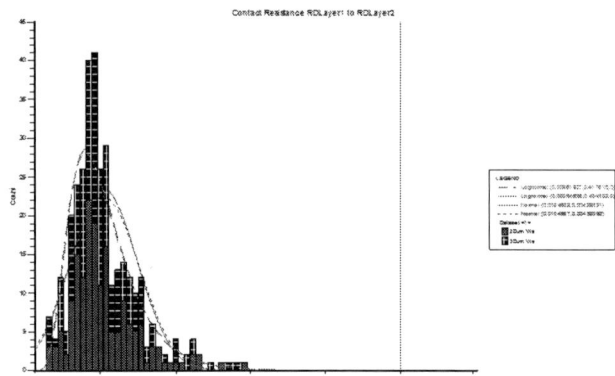

Figure 11 – Contact Resistance of First to Second RDL Layer for 20um and 30um via diameters

Together with this contact resistance structures also typical leakage structures in serpentine were included in the daisy chain test vehicle (**Figure 12**).

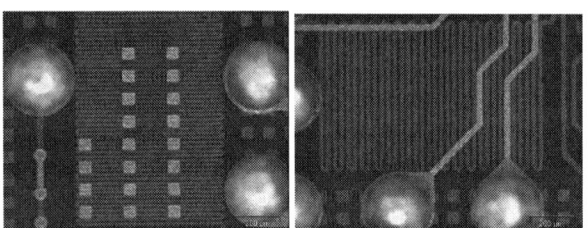

Figure 12 – Optical view of 10um and 15um Leakage structures in serpentine

Besides the designed structures integrated in this design NANIUM considering that new FE nodes technologies may require smaller pad sizes, included in daisy chain test vehicle narrow via structures to use them for process capability and manufacturability validation. In **Table 2** dimensions of those enhanced structures are defined.

Table 2 – Enhanced Via Dimensions for 2L RDL process capability and manufacturability validation

RDL Layer		Via Opening [μm]	Metal Overlay [μm]	Via On Via Offset (X/Y) [μm]
Die pad – to – RDL1	Via1	10	5	0/0
		15	5	0/0
RDL1 – to – RDL2	Via2	15	7.5	0/0
		20	10	0/0

By doing it NANIUM is validating its capability to place vias of 10μm and 15μm with metal overlay of 5μm in its eWLB designs.

After processing of test vehicle samples, these were submitted to detailed physical characterization (**Figure 13 and 14**) and to a complete reliability test plan including package and board level reliability testing (**Table 3**).

These activities are running at the moment and will be completed soon. The positive results obtained up to now show good capability to enable 2L RDL with eWLB at NANIUM.

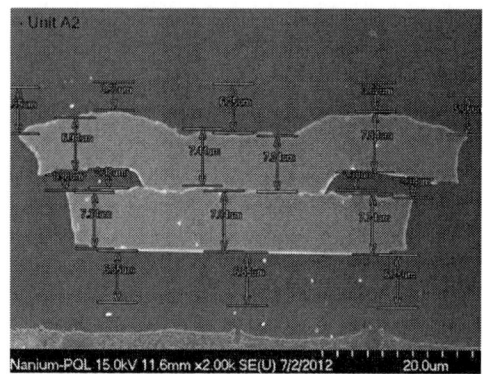

Figure 13 - Cross-Sections of Trace with 40μm with 20μm via in second dielectric, and 10μm Overlap in second redistribution layer

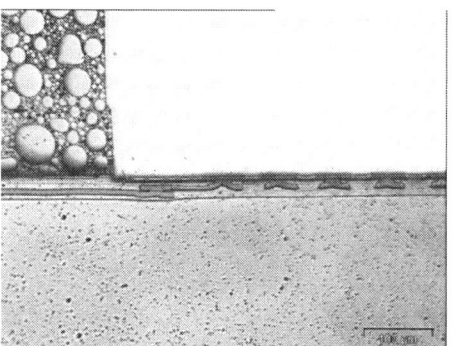

Figure 14 – Cross-section of 2L RDL test vehicle

Table 3 – Reliability test plan for development of 2L RDL

Reliability Test	Standard/ Spec	Pass Criteria	Final Assessment (Pass/Fail)
Moisture Sensitivity Level (MSL)	EIA/J-STD-020C (Level 1)	MSL1	Running
High Temperature Storage (HTS)	JESD22-A103 (Cond.C: Ta:175°C)	200hrs	PASS
Temperature Cycling (TC)	JESD22-A104 (-55°C / +125°C / 2cy/hr) Precond. L1; Tr:260°C	500x	Running
Temperature Cycling (TC)	JESD22-A104 (Cond G:-40°C / +125°C / 2cy/hr) Precond. L1; Tr:260°C	850x	Running
Temperature Cycling on Board (TCoB)	IPC 97-01 (-25°C / +100°C / 1cy/h)	1000x	PASS
Temperature Cycling on Board (TCoB)	IPC 97-01 (-40°C / +125°C / 1cy/h)	FF > 500 cycles	PASS
Drop Test (DT)	JESD22-B111	FF < 10% fails @ 20 drops	Running

SOLDER BALL PITCH 400UM

The reduction of ball pitch aims the increase in connection density of devices, which is crucial for the miniaturization trend on electronic devices. This ball pitch reduction is aligned with the increasing density of microelectronics, and with the <40nm foundry technologies. That is, it is necessary less Silicon area for the same chip functionality.

Since the reduction of ball pitch aims the higher interconnection density, it cannot derate the routing ability at RDL level, in either Fan-in or Fan-out packaging. Therefore, it must be followed by a correspondent reduction in trace width, as well as space between traces. **Figure 15** exemplifies this paradigm.

Using, as example, 20um line & space width in 500um pitch devices, it allows up to 4 traces between pads. If same line/space with would be kept in 400um pitch, only 2 lines would be possible, which would defeat the purpose of having higher connection density. To keep, at least, the same 4 traces, it is necessary to go down to 15um line/space width and further reduce ball pad area to ~70% of the value applied for 500um ball pad pitch.

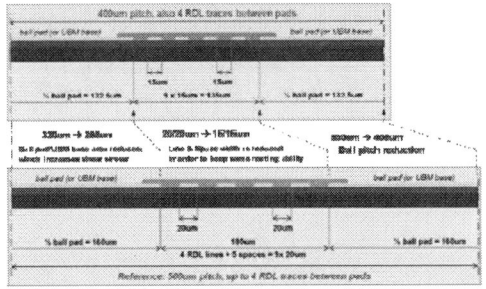

Figure 15 – Basic design considerations for routing capability of 4 RDL lines between ball pads when changing from 500 to 400um ball pitch

FINE RDL LINE AND SPACE

Thus, in order to allow finer ball pitches, ball pad area and line/space width must reduce. An alternative would be to increase the number of RDL layers, but the cost penalty of adding extra layers (which is a cost driver) may not compensate.
Ball pad area reduction brings higher shear stress level for the ball pad, so this places importance in materials and process to withstand such demand.

On its turn, the reduction of line space width, places higher challenges to lithography. Better optical resolution dielectrics and higher accuracy mask aligners are necessary.

NANIUM is actually working on this reduction, where customized daisy chain test vehicles are being tested with different line width and spaces. These vehicles include line widths and spaces from 20um to 13um and were submitted to a complete process monitoring and characterization measurement and inspection that included Leakage current measurements; CD measurements and optical inspections. Finally these vehicles are being submitted to physical characterization and THB testing that is running at the moment.

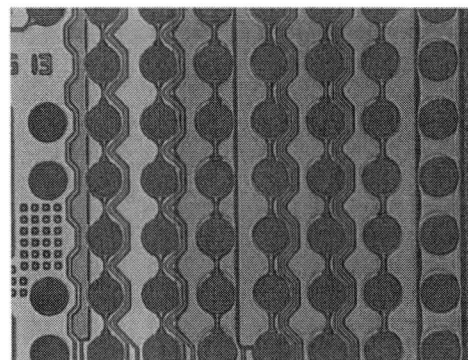

Figure 16 – Optical view of 15um / 13um meander for Leakage current measurements

REDUCTION OF DIE-TO-DIE DISTANCE AND DIE TO PACKAGE EDGE

With the aim to achieve the best valuable package to our customer, reduction of package form factor is mandatory. Benefiting from its reconstruction characteristics, eWLB can achieve outstanding reduced dimensions of SiP because of its capabilities for very small distances between dies and between dies and package edge (**Figure 17**).

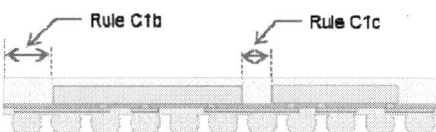

Figure 17 – Illustration of die-to-die distance and die-to-package-edge distance

NANIUM has been working on new developments on both characteristics.

Die-to-die distance of 200um was proven on a SiP [4] (**Figure 18**) and NANIUM is going now to test die-to-die distance of 50um with die placement accuracy +/-10um @3 sigma inside WLSiP package. Impact on reliability will be investigated.

Figure 18 – Reconstituted 300mm wafer with reduced die-to-die distance demonstrator

The reduction of the die-to-package-edge distance aims the increase of the multi-chip density inside a package. This was the motivation to investigate and decrease the previous value of 500μm to the minimum possible.

The multi-chip density inside a package is achieved not only by reducing the space between dies but also reducing the die-to-package-edge distance.

Experiments were carried out with main focus on the die-to-package-edge distance. Values tested were 50, 75 and 100µm for both dimensions 2 and 4 in **Figure 19**.

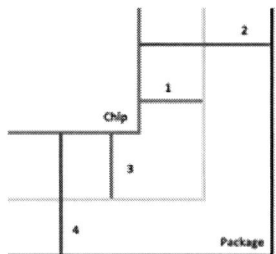

Figure 19 – Schematic of die-to-package-edge distance

Experiments were done by moving the die inside the package to one corner (**Figure 20**). All eWLB process was applied.

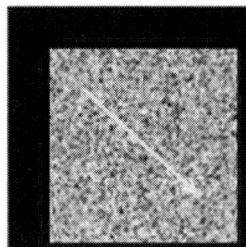

Figure 20 – Schematic of test vehicle for testing of reduced die-to-package-edge distance

In all samples submitted to cross-section, the minimum mold wall thickness observed was 30µm and did not reveal any package integrity problem. Therefore, we set 30µm as minimum mold wall to assure package integrity.

Figure 21 shows data collected on diced units for measurements 1 to 4 of figure 19, on targeting 75µm die-to-package-edge distance (data for 50µm and 100µm exhibit same distribution, so it is not shown.). Contributors for the observed spread and shift of the distributions go from die placement shift to dicing blade wearing.

Maximum deviation from the target to extrapolated distribution tail is 40µm. Based on the 30µm minimum mold wall, we can conclude that a minimum of 70µm die-to-package-edge distance is required.

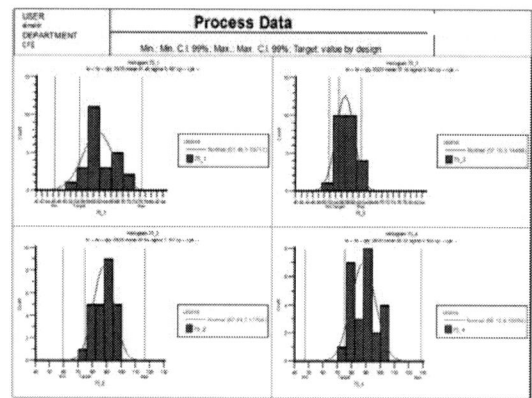

Figure 21 – Measurement of die-to-package edge distance for target value of 75um

Reliability results achieved so far (Pre-conditioning, TC & uHAST) are showing quite positive results (see **Table 4**)

Reliability Test	Standard/ Spec	Pass Criteria	Final Assessment (Pass/Fail)
Moisture Sensitivity Level (MSL)	EIA/J-STD-020C (Level 1)	MSL1	PASS
Temperature Cycling (TC)	JESD22-A104 (Cond G:-40°C / +125°C / 2cy/hr) Precond. L1; Tr:260°C)	850x	PASS
Unbiased HAST (uHAST)	JESD22-A118 Cond A: 130°C / 85% RH) Precond. L1; Tr:260°C)	96hrs.	PASS

Table 4 – Reliability test plan for development of die-to-package edge distance of 75um

Based on the obtained results and process tolerances, a nominal target for the die-to-package-edge distance was fixed in 75µm.

DEVELOPMENT OF 3D HETEROGENEOUS INTEGRATION SOLUTIONS

3DPLUS used DDR3 quad die JEDEC compatible memory modules to demonstrate their WDoD™ technology combined with eWLB provided by NANIUM [5].

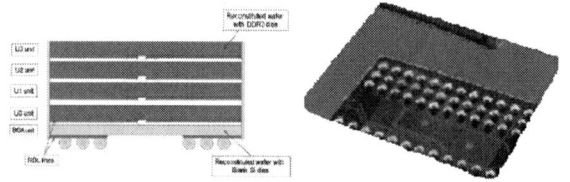

Figure 22 – Stacking solution of 4 Die DDR3 DRAM , using eWLB and WDoD technology

The interconnections to the BGA are made in RDL, on an additional blank silicon die. The first prototypes were built before NANIUM developed dual layer RDL process, and the limitation to a single layer RDL resulted in significant coupling effect between specific signals. For example, the value of the AC inductance matrix coupling coefficient between DQS or DQSB and DQ2 (see routing in the **Figure 23**) obtained by parasitic simulation is above 20%, while the value of this parameter for other signals with more favorable routing can be 10 times lower – 2% between DQ0 and DQ4, for instance. However, for the first technology demonstrators, the first priority was to demonstrate functionality and not achieving the highest DDR3 performance.

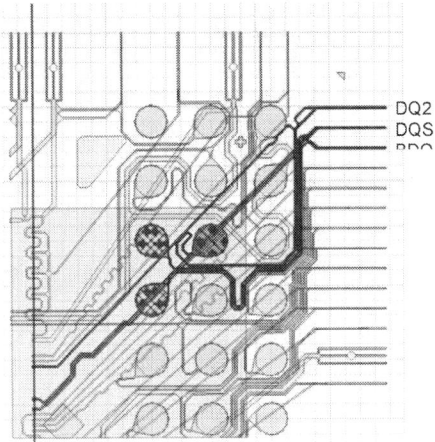

Figure 23 – Routing example of DQS and DQ2 signals

As a former Qimonda subsidiary, NANIUM is very experienced in testing SDRAM performance. 3D Plus DDR3 memory cube was tested for speed grades DDR3-1066F (7-7-7) and DDR3-1333H (9-9-9) in an Advantest T5501 memory tester, at room temp, -10°C and 95°C, with a test plan comprising the following blocks:

1. Contact and leakage test
2. IDD measurement at operating conditions
3. Functional test of the complete memory cell array at three different conditions: all timings relaxed / critical tRCD / 96 ms retention time.
4. ODT test
5. At-speed test with tight CL, tRP, tRCD and duty-cycle.
6. Verification of the interface timings per speed grade: setup/hold times, tDQSCK, tLZ, tQH, tDQSQ

The first prototypes that were functional, that is, that passed all the tests in blocks 1-4, all achieved 533 MHz, proving that this technology is adequate for building DDR3 modules. However, all the prototypes failed at least one test in the 667 MHz test sequence, thus not achieving frequencies above 533 MHz.

Electrical analysis showed evidence of significant crosstalk between DQ and DQS signals, as predicted by parasitic simulation of the package. The waveforms represented in **Figure 24** show DQS and DQSB signals when all DQ0-7 signals are fixed low. In this and the next pictures, only DQ5 and DQ7 are represented, but in every case the topology applied in all DQ0-7 is the same. In this picture, it is shown that at 667 MHz the signal already does not have time to reach its maximum value. The input capacitance of the 4 dies can justify this limitation in rise time, however the quality of the signal seems acceptable.

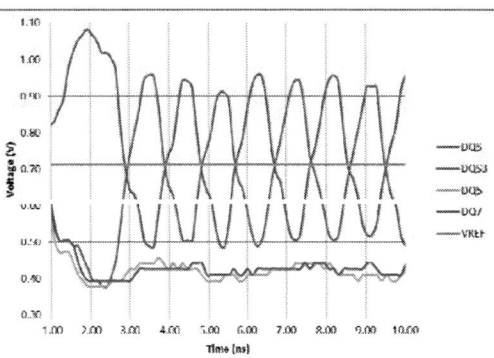

Figure 24 – Waveforms of DQS and DQSB signals when all DQ0-7 signals are fixed low

In the following two pictures (**Figure 25** and **Figure 26**), DQ0-7 are toggling simultaneously; in the first case, DQ0-7 follow the topology on DQS, while in the second case follow DQSB instead. Both pictures show the degradation in all the waveforms. Comparing the 2 pictures, it can be seen that the DQS/DQSB are influenced in opposite directions, depending on the DQ0-7 topology, confirming the coupling effect.

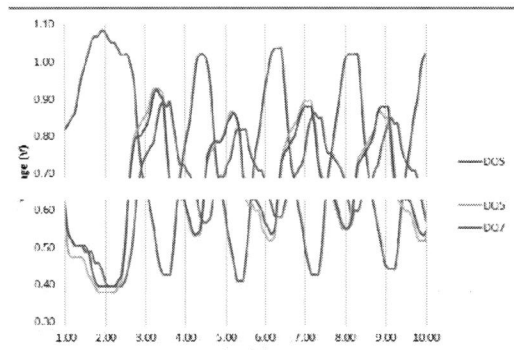

Figure 25 – Waveforms of DQ0-7 toggling simultaneously and following topology on DQS

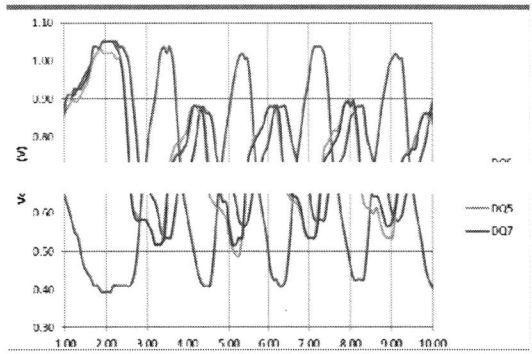

Figure 26 – Waveforms of DQ0-7 toggling simultaneously and following topology on DQSB

Although the dual die modules suffered from the same routing restrictions, and coupling, they could achieve 667 MHz; we believe this was because of the lower input capacitance.

151

In the next prototypes already being fabricated, the blank silicon die was replaced by a 4-layer PCB substrate, therefore there are expectations of improved electrical results.

As part of this joint development samples of this technology demonstrator were submitted to reliability testing with pass results (**Table 5**).

Reliability Test	Standard/ Spec	Pass Criteria	Final Assessment (Pass/Fail)
Moisture Sensitivity Level (MSL)	EIA/J-STD-020C (Level 3)	MSL3	PASS
High Temperature Storage (HTS)	JESD22-A103 (Cond.C: Ta:150°C)	1000hrs	PASS
Temperature Cycling (TC)	JESD22-A104 (Cond G:-40°C / +125°C / 2cy/hr) Precond. L3; Tr:260°C)	500x	PASS
Unbiased HAST (uHAST)	JESD22-A118 (Cond B: 110°C / 85% RH) Precond. L3; Tr:260°C)	264hrs	Running
Temperature Cycling on Board (TCoB)	Customer (-40°C / +125°C / 2cy/h)	FF > 500 cycles	Running

Table 5 – Reliability test plan for development of stacking solution of 4 Die DDR3 DRAM , using eWLB and WDoD technology

CONCLUSIONS

NANIUM has been one of the main user and driver of eWLB as FO-WLP technology.

On this path, NANIUM is continuously extending eWLB capabilities and enlarging its application field, by completing several new developments.

Examples of these developments are multi-layer RDL, fine RDL line width and space and reduced ball pitch, enabling of multi-die and reduced die-to-die-distance and die-to-package edge distance and finally 3D Heterogeneous integration.

As a result eWLB technology is gathering the adequate capabilities to be one of the main packaging technologies for SiP applications, now at wafer level – WLSiP.

ACKNOWLEDGEMENTS

The authors would like to thank and acknowledge the contribution of all colleagues from NANIUM that contributed to this paper and their continuous effort on the running activities.

Also the authors would like to recognize the confidence and openness of our customers that are working with us in some of these projects.

REFERENCES

[1] T. Meyer, G. Ofner, S. Bradl, M. Brunnbauer, R. Hagen. "Embedded Wafer Level Ball Grid Array (eWLB)", *10th Electronics Packaging Technology Conference*, December 2008, Singapore

[2] Yonggang Jin, *et al*, "Next Generation eWLB (embedded Wafer Level BGA) Packaging", *12th Electronics Packaging Technology Conference*, December 2010, Singapore

[3] Dr. Christian Val, et al, "Stacking of Full Rebuilt Wafers for SiP and Abandoned Sensors/applications", *European Microelectronics Packaging Conference*, 2009, Rimini, Italy

[4] J. Campos, "New Application for Fan-Out Wafer Level Packaging Technology", *8th International Wafer Level Packaging Conference*, October 2011, San Jose, California, USA

[5] Jerome Noiray, Dr Christian Val, Dr Pascal Couderc and T. Ferrara, "Stacking of known good rebuilt wafers for high performance memory application to high speed DDR3", MiNaPAD, April 2012, Grennoble, France

ADAPTIVE PATTERNING FOR PANELIZED PACKAGING

C. Scanlan, B. Rogers, T. Olson, C. Bishop, J. Kellar, and B.Y. Jung
Deca Technologies, Inc.
Tempe, AZ, USA
chris.scanlan@decatechnologies.com

ABSTRACT

Fan-Out Wafer-Level Packaging (FOWLP) or fan-out technology has held promise for a number of years; primarily as a means of packaging semiconductor devices with interconnect densities exceeding the capabilities of standard Wafer Level Chip Scale Packaging (WLCSP). With FOWLP technology, die are embedded in a molded panel, and I/Os are then redistributed over the larger effective surface area using conventional WLCSP techniques. The packages are then singulated and attached directly to a printed circuit board (PCB) or low-cost substrate. This technology provides one of the smallest and lightest possible package form factors; enables more I/Os for a given pitch with excellent electrical properties; and eliminates the need for custom substrates used in flip chip or wirebond Ball Grid Array (BGA) packages. Despite its promise, widespread adoption of FOWLP packaging has been limited largely by cost and yield issues. The requirement for high die placement accuracy when forming the molded panel restricts throughput at the die pick-and-place operation, leading to high process costs. Die drift, or movement during panel molding, limits via and RDL design rules and ultimately can result in yield loss when the drift is excessive. Managing or overcoming die offset is one of the keys to making FOWLP competitive with other package formats.

This paper describes an approach to FOWLP that allows die offset to increase by an order of magnitude compared with conventional methods. Using a novel Adaptive Patterning* technology, real-time designs are created for each package within each panel during the manufacturing process. After panelization, the position of each die within each molded panel is precisely measured. Information is fed into a proprietary auto-routing design tool on a per panel basis. The resulting pattern layers are then issued to a lithography system which dynamically implements the unique design on a per panel basis. Dynamic layers include various design features such as vias or redistribution layers (RDL).

The ability of adaptive patterning to correct for deviations in die location can result in both improved yield and higher panelization throughput, thereby enabling the industry to finally realize the cost, flexibility, and form factor benefits of FOWLP. In the paper, adaptive patterning examples will be presented and the benefits and limitations of the technology will be discussed.

Key words: adaptive patterning, fan-out wafer-level packaging (FOWLP), panelized packaging, WLP

INTRODUCTION

The handheld consumer electronics space, where portability and increasing functionality are strong drivers, continues to motivate the transition to packaging approaches that provide small size, high performance, and low cost. Wafer Level Chip Scale Packaging (WLCSP), which offers the smallest packaging form factor, has often been a preferred option for addressing the handheld market. In WLCSP, chip I/Os are generally fanned-in across the die surface using polymer and redistribution line (RDL) buildup layers to produce an area array, and large solder bumps are formed at the terminals by ball drop or plating. These additive processes allow the chip to be attached directly to a PCB with high reliability. However, two progressions in front-end chip manufacturing offer challenges to packaging in a WLCSP format: (1) die shrink, enabled by advancing semiconductor technology nodes, makes it increasingly difficult to fit all of the large solder ball I/Os on the die surface; and (2) increasing chip functionality produces a need for more I/Os, also making WLCSP packaging more difficult. One approach to extending WLCSP is to shrink the size of the I/Os or solder bumps on the chip surface so that more can fit within the chip area. However, this approach is generally limited by a lack of assembly infrastructure and higher assembly costs for dealing with the smaller or tighter pitch I/Os. [1]

Fan-out, or FOWLP has been offered for a number of years as an alternative for addressing constraints to WLCSP [2,3]. In this technology, chips are singulated and then embedded in a panel. A common method for forming this panel is to place the chips face-down on a carrier at a desired pitch and then mold over them using compression or print molding. The molded panel is subsequently separated from the carrier. The panel is often formed in the shape of a wafer, so that standard wafer processing techniques can be used to create buildup layers on the panel surface. The extra panel surface around the chip allows I/Os to be both fanned in over the chip and fanned out across the mold compound, thus accommodating a larger number of I/Os. After buildup layer processing and solder ball attachment, the packages undergo backgrinding, laser marking, and singulation, just like WLCSPs. The resulting package is often just slightly larger than the chip and just large enough to accommodate the I/Os. Like WLCSPs, the package is ready to be mounted directly to a PCB.

The potential benefits of FOWLP are numerous. This technology provides the smallest and lightest possible form factor for packaging small, high I/O chips that cannot be packaged as WLCSPs. As with WLCSPs, the device-to-board connections through thick copper routing layers and large solder balls offer excellent electrical properties and performance. When hitting acceptable cost targets, FOWLP can potentially displace other forms of packaging, such as flip chip or wirebond BGAs. In those cases, it generally brings a size advantage and eliminates the need for custom substrates, significantly simplifying the supply chain. Finally, FOWLP enables the connection of two or more chips in the fan-out routing layer, facilitating multichip and system-in-package (SIP) applications [4-7].

Cost, yield and reliability issues have effectively limited the widespread adoption of FOWLP despite its promise. Placing singulated chips on the carrier to form the molded panel requires high placement accuracy. Any misplacements can lead to pattern overlay difficulties in the buildup process on the reconstituted panel. The requirement for high placement accuracy restricts throughput at the pick-and-place operation, leading to high process costs. During the molding operation and mold cure, die drift or movement can occur. This die drift can further complicate pattern overlay matching in the buildup process on the panel and can result in yield loss when the drift is excessive. In addition to die offset issues, several other challenges must be addressed for successful FOWLP packaging. The molded panels tend to warp during the buildup process; limiting or minimizing this warpage is important to enabling the use of standard wafer processing equipment. A low-cure polymer is required for the buildup layers over the mold compound. Identifying a low-cure polymer with good mechanical properties is important to ensuring package reliability. Finally, the coefficient of thermal expansion (CTE) mismatch between the mold compound and the silicon can lead to reliability issues and must be carefully managed.

Addressing or overcoming die offset and resulting overlay issues is one of the keys to making FOWLP competitive with other package formats. This paper discusses a novel approach called Adaptive Patterning, which allows die offset to increase by an order of magnitude compared with conventional methods. The ability of adaptive patterning to correct for deviations in die location can result in both improved yield and higher panelization throughput, thereby enabling cost-effective FOWLP packaging.

BACKGROUND
To illustrate the challenges in die position control, bond pads at a 45µm pitch, representative of an advanced technology node, are shown in Fig.1. A 20µm via is shown centered in the bond pad opening (BPO), representing the first layer in the fan-out buildup. The bond pad is assumed to be 40µm, with a BPO of 38µm. At these dimensions, a die shift of no more than 9µm in X and 9µm in Y can be tolerated before the via is no longer making full contact to the bond pad, as shown in Fig. 2.

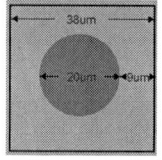 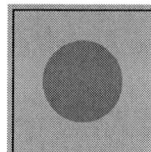

Fig. 1. Illustration of perfect die placement for an advanced technology node: 20µm via centered in bond pad openings (BPO) of 38µm on a 45µm pitch.

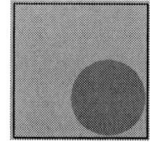 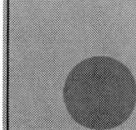

Fig. 2. Die shift by -9µm in X and 9µm in Y: 20µm via just makes full contact with BPO

Die rotation errors also present a challenge, as is illustrated in Fig. 3. For a 2mm x 2mm die, rotation placement errors as small as 0.5 degrees can result in a 9mm offset in the die corners, essentially using up the maximum 9mm shift budget of the previous example. As shown in Fig. 3, the situation gets significantly worse with larger die sizes.

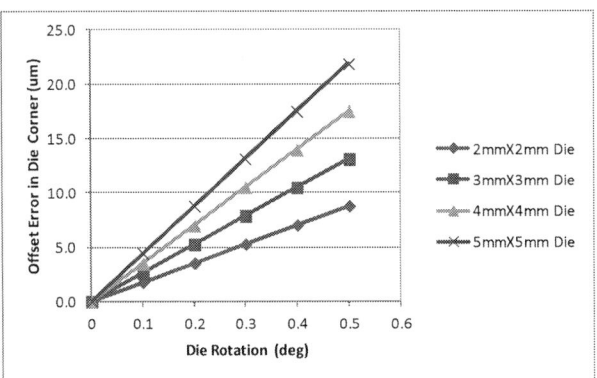

Fig. 3. Contribution of die rotation error to offset error in die corner vs. die rotation placement errors.

This demands excellent die placement and rotation control in the pick-and-place operation used to form the panel. Generally, die placement accuracy of < 10μm and rotational accuracy of < 0.1deg are required. Thus, relatively slow speeds must be utilized at the pick-and-place operation, adding significantly to the overall cost of FOWLP packages.

Die location can be further adversely affected by the panel molding operation. Die drift or shifts in position are commonly encountered during panel molding. Expected shifts can be modeled and compensated for in the pitch used at die placement. However, random shifts can result in yield loss, when the die placement error plus the random shift error exceeds the maximum shift budget for successful overlay of the buildup layers.

One strategy for dealing with die mislocation is to utilize design rules that accommodate substantial shifts and still result in yielding parts. An example would be limiting FOWLP to devices with large bond pad pitch and bond pad openings, so that the via and RDL layers in the buildup have a good chance of making contact to the BPOs. This approach, however, severely limits the number and types of chips to which FOWLP can be applied.

An alternative approach, possibly utilizing stepper technology, is to use local alignment schemes to better align the buildup layers to the die pads [8]. This has been demonstrated to result in improved alignment to mislocated die, and thus improved overall yield. However, this approach still requires a high degree of die placement and die drift control. In the extreme case, steppers can be used to perform die-by-die alignment but with a significant throughput penalty, making this approach impractical for a production environment.

ADAPTIVE PATTERNING METHOD

A new manufacturing method called adaptive patterning has been developed to correct for die shifts inherent in any

chips-first, panelized packaging process. The first application of this technique is FOWLP, but it can also be applied to embedded die in substrate or other chips-first processes. The adaptive patterning method draws from techniques used in traditional wirebond packaging. In standard wirebond package assembly processes, die are placed using a high-speed die placement machine and rigidly attached to a package substrate, typically a strip containing an array of units. Subsequently, a wire bonder inspects the strip to measure the actual location of the die and dynamically adapts the wirebond program for each individual chip, allowing the wire bonder to accurately connect the bond pads on the chip to the corresponding bond fingers on the substrate or leadframe. This process of dynamically adapting the chip-to-package interconnect pattern has previously not been technically feasible and economically viable for fan-out or embedded die processes wherein the die is placed prior to forming the routing layers on the package.

The adaptive patterning process developed for FOWLP works by dynamically adjusting a portion of the interconnect structure to accurately connect to the bond pads or other structures on each individual die in the molded panel. The process is summarized in Fig. 4. First the die are thinned and singulated from the native wafer. Die are placed onto a temporary carrier and a compression molding process is used to encapsulate the die, forming a panel. While the initial implementation uses a form factor similar to a 300mm semiconductor wafer, the method can be readily extended to larger panel formats of any shape. Next, an optical scanner inspects features on each die to determine actual position and rotation of every die on the panel with respect to a panel frame of reference. Using the actual die position data for each die on the panel, a proprietary design tool adjusts the fan-out unit design for each package on the panel so that the first via layer and RDL pattern are properly aligned to bond pad features on the die. The individual package designs are combined to form a drawing of the full panel for each of the layers that need to be adjusted, typically including the first via layer on the panel and the first RDL layer. The design files for each panel are imported to a lithography machine, which uses the design data to dynamically apply a custom, adaptive pattern to each panel.

Multiple strategies for adapting the pattern were considered. The simplest method would be to adjust only the first dielectric via layer to the corresponding contact pads on the panel.

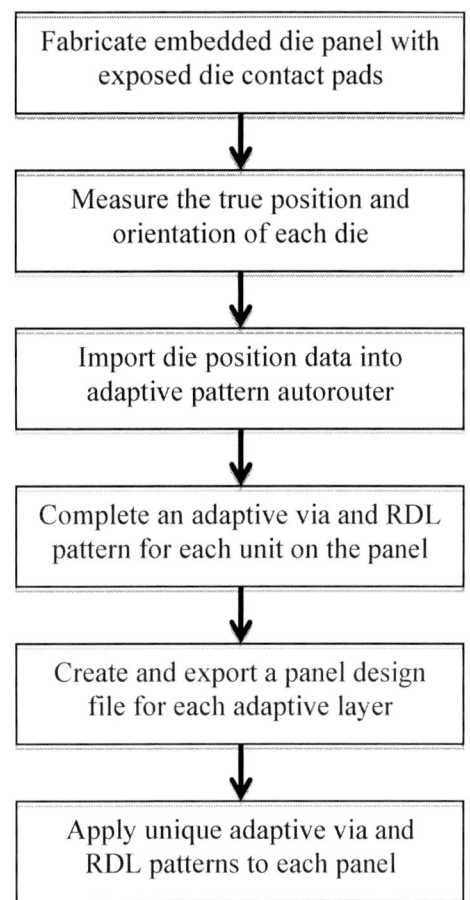

```
┌─────────────────────────────────┐
│  Fabricate embedded die panel with │
│      exposed die contact pads      │
└─────────────────────────────────┘
                 │
                 ▼
┌─────────────────────────────────┐
│   Measure the true position and   │
│       orientation of each die      │
└─────────────────────────────────┘
                 │
                 ▼
┌─────────────────────────────────┐
│     Import die position data into  │
│      adaptive pattern autorouter   │
└─────────────────────────────────┘
                 │
                 ▼
┌─────────────────────────────────┐
│   Complete an adaptive via and RDL │
│      pattern for each unit on the panel │
└─────────────────────────────────┘
                 │
                 ▼
┌─────────────────────────────────┐
│    Create and export a panel design │
│      file for each adaptive layer   │
└─────────────────────────────────┘
                 │
                 ▼
┌─────────────────────────────────┐
│    Apply unique adaptive via and   │
│      RDL patterns to each panel     │
└─────────────────────────────────┘
```

Fig. 4. Adaptive Patterning flow.

The RDL capture pad overlying the via can be enlarged to accommodate shifts in the via position with respect to a static RDL pattern. This approach is limited to larger pad pitch applications and is therefore not extendable to advanced technology semiconductor nodes.

An alternate approach is to adjust the position of the first via layer and the entire RDL pattern for each unit in order to align to the actual position of the die on the panel. This technique eliminates the via-to-RDL capture pad offset issue described above, so it has the potential to support designs with finer pad pitch. However, for a package with a single fan-out routing layer, the overlay problem is moved to the next via layer underlying the UBM. It is not desirable to allow the position of the UBM pattern to shift with respect to the edge of the package – it must be held constant. Therefore, if the entire RDL pattern shifts with respect to the fixed UBM pattern, the shift must be accommodated

either by increasing the size of the underlying capture pad on the RDL layer or by reducing the diameter of the via connecting the UBM to the RDL layer. In addition, this technique will not work for multi-chip modules wherein die will shift with respect to each other within the same unit design. This technique is therefore limited to smaller, low routing density, single chip packages.

A variation on this technique has been considered. It is possible to create a discrete number of fixed RDL patterns during the initial design process, each pattern being applicable to a subset of the possible die shifts defined by the process capability of the panelization process. In this case, the best fit design is selected for each unit on the panel, and a minor offset in x-y and theta is applied.

The ultimate flexibility could be achieved by calculating a complete custom RDL pattern for each unit on the panel. This method has the potential to address both die-to-RDL and RDL-to-UBM alignment without compromising on overlay design rules. Although we have demonstrated feasibility of this approach, there are some potential disadvantages. The algorithms required to complete a full package auto route are quite complex and less capable for reliably replicating complex patterns such as inductors and large power and ground planes.

ADAPTIVE PATTERNING USING PRESTRATUM

The most flexible and consistent adaptive patterning method is a hybrid approach, in which a portion of the first RDL remains fixed with respect to the package outline and BGA array. This fixed partial pattern, or prestratum, includes capture pads for subsequent pattern layers including dielectric vias and UBM structures. However, a small portion of the RDL layer in close proximity to the die contact pads is omitted from the prestratum pattern. After scanning the panel to measure the actual position and orientation of each unit, the design of each unit on the panel is completed to connect the prestratum pattern to the die contact pads and their corresponding dielectric vias. The adaptive region in which the RDL traces are allowed to dynamically change is typically on the order of 100μm to 200μm surrounding the contact pads on the chip.

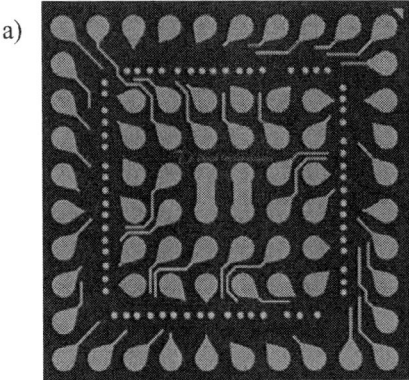

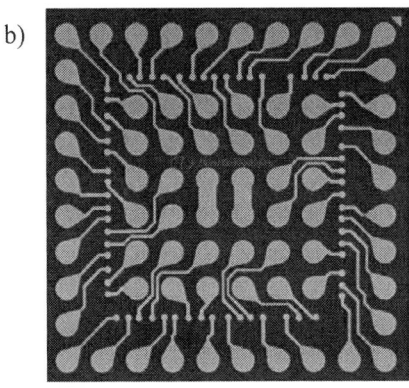

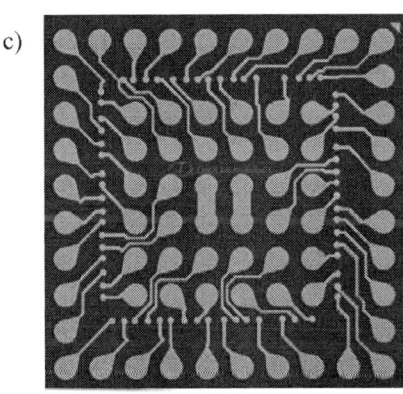

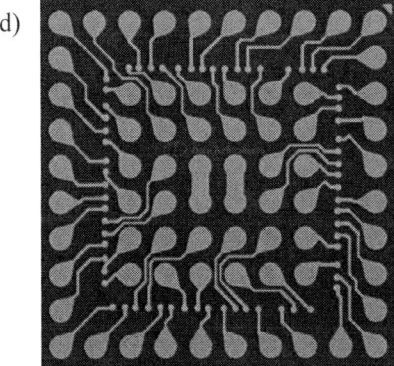

Fig. 5. Adaptive pattern design using prestratum. a) prestratum design with nominal position of die contact pads b) completed design for the nominal die position c) completed design with die shift (x, y, theta) of –50μm, -

50μm, 0.5° d) completed design with die shift (x, y, theta) of 50μm, 50μm, -0.5°

Figure 5a shows an example of a fan-out RDL layer with only the prestratum pattern routed. This portion of the design is created in a traditional layout tool, such as Cadence. In order to determine the optimum prestratum design and to ensure routability of the design for all possible die shifts, a Monte Carlo analysis is performed. This is done using a proprietary design tool by applying an array of die shifts in x, y, and theta that represents the expected range of die shift resulting from the panelization process. Any design rule violations or routing errors can be characterized quickly in the design environment and corrected by adjusting the prestratum design prior to prototyping.

Figure 5b shows the completed RDL design wherein the prestratum pattern has been connected to the die bond positions on the chip. The portion of the routing is highlighted. This figure represents the nominal design, i.e. the case wherein the die shift and rotation is nil.

Figure 5c shows the completed RDL design for the case where the die is shifted from the nominal position by 50μm in x, 50μm in y and -0.5° in theta. Note that the prestatum pattern remains constant, but the autorouter changes the adaptive region of the RDL layer to ensure contact is made with the die contact pads. Figure 5d shows a completed design for the case where die shift is -50μm in x, -50μm in y and 0.5° in theta with respect to nominal. Again, the prestratum is held constant and the adaptive region of the RDL pattern is adjusted by the autorouter to make contact with the die contact pads. Monte Carlo analysis was completed to validate that routing is feasible with shifts in X and Y of up to +/- 80μm and rotation up to 0.5 degrees.

Note that in all of the adaptive patterning methods described above, the first dielectric via patterns are adapted to align precisely with the contact pads on each chip in the panel.

FUNCTIONAL DEVICE EXAMPLE
The adaptive patterning method has been demonstrated on a FOWLP containing a functional mixed signal IC with die size of 2.9mm x 4.47mm. The package has a body size of 4.65mm x 4.9mm and a 92 ball BGA at 0.4mm pitch. The package contains a single fan-out RDL layer, and an in-line arrangement of die contact pads at 102μm pitch. The RDL prestratum design for the package is shown in Fig. 6. Note that the prestratum design contains a fixed ground pattern at the center of the package.

Fig. 6: RDL prestratum design containing a fixed ground plane structure.

The FOWLP was constructed on a 300mm diameter molded panel using the method described in Fig. 4. A first photopolymer layer was applied using adaptive patterning to align the via pattern to the contact pads on each unit. A unique RDL pattern was created for each unit on the panel to connect the prestratum traces shown in Fig. 6 to the associated polymer via overlying the die contact pads. An example of a finished package is shown in Fig. 7. The device shown exhibits die offset of 10μm in X, 20μm in Y, with rotation error of 0.3 deg, adding effectively 7μm of error to X and 11μm of error to Y in the corner region. Total misplacement in the corner region is ~ 17μm in X and 31μm in Y. The adaptive RDL traces effectively align to the vias resulting in an electrically functional unit.

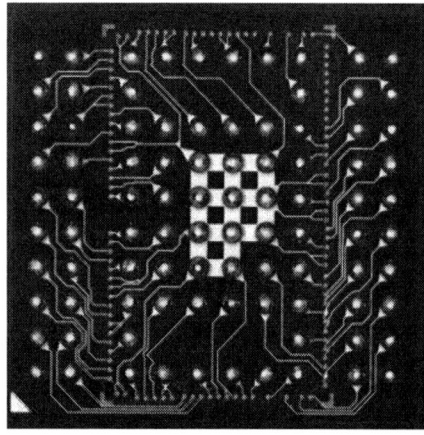

Fig. 7: Fan-out package with adaptive pattern: 2.9mm X 4.47mm die in a fully molded 4.65mmX4.9mm package; 92 balls, 0.4mm pitch.

CONCLUSIONS

Chips-first panelized packaging technologies have been in development for more than 2 decades, but have been limited by the ability of traditional patterning methods to accurately align to the die, which tend to shift within the panel. A method of adaptive patterning for FOWLP and embedded die packages has been demonstrated that solves this problem. Adaptive patterning allows die offset from the panelization process to increase by an order of magnitude compared with conventional methods. The prestratum design methodology described in this paper offers the flexibility to adapt to die position shifts in assembly while maintaining a mostly constant, deterministic routing pattern. The ability of adaptive patterning to correct for deviations in die location can result in both improved yield and higher die placement throughput, removing the key barriers to broad adoption of FOWLP and embedded die packaging.

*Adaptive Patterning™ by Deca Technologies, Inc.

REFERENCES

[1] Anderson, R., et. al. "Advances in WLCSP Technologies for Growing Market Needs," *IWLPC Proceedings*, Oct. 2009.

2] Brunnbauer, M. et. al., "Embedded Wafer Level Ball Grid Array (eWLB)," *Electronics Packaging Technology Conference 8th Proceedings*, Dec. 2006.

[3] Keser, B. et. al., "The Redistributed Chip Package: A Breakthrough for Advanced Packaging," *2007 Electronic Components and Technology Conference*, pp. 286-291,

[4] J. Sabatini, "GE's High-Density Overlay Technology," *Surface-Mount Technology*. Vol. 6, No. 3. Mar. 1992, pp. 18-19.

[5] Daum, W. et. al., "Overlay High-Density Interconnect: A Chips-First Multichip Module Technology," *Computer*, vol. 26, no. 4, April 1993, pp. 23-29.

[6] Meyer, T. et. al., "eWLB System in Package – Possbilities and Requirements," *IWLPC Proceedings*, Oct. 2010, pp. 160-166.

[7] Kang, I.S, et. al., "Wafer Level Embedded System in Package (WL-ESiP) for 3D SiP Solution," *IWLPC Proceedings*, Oct. 2010, pp. 153-159.

[8] Hsieh, R.L. et. al., "Lithography Challenges and Considerations for Emerging Fan-Out Wafer Level Packaging Applications," *IWLPC Proceedings*, Oct. 2009.

OPTICAL PROFILOMETRY OF SUBSTRATE BOW REDUCTION USING TEMPORARY ADHESIVES

[a]Paul Flynn and [b]John Moore
[a]FRT of America, LLC, [b]Daetec, LLC
[a]San Jose, CA, [b]Camarillo, CA
[a]pflynn@frtofamerica.com, [b]jmoore@daetec.com

ABSTRACT

Incentives of form factor, performance, and cost reduction are urging the integration of three-dimensional packaging of integrated circuits (3DIC) and promoting thinner substrates. To maintain this trend, new ways of rapid metrology measurement with solutions to achieve substrate bow reduction are needed. FRT's optical profilometry systems offer high resolution and rapid scanning to map a substrate in minutes [1]. When combined with a removable temporary bonding solutions, simple and low-cost options exist for bow reduction to the end user [2].

Key words: profilometry, bow, warp, TTV

INTRODUCTION

It is difficult to predict what variability exists due to thinning semiconductor wafers and die; however, this irregularity commonly increases as thickness is reduced [6]. For example, bow is observed in a LED wafer when its internal stress exceeds the intrinsic strength of sapphire when thinned to 100um (Fig. 1).

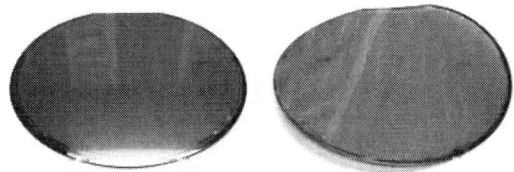

Full thickness (~700um)   Thinned (~100um)

Fig. 1. Sapphire LED substrate at full thickness (left) and thinned (right) with the presence of bow.

Substrate stress may originate from metal thickness, component design and layout, and thinning. These may be unavoidable, leading to bow and warpage, which contributes to yield loss in 3DIC. Measurement of these characteristics must occur and where necessary, be supported by temporary carriers.

Optical Profilometry

Substrate bow [3], warp [4], total thickness variation (TTV), and flatness [5] are measured to SEMI Standards with either a single sensor or dual, opposed sensor configuration. The entire substrate or wafer can be mapped or measured by a series of profiles. Non-contact optical profilometry is the preferred method of choice for high resolution, speed, and reliability. FRT's capabilities extend to 450 mm wafers and

larger substrates for determining 3D measurements locally or across the entire sample. Detailed images and software driven statistics enable rapid identification of the parameter of interest, enabling critical decisions when searching for solutions that involve handling expensive thinned substrates (Fig. 2).

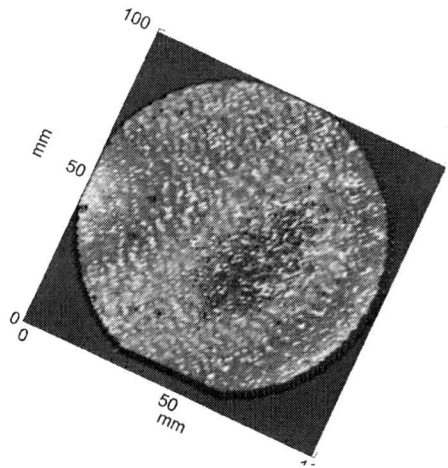

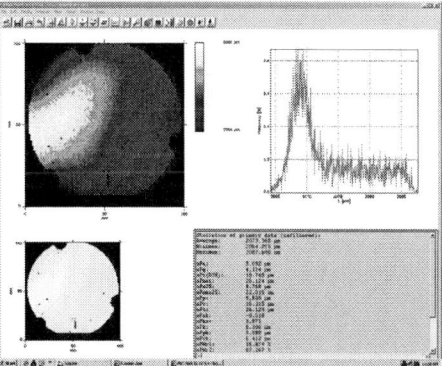

Fig.2. Flatness and TTV scan of a 100 mm wafer.

The equipment sensor resolution can be set in the z range to 3 nm and 1 um in the x-y direction with a z-working distance to 5 mm. The optical sensor uses a positioning camera, which works as an OM to define the scanning area. Included on FRT's MicroProf® series, they allow rapid and accurate topography measurements at the R&D level or with automation on sample sizes up to 600 mm x 600 mm (Fig. 3).

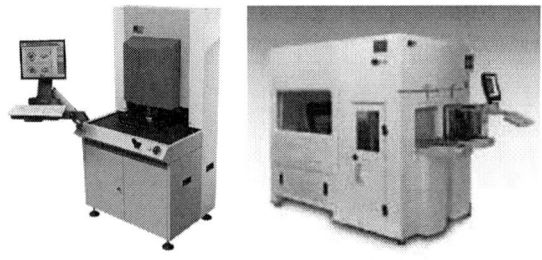

Manual Operation   Automated – 300mm

Fig. 3. FRT's MicroProf® series of profilometers.

Thin Substrate Handling

Of the common wafer support practices, an adhesive bonded carrier is the most reliable to protect thin substrates and exhibit chemical and thermal resistance necessary for backside work (Table 1).

Table 1. Options for thin substrate handling.

Method of Handling	Substrate Thick (um)	Chem & Therm Resistant	Single or Batch Process	Backside Process Support
Tape	>50	No	Both	No
Vacuum Chuck	>50	No	Single	No
Adhesive Bonded Carrier	<25	Yes	Both	Yes

Temporary Adhesives

All of the current commercialized adhesive technologies use a carrier for temporary support. The carrier enables good surface planarity, low TTV, and reduces both internal stress and wafer bow during grinding [6-9]. Liquid spin-on forms of adhesives offer easy control of TTV with acceptable thinning uniformity achieved with this value at ≤0.5% [10-11]. The adhesive supports backside processing, including TSVs, metallization, and dicing. A common feature of temporary adhesives with carriers includes two active stages, namely, bonding and de-bonding (Fig. 4).

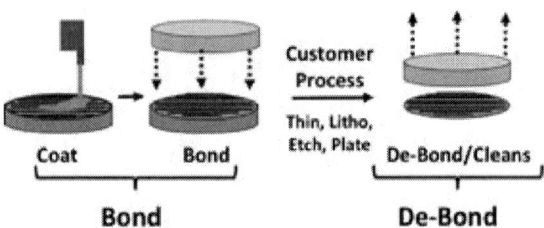

Fig. 4. Two active stages to the use of any temporary adhesive and carrier, bonding and de-bonding. Cleaning is included in the de-bonding practice.

Table 2. Commercial temporary adhesives and their process parameters.

Firm	Chemistry	DeBond Method	Batch or SW	Cleans
BSI	Rubber	Chem. diffusion, therm. slide, peel	SW	Solvent
3M	Acrylic	UV and/or ablate & peel	SW	Solvent
TMAT, Dow Corning	Silicone	Peel	SW	Solvent
DuPont	Polyimide	Ablate/peel	SW	Solvent
TOK	Acrylic/ Styrenic	Ablate &/or chem. diffusion	Batch	Solvent
Daetec	Various AQ	Chem. diffusion	Batch	Detergent

Products used in the marketplace include the following: a) rubber/olefinic [12-13], b) acrylic [14], c) silicone [15], d) polyimide, and e) rosin-urethane [16]. Although these chemistries vary, their application is similar by direct wafer coating and carrier bonding (Fig. 4). The main variance in performance and complexity is in their de-bonding (Table 2 & Fig. 5).

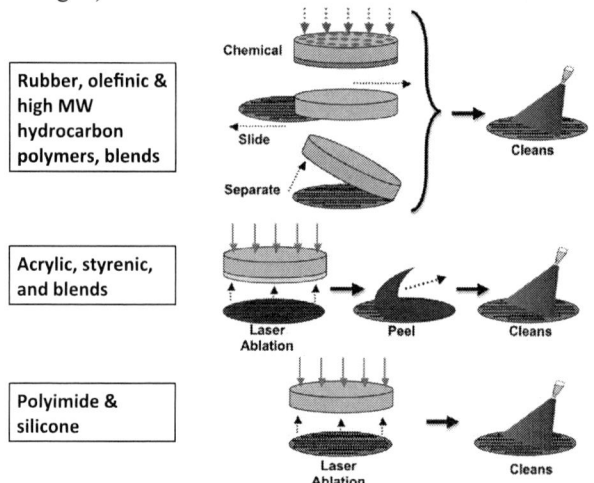

Fig. 5. Temporary adhesives commercially available, exhibiting variable de-bond performance and complexity.

EXPERIMENTAL

Readily available thinned interposer die are acquired by several suppliers in the industry (Fig. 6). Temporary bonding adhesives and other developmental products are readily available [2]. DaeCoat CD300, CS300, and FS300 with detergent DaeClean DP-108 and SL3200, SL1750 cleaning agents. Coatings are produced on a Brewer Science, Inc. CB-100 spin-coater, while spray and encapsulation uses custom tooling. Configuration of interposer die applied to varying carriers with adhesive and measured by optical profilometry (Fig. 7). Metrology data is generated by a FRT MicroProf® optical profiler [1]. Measurement conducted on the diagonal end-to-end, producing thickness profiles with statistics (Fig. 8). Modified thermogravimetric test methodology for outgas is conducted by typical laboratory scales (+/- 0.1mg, Fig. 9). UV cure equipment includes the Intelli-Ray 400 microprocessor controlled light curing system (Uvitron

International, www.uvitron.com). Measurement of a thermal deviation from a planar surface by shadow moire', similar equipment to the commercial manufacture (www.akrometrix.com).

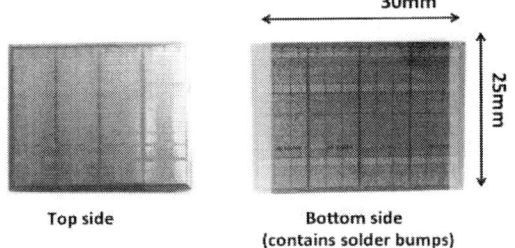

Fig. 6. Interposer die dimensions.

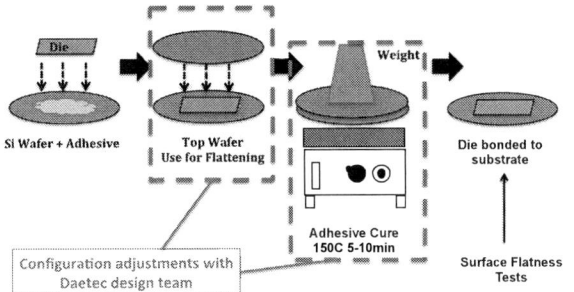

Fig.7. Process flow in bonding interposer die and flatness measurement.

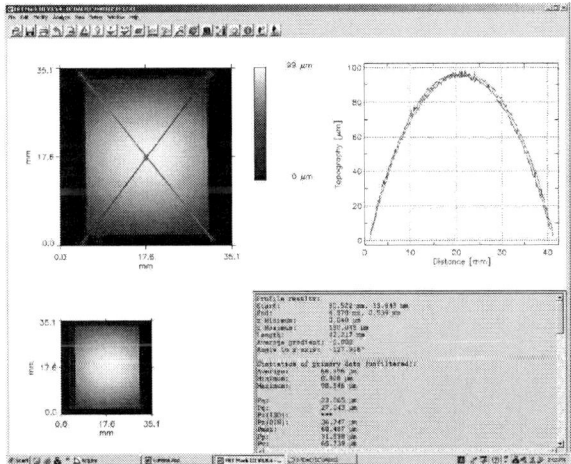

Fig. 8. Optical profilometer output with statistics.

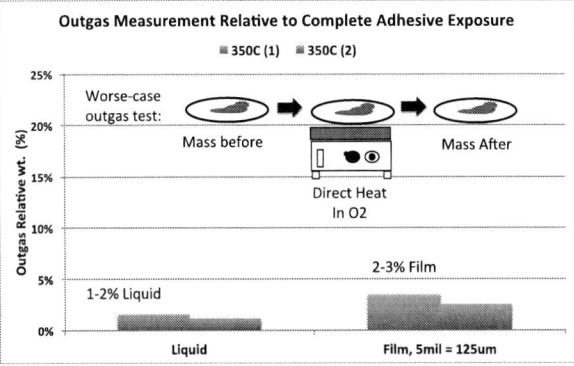

Fig. 9. Outgas data on thermal resistant adhesives.

RESULTS

Thinned interposer die are measured by optical profilometry and are shown to exhibit a bow of >100um as deviation from a planar surface. In the following example, interposer die #5 is observed at the initial condition (Fig. 10), coating on bumped side (Fig. 11), mounted to a carrier (Fig. 12), and then observed for bow in the affixed position to the carrier (Fig. 13).

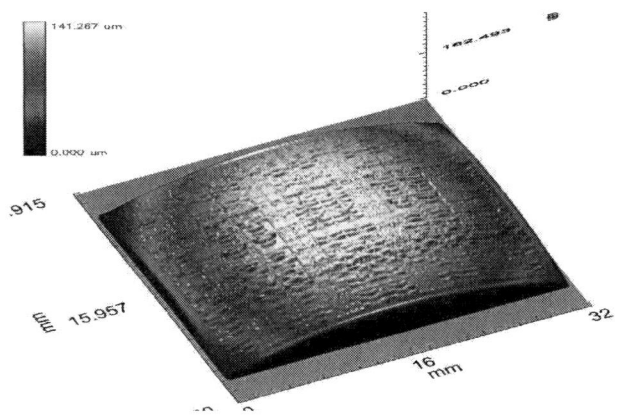

Fig. 10. Initial optical profilometry of interposer die #5, indicating bow extending to >140um.

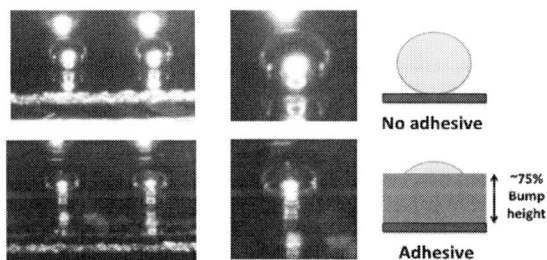

Fig. 11. Adhesive planarization of bumps, $\geq$75%.

Fig. 12. Bonding interposer die to different carriers as designed internally according to porosity and capillary action during debond.

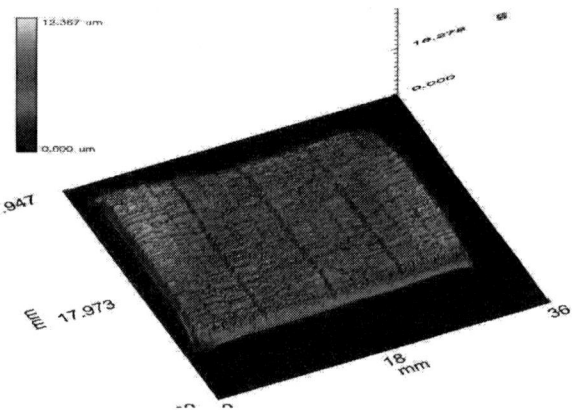

Fig. 13. Post optical profilometry of interposer die #5, affixed by temporary adhesive to carrier, indicating bow reduction to 12um (~90% reduction).

Using multiple adhesive configurations with liquid UV/cure and film forms, interposer die are attached to a range of carrier substrates as solid, semi-porous, and porous (Fig. 10). Once affixed to the carriers, optical profilometry is performed and reporting the maximum and average relative thickness from center (Fig. 14) and percent relative reduction (Fig. 15). Shadow moire' thermal topography results of several interposer die through reflow temperature of 260 °C, achieving a deviation satisfactory to the process (Table 3).

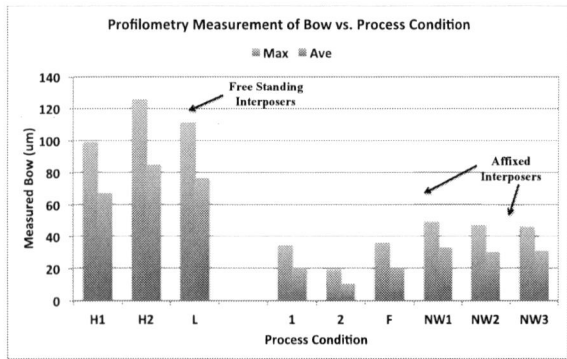

Fig.14. Bow measurement of interposer die.

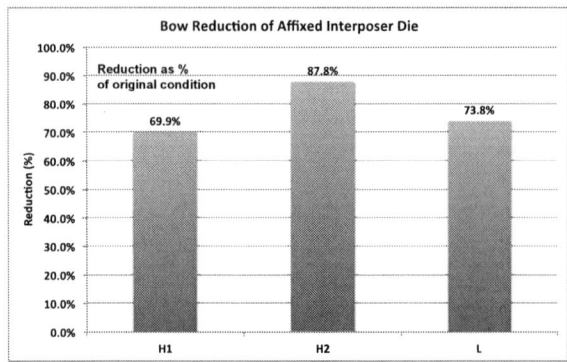

Fig.15. Bow reduction as percent of original condition.

Table 3. Shadow moire' test condition of adhesive.

Sample #	Adhesive type	Shadow moire'
407	DaeCoat FS300	Pass
#1	DaeCoat CS300	Pass
#2	DaeCoat CS300	Pass
Un-supported	- N/A -	Fail

DISCUSSION

Bow reduction by temporary affixing interposer die to carrier substrates is the beginning to a process development for down stream further 3DIC processing. One such temporary adhesive for coating and planarizing of large features is DaeCoat CS300. Once cured, it can then be affixed to the carrier by a film version of the same chemistry, Daecoat FS300. A process with porous substrates is given (Figs. 16-17).

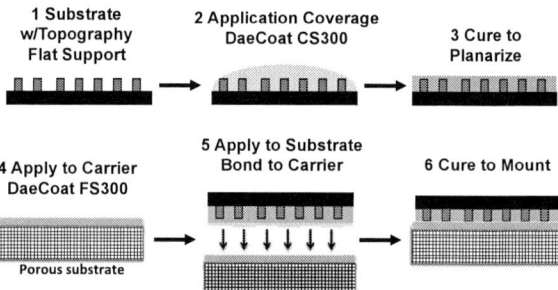

Fig. 16. Process flow for affixing interposer die.

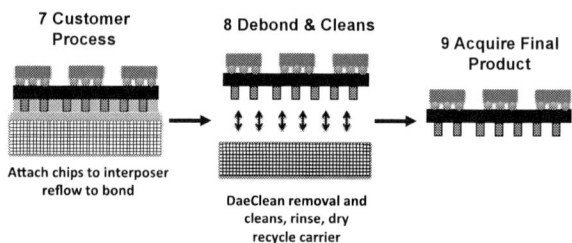

Fig.17. Post-processing of affixed interposer die, demount, and cleans

The tested temporary adhesives in this study are easily removed (cleaned) from the surface, leaving the substrate in a pristine condition. Usual cleans practices involve the application of solvents or aqueous mixtures followed by an alcohol or water rinse. For simple de-bond, it is recommended to use of perforated carriers or capillary diffusion substrates in a batch process.

To support this approach, a new generation of porous substrates is being explored. These substrates are designed to allow greater fluid contact to the adhesive and aid in batch de-bonding. Of the carriers under review, some exhibit smooth surfaces with Rq values of 1um or less and flatness within 10-25um (Fig. 18).

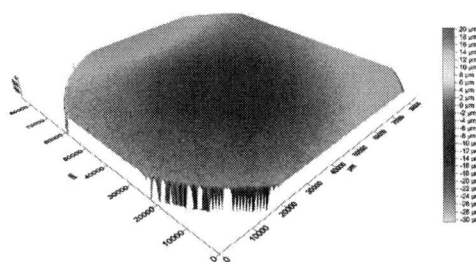

Fig 18. Flatness of a porous carrier, used to demonstrate batch de-bonding processes.

For several decades, temporary adhesives have been used in mounting wafers to carriers, thinning, backside processing, and finishing by a batch de-bonding process [10-12]. Bonded wafers are assembled in a cassette and then immersed into the cleaning solution. Penetration occurs via conduction channels at the side and through a perforated carrier, allowing breakdown of the adhesive. The adhesive gives way and demounting occurs with simultaneous cleaning in the same bath. Customers have designed their own fixtures to separate the carrier from their product without yield loss.

Where wafers are used, it is preferred that de-bonding and cleaning occur while the thinned wafer is supported on a tape (film frame). This practice requires cleaning and process to be fully compatible (safe) with the chemistry of the tape (Fig. 19).

| Film Attachment | Wafer Cleans | Dicing |
| Carrier Demount | Safe for Tape | |

Fig. 19. Process flow for cleans while product wafer is supported on tape (film frame).

Most acrylic-type wafer taping media exhibit limited compatibility to most organic solvents used to clean/remove temporary adhesives, especially those described earlier (Table 2 & Fig. 5). For this reason, it is encouraged to use of aqueous-soluble adhesives, to achieve full compatibility with the film frame without the need of additional equipment would be necessary.

The primary reasons in using aqueous soluble adhesives include process simplification, material compatibility, cost reduction, and environmental safety. Whether it be cleaning with water or detergents, the chemistries are non-flammable, non-toxic, and do not generate evaporative material to trigger air permit requirements. Subtleties exist in aqueous cleans, and many believe it to be more challenging to control than organic solvents. Effective aqueous systems are built with additives that prevent irregularities during processing. Detergents are mixed with purified water. Ingredients in the detergent mix with species from the adhesive to prevent redeposition, inhibit corrosion, and stop scale build-up. These so-called detergents are complex and

offer a balance in chemistry to deliver performance at the selectivity that is desired by the process.

CONCLUSIONS

This paper presents data and process suggestions to reduce the occurrence of bow and warpage in thin substrates. By utilizing optical profilometry as a measurement tool with the use of temporary adhesives, a means to reduce bow by nearly 90% is shown to be feasible. This equipment and material options allow subsequent 3DIC processes to occur without yield loss. Using an aqueous soluble adhesive, the process can be planned for batch processing using simple and low cost detergents.

ACKNOWLEDGEMENT

The authors would like to thank the staff of our companies, and especially to Mr. Alexander Brewer for his support.

REFERENCES

[1] FRT of America, www.frtofamerica.com
[2] Diversified Applications Engineering Technologies (Daetec), www.daetec.com.
[3] SEMI MF 534 Test Method for Bow of Silicon Wafers.
[4] SEMI MF 657 Standard Test Method for Measuring Warp on Silicon Wafers by Noncontact Scanning.
[5] SEMI MF1530 Standard Test Method for Measuring Flatness, Thickness, and Thickness Variation on Silicon Wafers by Automated Noncontact Scanning.
[6] C. Orlando, T. Goodrich, and E. Gosselin, Backside Mounting Procedures for Semiconductor Wafer Processing, *Proceedings for GaAs MANTECH,* 2001, pp. 189-191.
[7] J.Moore, Materials and Conditions Used to Optimize Thinning, Processing, and Dismounting GaAs Wafers, *SEMICON-WEST Technical Symposium (STS): Innov. Semi. Mftg.,* July 2001, pp. 339-348.
[8] I. Blech and D. Dang, "Silicon Wafer Deformation After Backside Grinding, *Solid State Technology,* August 1994, pp. 74-76.
[9] C. McHatton and C. Gumbert, Eliminating Backgrind Defects With Wet Chemical Etching, *Solid State Technology,* November 1998, pp. 85-90.

[10] D. Mould, and J. Moore, A New Alternative for Temporary Wafer Mounting, *Proceedings for GaAs MANTECH,* April 2002, pp.109-112.

[11] J. Moore, A. Smith, D. Nguyen, and S. Kulkarni, High Temperature Resistant Adhesive for Wafer Thinning and Backside Processing, 2004 *Proceedings for GaAs MANTECH,* May 2004, pp. 175-178.

[12] A. Smith, J. Moore, and B. Hosse, A Chemical and Thermal Resistant Wafer Bonding Adhesive for Simplifying Wafer Backside Processing, *Proceedings for GaAs MANTECH,* April 2006, pp.269-271.

[13] U.S. Patent No. 7,678,861, J. Moore and M. Fowler, March 16, 2010.

[14] U.S. Patent Applications 2009/0017248 A1 (2009), *Larson et al.*; 2009/0017323 A1 (2009), *Webb et al.*; and International Application WO 2008/008931 A1 (2008), *Webb et al.*

[15] U.S. Patent No. 7,232,770, J. Moore and A. Smith, June 19, 2007.

[16] U.S. Patent Nos. 6,869,894 and 7,098,152, March 22, 2005 and Aug. 29, 2006, J. Moore.

WAFER SPRAY COATING FOR PRE-APPLIED UNDERFILL

Akira Morita and James Klocke
Nordson ASYMTEK
Carlsbad, CA
akira.morita@nordsonasymtek.com; james.klocke@nordsonasymtek.com

ABSTRACT

Recently, the process of pre-applying underfill onto a wafer is getting attention in 3D packaging. The pre-applied underfill material is placed on the bumped side of the wafer before dicing. Then the dies (with underfill) are stacked on another chip or wafer after dicing. There are a few ways to apply underfill onto the wafer: vacuum lamination, spin coating, and spray coating. Each of them has different challenges.

- Vacuum lamination needs expensive vacuum lamination equipment.
- Spin coating has material wastage.
- Spray coating must make consistent coating thickness over wafer.

This paper will demonstrate that using our experience with conformal coating and jet dispensing, we have evaluated wafer spray coating for pre-applied underfill application. This paper will present how we address the thickness consistency and wastage reduction challenges when using wafer spray coating.

Key words: 3D package, underfill, pre-applied, wafer coating, spray coating, material wastage.

INTRODUCTION

3D packages with through-silicon vias (TSV) have been enjoying a strong momentum recently for use in many semiconductor applications such as memory, MPU, application processor, and FPGA in mobile, data center and telecom applications. This is because these semiconductors need more bandwidth, and have a smaller form factor that saves power for the end applications.

Underfill for 3D packages includes two types: underfill between stacked silicon dies for micro-bump bonding, and underfill between silicon die and organic substrate for conventional flip chip bonding. Both types of underfill are usually required for a 3D package. The same technologies such as capillary underfill for a single flip chip package are applicable to the underfill between silicon die and organic substrate. The underfill between stacked dies needs a new technology in terms of both material and application methods mainly because of the tight geometries, even though there is no CTE mismatch between stacked silicon dies.

In 3D packaging, very tight bump pitch and bump height between stacked dies are required, for example, 25um bump height and 40um pitch. This geometry makes the conventional underfill process (capillary underfill) difficult in terms of underfill penetration speed under dies and increases the possibility of undesirable void creation.

There are three major technologies to apply underfill: capillary underfill, molded underfill and pre-applied underfill. Capillary underfill has been the dominant technology since flip chip technology began. Molded underfill has emerged recently for high-volume production especially for mobile devices because it combines the two processes of over-molding and underfill. Because capillary and molded underfill techniques apply underfill after die bonding, they are significantly challenged by the tight bump pitch and height between stacked dies because of slow penetration speed (even no penetration) and void possibility.

On the other hand, pre-applied underfill is not challenged by tight bump pitch/height because it is applied before die bonding. Therefore, pre-applied underfill is getting attention in the market. Pre-applied underfill on a wafer is especially attractive because bumps on die don't have to go through an underfill layer to reach the pads. In underfill on backside of the dies (non-bumped side), bumps on dies have to go through the underfill. And application on wafer is more productive than application on chip by chip.

REQUIREMENTS FOR PRE-APPLIED UNDERFILL

For a long time, pre-applied underfill has been used for flip chip packaging as long as the die size is small (such as less than 5mm) and the bump number is fewer than ten. These conditions can avoid pre-applied underfill challenges relatively easily: trapping fillers between bumps and pads, and trapping air in the underfill (voids) during the bonding process. But many existing and potential 3D package devices such as FPGA, MPU, memory and application processors have larger die size and significantly more bumps (i.e. >10,000 bumps) compared to the traditional devices adopting pre-applied underfill.

Underfill material suppliers suggest many various formulas and bonding/curing processes including temperature profiles and pressure for preventing the two challenges. Additionally, underfill coating shape on wafer such as thickness, topology and tolerance have been defined gradually with coordination between underfill material formulators and equipment manufacturers. Typical shape requirements include the following.

- Flat underfill surface
- Bump tops appear on underfill surface

- Thickness +/-10% tolerance
  (Typical thickness is 20um ~ 30um)

Figure 1 shows three shapes of underfill. Underfill height 2 is the most popular requirement currently. Underfill height 1 could be acceptable because of other process and material conditions. But Underfill height 3, which has hollow surfaces between bumps, is unacceptable because the hollow surfaces trap air in bonding.

The underfill thickness is mainly determined by the underfill volume requirement between dies. If the pad side surface on another die (counter part of die with underfill) is irregular because of circuitry, the required volume would change.

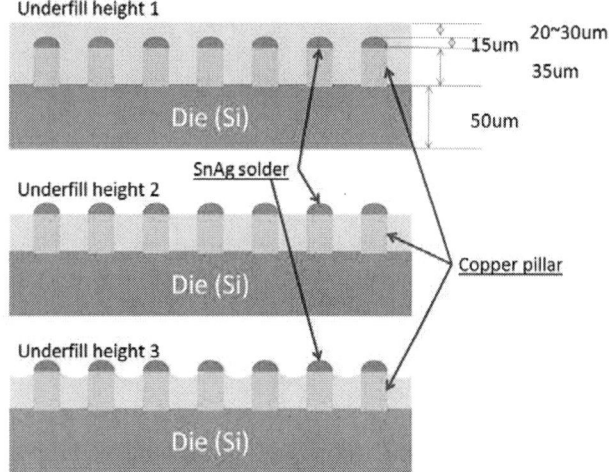

Figure 1. Pre-applied underfill shapes with different volume

These shape requirements are important to consider when selecting an underfill application technology.

APPLYING PRE-APPLIED UNDERFILL
Vacuum lamination, spin coating, and spray coating are processes used to apply uniform thin films or coating to flat substrates.

Vacuum lamination
Vacuum lamination has several different processes by equipment suppliers. The following process [1] is an example and Figure 2 illustrates the process flow below.

1. The substrate and film base pre-applied underfill or non-conductive film (NCF) are loaded into diaphragm vacuum chamber.
2. Vacuuming starts from substrate side and pressurization starts from NCF side.
3. Diaphragm makes NCF stick to substrate by pressure, and heat is applied to substrate if necessary.
4. Vacuuming and pressurization are reversed to release and unload the substrate.

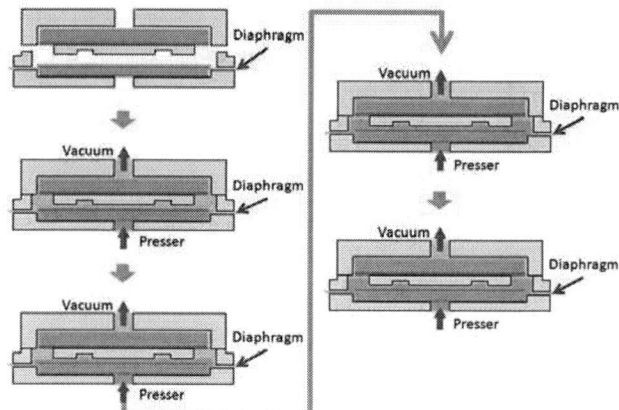

Figure 2. Vacuum lamination process flow chart

Spin coating
Spin coating has been established as described below, and is illustrated in Figure 3.

1. The substrate is loaded on a rotator.
2. An excess amount of a solution is placed on the substrate, which is then rotated at high speed in order to spread the fluid by centrifugal force.
3. Rotation continues while the fluid spins off the edges of the substrate, until the desired thickness of the film is achieved.

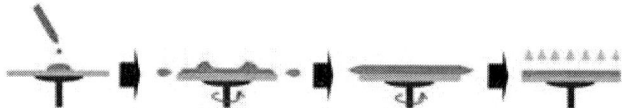

Figure 3. Spin-coating process flow chart [2]

Spray coating
Spray coating has several different processes. The following process was used for this study, and its flow chart is in Figure 4 below.

1. The substrate is loaded into a spray coating machine.
2. The spray applicator starts coating the substrate while the system coordinates fluid flow rate, substrate rotation rate and coating thickness.
3. The substrate is unloaded after coating is complete.

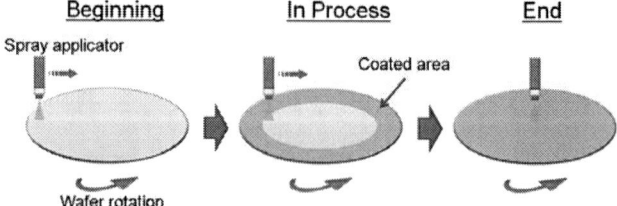

Figure 4. Spray coating process flow chart

Pre-applied underfill by spray coating is shown in Figure 5.

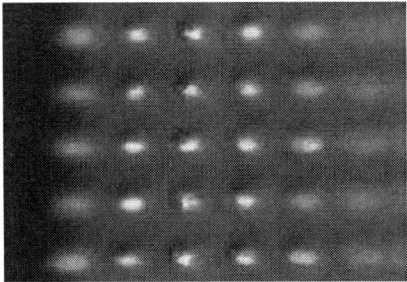

Figure 5. Pre-applied underfill by spray coating

APPLYING PROCESS PROS AND CONS
Vacuum lamination pros and cons
The process of vacuum lamination has been applied to various film applications such as die attach film and back grinding tape. It has advantages such as thickness uniformity and flatness after lamination. These advantages are valuable benefits for pre-applied underfill requirements. On the other hand, bumps need to go through non-conductive film (NCF) so that bump tops appear on the NCF surface. As a result, the underfill surfaces between bumps have s creases or deformation. The deformation can be flattened by heat and/or an adjustment to the viscosity of the underfill used in the vacuum lamination process. Another way to prevent the deformation is to make holes on the NCF, matched with bump locations on the wafer in advance. In this case, the vacuum lamination machine needs to make the alignment between bumped wafer and NCF for the holes to match. [3]

This process needs pressure to make NCF stick to the bumped wafer. This pressure could break interconnection layers (BEOL) if the layers are low-k material. Vacuum lamination is suitable to the entire wafer. It is difficult to laminate NCF on just specific areas on wafer. Under certain conditions, such as 2.5D applications where the wafer is used as a silicone interposer, only selective areas need pre-applied underfill. In this application, pre-applied underfill is required on the wafer interposer and not on the die. Thus, there are areas on the interposer that do not require underfill and must be selectively removed. Vacuum lamination has limitations for this request.

Vacuum lamination needs relatively expensive equipment, more than $250,000, because it requires vacuum chamber and hole alignment. And equipment size also is larger than others: more than 3m wide.

Spin coating pros and cons
Spin coating has been used for many years to apply thin layers of photoresist to semiconductor wafers. This is a highly reliable and well-studied method of applying coatings. With adjustments to the fluid viscosity and control of the spin speed, a recipe can be developed to predict the coating thickness. More complicated recipes can be developed with multiple spin speeds to overcome problems with single-speed coating. In some cases spin coaters can be

housed inside sealed chambers to control the evaporation rate of solvents in the coating materials, allowing better coverage of non-flat topographical features, such as walls and edges on MEMS devices.

Spin coaters use a variety of spray heads to apply photoresist and other materials to wafers. Fan- and cone-shaped sprays are often used to dispense an even coating on the wafer. In addition, ultrasonic spray heads have been employed to create a mist over the wafers. However they almost always use low-viscosity fluids and the speed of the wafer spinner to control the coating thickness on the wafer.

One major disadvantage to spin coating of wafers is the amount of wasted material. Coating material is flooded onto the center of the wafer, and it is thrown laterally at high speed to obtain a desired thickness. With 200mm diameter wafers it is estimated that approximately 48% of the materials applied to the wafer ends up wasted (as it spins off the wafer). With larger, 300mm wafers the waste is in the 75% range. This is one of the reasons that ultrasonic misting heads are used in spin coaters, to minimize wasted fluid. Although ultrasonic heads produce fine spray mists with a tight distribution of particles in the range of 25um, they need low-viscosity materials to function effectively. When using inexpensive fluids, the waste is not so much of a problem, but when materials cost over $1000 per liter, it becomes very expensive to coat wafers with poor transfer efficiencies.

Other problems that result from the spin process, such as edge beads, can cause downstream problems. The bead is a product of the spinning of the fluid and surface tension effects at the edge of a wafer. Equipment manufactures add separate spray attachments to the wafer spinners to remove the edge beads, using a small spray of solvent. This adds complexity and cost to the process, and a solvent/resist waste stream. One good feature of spin coating is its ability to coat over minor surface problems cause by dried water spots. The shearing action can actually move particles or other forms of contamination. However coatings may pull back if there is an area on the wafer where the fluid does not wet to or a small pit or projection (from the surface) exists.

Because spin coating needs low-viscosity fluids, it is not uncommon for users to thin high-viscosity fluids with solvent and apply several coatings on a wafer to get to a desired thickness. Dilution of materials can limit the fluid formulator's options. There are some cases when the material formulators would prefer to use high-molecular-weight adhesives, which have higher bond strength. If the material is thinned with solvent, the adhesive strength is lowered. With adhesive thickness on the order of 10um there is very little adhesive to start with, and to lower the adhesive strength to facilitate the coating equipment is not desirable.

Spray coating pros and cons

Spray coating is quite popular and used in many various industries. In the electronics industry, conformal coating on printed circuit board assemblies (PCBA) by spray applicator with silicone, urethane and other materials has been established for higher reliability electronics, protecting moisture from penetrating the PCBA. Also, flux coating is a major spray coating application for flip chip packaging and CSP/BGA assembly on board.

Spray technology intrinsically gives atomized liquid direction: thus selective coating (coating selective areas of a substrate) is possible. The edge definition of selective coating varies, depending on spray applicator technologies and liquid materials. Tight edge-definition ranges could be a few millimeters, which is the distance between no-coating area edge and coating area edge with pre-determined coating thickness. Therefore, a spray applicator can coat just the wafer area without much wastage. This is a clear advantage over spin-coating for pre-applied underfill. And this selective coating could be used to coat a specific area on a wafer, while spin-coating and vacuum lamination are limited in selectivity.

Spray applicators can adjust atomizing parameters such as fluid pressure and nozzle size to accommodate a wide range of fluid viscosities to make consistent coating. This is another advantage over spin-coating, which has just rotation speed for its major adjustment. And higher viscosity range is also another challenge. The rotation/spray process has been reported by several authors [4] but not using high-viscosity fluids.

However, thinner coating thickness and consistency of thickness are major challenges for spray coating against pre-applied underfill requirements because typical thickness and consistency requirements are 20um ~ 30um +/- 10%.

By reducing to a fairly low rotation speed, such as less than 1000 rpm, any edge build up, which happens in spin-coating, would be eliminated as well as any shadowing effect if you have a bump structure on the wafer. An atomizing spray head is used to create adhesive particles from the highly viscous fluid. This process does not flood the wafer, or use the spin/shear effect to thin the fluid layer. To coat the wafer a controlled arc motion path for the spray head is used. This results in an even coating over a 300mm diameter wafer, to within a few microns uniformity. Coverage is uniform to the edge of a wafer within 1mm of edge. See Figure 4 for a process flow for spray coating.

This study focuses on the following points to evaluate the technical feasibility of wafer spray coating equipment.
- Coating thickness on wafer: 20um
- Coating consistency: +/-10%
- Material wastage: less than 20%

APPARATUS FOR ROTATION/SPRAY PROCESS
Spray coating equipment for this study includes:

- An X-Y robotic dispenser system
- An air atomizing spray applicator
- Speed-programmable wafer rotator
- An integrated balance scale

The coating set-up for rotation/spray process consists of a speed-programmable wafer rotator and chuck. These are situated inside the X-Y robotic dispenser that is using the spray applicator. The air spray applicator uses a cone-in-seat valve adjustment system to vary the fluid flow. Typical application settings are;
- Fluid pressures range 2 to 3 PSI (0.14 to 0.21 bar) to >30 PSI (>2 bar)
- Atomization pressures range 10 to 50 PSI (0.69 to 3.4 bar)
- Nozzle diameter range 0.003 to 0.010 in. (0.0075 to 0.025 mm).

The rate of fluid flow is controlled by using a patented mass flow controller device that is integrated into this system. Through testing, the mass flow characteristics of the applicator can be determined by setting a flow rate from the applicator and characterizing this with a film thickness. Figure 6 shows the mass flow calibration principle.

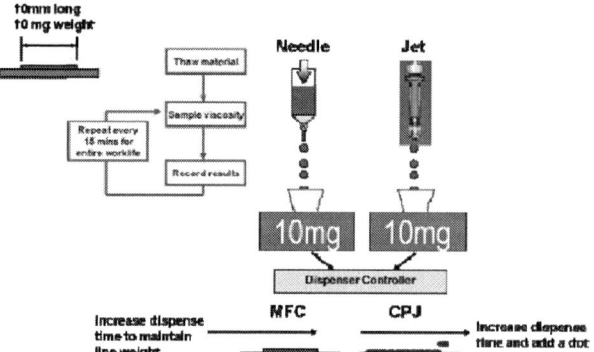

Figure 6 Mass Flow Calibration principle

A particular issue with rotation/spray to consider is that the linear speed of a spinning wafer at the circumference is large and zero at the center. Therefore the control of the spray applicator movement has to compensate for this problem. At the same time, if uniform coating to the edge of the wafer is critical, the spray applicator has to be able to coat to the very edge of the wafer, with minimal waste of fluid at the edge. Figure 7 shows a 300mm wafer that has been coated using this process.

Figure 7. 300mm wafer that has been coated

COATING THICKNESS MEASUREMENT METHOD
Coating thickness is measured by profilometer. For accuracy, the bare wafer surface was revealed by scratching off the coated area. Then a profilometer measured the distance between coated area and bare wafer area, see Figure 8.Measurement results by profilometer for 20um thickness are shown Figure 9.

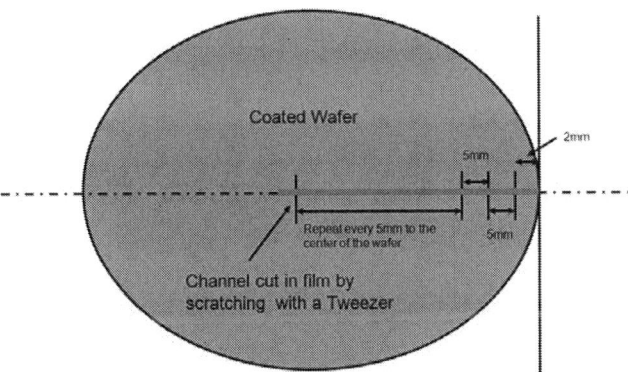

Figure 8. Coating thickness measurement method diagram

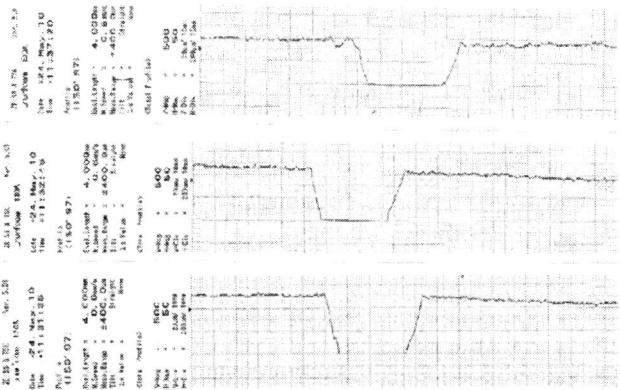

Figure 9. Measurement results by profilometer for 20um thickness

RESULTS
With a 300-mm diameter wafer, the volume of fluid on the wafer was typically 1 gram of material. Several different spray applicator motions were evaluated but the preferred motion was an arch across the wafer. Depending on the film thickness accuracy required, the coating time for a 300mm wafer was as low as 13 seconds. But if an accuracy of +/- 10% was required on a 20um thick film, the coating time was as long as 120 seconds.

Spray coating thickness was measured for a wafer using a 20um thickness target. The thickness distribution (Figure10) shows the coating consistency is within 20um +/-10%. Coating coverage is uniform to the edge of a wafer within 1mm of edge.

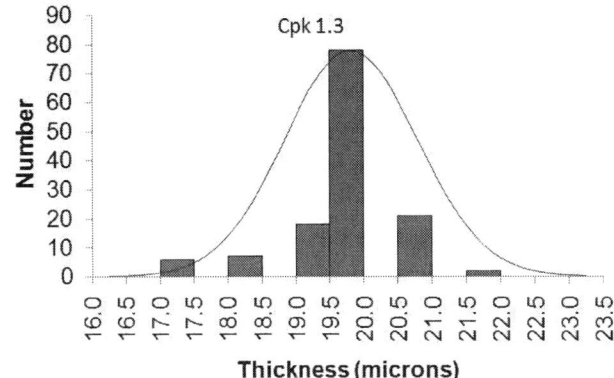

Figure 10. Spray coating thickness distribution for wafers.

Material wastage was measured for 4 wafer samples; results are shown in Table 1. All of the samples show less than 20% of material wastage.

Table 1. Material wastage results

Sample #	Blank Wafer wt. [gm]	Dis. Wafer wt. [gm]	Initial Syringe Wt. [gm]	After dis. Syringe wt. [gm]	Wafer net dis wt. [gm]	Syringe net dis. wt. [gm]	Wastage [gm]	Wastage [%]	Dist. From edge [mm]
1	52.7174	53.5882	36.0588		0.8708	1.0641	0.1932	18.2	0
2	51.2733	52.1463			0.8730	1.0641	0.1911	18.0	0
3	51.5687	52.4431			0.8744	1.0641	0.1897	17.8	0
4	51.6906	52.5685		31.8026	0.8780	1.0641	0.1861	17.5	0

Material wastage was measured by comparison between syringe weight change and wafer weight change before and after coating.

CONCLUSIONS
3D package requirements dictate the need for thin and consistent coating of pre-applied underfill materials. Pre-applied underfill volumes require 20um ~ 30um film thicknesses, with a +/-10% tolerance (on the thickness). Material wastage should be minimized to save costs. The process must be flexible (high volume/low mix and low volume/high mix) with regard to selectively placing the underfill material only where it is needed, and the coating process equipment should be minimally invasive (small footprint) to the production floor.

ACKNOWLEDGMENTS
The authors would like to thank Mr. Masaaki Hoshino and Ms. Satomi Kawamoto of NAMICS Corporation, Mr.

Hironori Kurauchi and Mr. Hironari Mori of Sumitomo Bakelite, and Ms. Heakyoung Park and others from Nordson ASYMTEK for their help.

REFERENCES

1. Company website for MEIKI, http://www.meiki-ss.co.jp/mac/index.html.

2. Spin-coating entry for Wikipedia-Netherlands http://nl.wikipedia.org/wiki/Bestand:SolGel_SpinCoating.jpg

3. Eric Huenger, et al., "Development of a Low Temperature Curing Aqueous Base Developable Photoimageable Dielectric for Wafer Level Packaging," IMAPS Device Packaging, March 2012.

4. Mark Whitmore, Jeff Schake, "Screen and Stencil Printing for Wafer Backside Coating," 33rd International Electronics Manufacturing Technology Conference, 2008.

ROOM TEMPERATURE DEBONDING – AN ENABLING TECHNOLOGY FOR TSV AND 3D INTEGRATION

Garrett Oakes[2], Thorsten Matthias[1], Eric Pabo[2], Jürgen Burggraf[1], Daniel Burgstaller[1], Markus Wimplinger[1] and Paul Lindner[1]

EV Group
Florian/Inn[1], Austria, Tempe AZ[2]
G.Oakes@EVGroup.com

ABSTRACT

Thin wafer processing is a critical technology for TSV manufacturing and 3D integration. Thin wafer processing allows reducing the aspect ratio of the vias, thereby reducing the total processing cost and enables ultra-thin packages for handheld applications. Temporary bonding to a rigid support carrier and debonding after backside processing have been used for thin wafer handling/processing for many years. However, so far all the debonding methods imposed severe limitations on the manufacturability. For light induced debonding the carrier had to be transparent and for solvent based debonding the carrier had to be perforated. For thermally induced debonding, "slide-off debonding" the debonding temperature had to be below the reflow temperature of the solder bumps, which limited the maximal process temperature of the adhesive.

In this paper we describe a new debonding method at room temperature. This new technology decouples the debonding process from the adhesive properties, which creates a de facto material independent debonding standard. As the debonding process does not rely on the adhesive properties a major boundary for adhesive engineering has been removed. The debonding method is compatible with bumps or pillars in the bond interface as well as on the backside of the wafer stack. No force is applied on the bumps during debonding which results in very high yields.

Keywords: TSV, 3D Integration, Room Temperature Debond

INTRODUCTION

3D stacked ICs (3Ds-IC) have been a hot topic for several years, but recent announcements from leading image sensor and memory manufacturers show that 3Ds-ICs finally move into high volume manufacturing. The main difference between a standard 2D wafer fab and a 3Ds-IC wafer fab is the ability to process both sides of an ultra-thin wafer and to manufacture through silicon vias (TSVs). Wide I/O DRAM is currently targeting 20 μm thin wafers. The most obvious reason for thin wafers is the reduced form factor, which is especially important for handheld devices. However, probably even more important is that thinner wafers enable significant cost reduction for TSVs. The silicon real estate consumed by the TSVs has to be minimized in order that the final device provides a performance advantage compared to traditional 2D devices. The only way to reduce area consumption by the TSVs is to reduce their diameter. For a given wafer thickness the reduction of TSV diameter increases the TSV aspect ratio. However, the cost and cycle time of the main TSV manufacturing process steps etching, barrier/seed layer deposition and plating increases significantly with higher aspect ratio. Thinner wafers enable smaller TSV diameters and lower TSV aspect ratios and thereby enable lower cost for TSV manufacturing [1]

The implementation of thin wafer processing in high volume memory manufacturing has brought a significant change of the requirements. In the past the early adopters of thin wafer processing in the fields of power electronics and compound semiconductors designed the backside process flow around the ability to handle and process a thin wafer. Today stacked memory applications the compatibility with standard processes at highest yield is a must. The thin wafers today usually have microbumps on both sides. To ensure high yield for thermo-compression microbump bonding, the thin wafers have to fulfill wafer fab cleanliness requirements after debonding. In a nutshell the industry demands standardized processes for thin wafer handling. The revolutionary ZoneBOND™ technology achieves just that –standardized and material independent processes and equipment.

THIN WAFER PROCESSING BY TEMPORARY BONDING TO A CARRIER

Processing of thin wafers is enabled by temporarily bonding the device wafer to a rigid carrier wafer. Figure 1 shows the principle of thin wafer processing. After front-side processing of the device wafer is completed, the device wafer is bonded face down to a rigid carrier wafer. A temporary adhesive is used for wafer bonding. Silicon or glass wafers are the most common choices as carrier. The carrier wafer gives mechanical support for the device wafer during backgrinding and all subsequent backside processing steps. After the entire backside processing of the thin device wafer is completed, the thin wafer is debonded from the carrier and mounted on a dicing tape on a film frame.

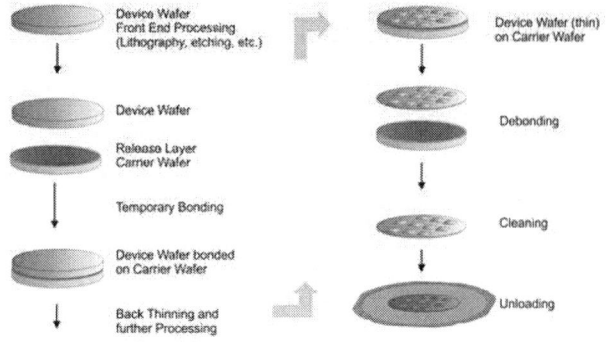

Figure 1: Principle of thin wafer processing using temporary bonding to a carrier wafer.

DEBONDING – THE KEY FOR RELIABALE THIN WAFER PROCESSING

In the past the debonding method, the adhesive properties and the carrier properties were closely linked to each other. This link between debonding method, adhesive and carrier imposed severe limitations on the manufacturability. Laser induced debonding required a glass carrier. For solvent release adhesives the carrier had to be perforated in order to provide access for the solvent to the bond interface. The perforation of the carrier causes the thin Si wafer to flex above the holes during backgrinding and polishing, which results in thickness non-uniformities ("dimples"). Combining thermoplastic materials with perforated carriers can lead to adhesive squeeze out through the carrier perforation during wafer bonding and other processes at elevated temperature. For thermally induced slide-off debonding a thermoplastic adhesive is used. Heating the adhesive to the debonding temperature reduces the viscosity of the material, which allows to slide off the thin wafer without mechanical stress. Of course the debonding temperature has to be lower than the reflow temperature of the bumps. The higher the temperature the lower is the adhesive viscosity and thereby the ability of the adhesive to withstand the internal stress of the thin wafer. Therefore for most slide-off adhesives the maximum operating temperature is only 50 – 100 °C higher than the debonding temperature. So in the past the debonding method defined the choice of carrier and adhesive. The debonding method imposed severe limitations on the adhesive properties.

A revolutionary new temporary bonding and debonding technology, ZoneBOND™, breaks this link between debonding method and adhesive properties. With ZoneBOND™ technology the debonding process is not at all a function of the adhesive any more – debonding has become a function of the carrier. Figure 2 shows the principle of the ZoneBOND™ carrier. The ZoneBOND™ carrier has two zones, which differentiate by the degree of adhesion between the adhesive and the carrier. The adhesion in the center zone is reduced, whereas full adhesion is at work in the edge zone. It is important to note that the surface of the device wafer does not have to be treated at all for ZoneBOND™, which makes the technology compatible

with any kind of surface passivation. This is especially important with regards to assembly after thin wafer processing. Debonding methods which rely on surface modifications of the device wafer have the inherent risk of causing adhesion problems with the underfill material during die bonding.

ZoneBOND™

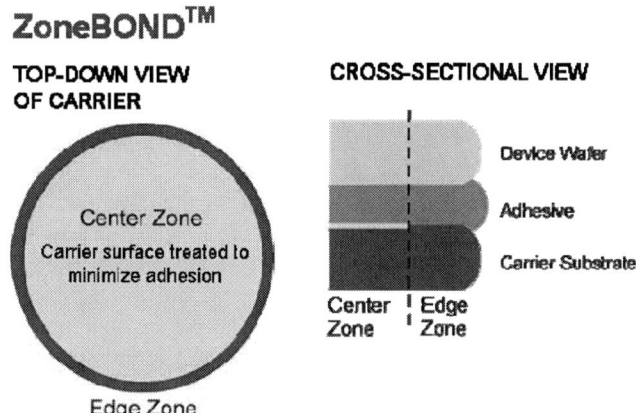

Figure 2: Left: top down view of a ZoneBOND™ carrier, Right: cross section of a ZoneBOND™ carrier; the carrier surface of the center zone is treated such that the adhesion is reduced.

Figure 3 shows the debonding principle using a ZoneBOND™ carrier. During the first step, the Edge Zone Release (EZR®), the adhesive in the edge zone is dissolved. The center zone with the reduced adhesion is now the only connection between the thin device wafer and the carrier. The device wafer is separated from the carrier during the Edge Zone Debond (EZD®) step with a pure mechanical separation at room temperature. It is important to note that the actual separation happens between the adhesive layer and the carrier wafer. This means that the debonding is totally independent from the top passivation layer of the device wafer. The debonding is also totally independent from topography in the bond interface: spherical bumps or pillars; 35µm or 80µm feature size, even for stacked dies on an interposer – as the debonding process happens at the boundary between adhesive layer and carrier the topography does not have any impact on the debonding process. During the EZD® step the bumps in the bond interface are embedded in the adhesive layer. No vertical or shear force is applied to the bumps during debonding, which eliminates the risk of bump damage. After debonding the device wafer is cleaned in a dedicated thin wafer cleaning module. Thermoplastic adhesives are removed residue free by solvent stripping. Solvent cleaning process has the advantage that it works well with spherical bumps – no force is applied to the bumps during adhesive layer removal.

1st Step : Edge Zone Release (EZR®)

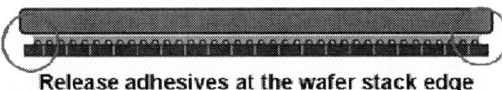

Release adhesives at the wafer stack edge

2nd Step : Edge Zone Debond (EZD®)

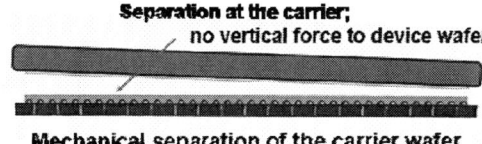

Mechanical separation of the carrier wafer

Figure 3: ZoneBOND™ Debonding principle: during the Edge Zone Release (EZR®) step the adhesive edge zone is dissolved; the thin device wafer is separated from the carrier during the Edge Zone Debond (EZD®) step.

Figure 4 shows the entire debonding process flow for ZoneBOND™ as it is implemented in the EVG850DB production debonding system. The bonded wafer stacks are delivered in a FOUP to the system. First the Edge Zone Release step is performed as single wafer process. As a result of the EZR process a thin free standing wafer edge is created. Putting the wafer stack after the EZR process into a FOUP can result in edge chipping. Combining EZR and EZD in one system is a necessary step for high yield debonding. Then the wafer stack is mounted on a film frame. Performing the edge zone release process prior to film frame mounting allows using dicing tapes, which are not compatible with the solvents used for EZR. This gives foundries and OSATs full freedom of choice for the dicing tape. After debonding the device wafer has to be cleaned. During cleaning the dicing tape is protected from exposure to the cleaning chemistry. The thin wafers on film frame are unloaded into a cassette.

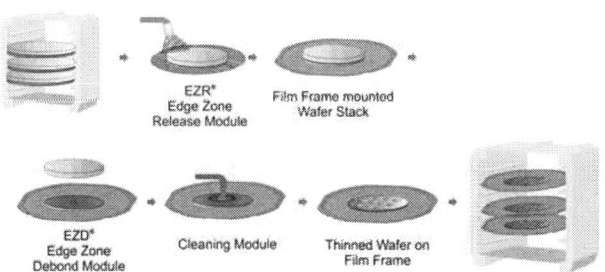

Figure 4: Full ZoneBOND™ debonding process sequence

The analysis of the process sequence shows that the debonding is completely independent from the adhesives properties. This opens up a complete new field for adhesive engineering. Taking away the necessity to design the debonding process into the adhesive enables a higher focus on the other success criteria for a temporary adhesive. The adhesive has to be solid at low temperature to enable backgrinding without dimples. It has to withstand the thermal and chemical exposure during backside processing. After debonding the adhesive has to be cleanable according to the requirements of a wafer fab. There must not be any

residue or modification to the device wafer surface in order to allow standard underfill processes during assembly. ZoneBOND™ debonding works with all kind of materials: dedicated thermal release adhesives, dedicated laser release materials as well as dedicated solvent release materials. Debonding is now a standardized process independent from the specific temporary adhesive. This standardization is a major milestone in thin wafer handling.

For the introduction of a new manufacturing technology like thin wafer handling another important milestone is a versatile supply chain. Many semiconductor manufacturers have a multiple source policy for all materials in the fab. EV Group provides the ZoneBOND™ Open Platform. Our customers get complete freedom of choice in regards to the temporary adhesive. Any type of adhesive from any supplier can be used. Thereby ZoneBOND™ technology provides investment protection. Figure 5 shows the temperature capability of various high temperature materials. We are positive that in the future there will be exciting new developments in temporary adhesives – the standardization of the debonding process ensures that today's equipment will work with tomorrow's adhesives.

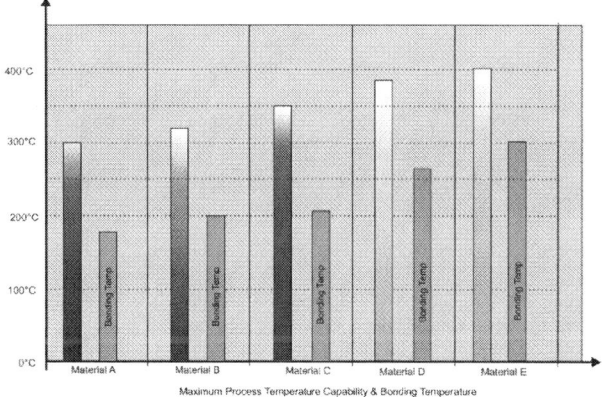

Figure 5: Various adhesives with high temperature capability

The availability of carrier wafers is another important aspect for a reliable and versatile supply chain. EV Group enables its customers to manufacture the ZoneBOND™ carrier by themselves. For pilot line and low volume production the carrier preparation modules are integrated into the EVG850TB temporary bonding system. For high volume manufacturing a dedicated ZoneBOND™ carrier manufacturing system is provided. Standard silicon or glass wafers can be used as carrier wafers which gives long term cost clarity. Temporary bonding and debonding with ZoneBOND™ technology has been implemented on the EVG850 platform (fig. 6). Up to nine process modules ensure optimal throughput matching of the various process steps. Wafer logistics is an important aspect as wafer bonding is different from any other fab process. Two incoming wafers are bonded and typically unloaded into a dedicated receiving cassette. This means that thre FOUPs are used just for one processing cycle. For continuous mode of operation it is necessary to have six FOUPs on the tool, which has been implemented with a local FOUP storage system. The same applies to debonding - three cassettes (1x incoming wafer stack, receiving cassette for the carriers, receiving cassette for thin wafer on film frame) are used just for one processing cycle. In a previous article we described the importance of thickness variation control during temporary bonding. An integrated metrology module ensures seamless integration with the TSV reveal processes after thinning [2]. The wafer-to-carrier alignment during temporary bonding achieves better than 50 μm (3σ) alignment accuracy. This prevents downstream problems with the thin wafer edge, but also ensures uniform backside process results.

[2] M. Wimplinger et al., In-line Infrared Metrology for High-Volume Temporary Bonding Applications, Chip Scale Review, August 2011

Figure 6: EVG850 production platform for temporary bonding and debonding

CONCLUSION

ZoneBOND™ is a revolutionary breakthrough in thin wafer processing. It enables room temperature debonding method, which is independent from the properties of the temporary adhesive. Thereby it enables a standardization of the debonding process and debonding equipment as it is material independent. The ZoneBOND™ Open Platform enables a versatile supply chain with multiple adhesive suppliers. Manufacturing of the carrier wafer is integrated in the high throughput EVG850 temporary bonding/debonding production platform.

ACKNOWLEDGEMENT

LowTemp®, EZR® and EZD® are registered trademarks of EV Group, St. Florian am Inn, Austria

ZoneBOND™ is a registered trademark of Brewer Science, Inc., Rolla, MO, USA

REFERENCES

[1] P. Siblerud, Cost effective TSV integration, Proc. Pan Pacific Microelectronic Symposium 2010

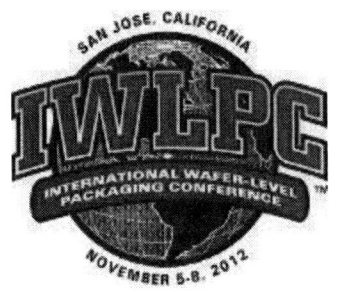

Heterogeneous Packaging for MEMS

Matt Apanius, Desich SMART Center

Desich SMART Center

Overview

- A MEMS packaging development foundry that creates system integration solutions for customers developing MEMS-based products

- Advanced and standard packaging capabilities include back-end wafer processing, package assembly, performance testing, accelerated reliability testing, and package design

- Key MEMS packaging and testing services in one location is a cost-effective customer solution that reduces lead times

Package Assembly Basics

Process steps that take place after wafer fabrication

Wafer Prep

Wafer Dice

Die Attach

Wire Bond

Encapsulation

- Utilization of standard package assembly processes to get the job done

- MEMS devices integrated at the wafer level still need to be connected to a circuit board at some point

- Therefore, the interconnect interface (the package) still needs to exist

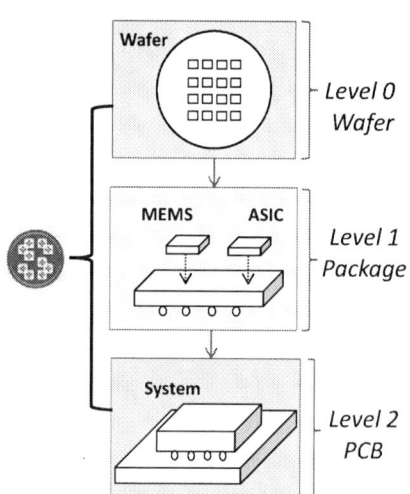

Heterogeneous Integration

Also known as System-in-Package...

Side by Side Placement	Stacked Structure	Embedded Structure

• Wire bond
• Flip chip
 	
+ passive components
	• PoP
	• Stacked die
	  –> WB + FC
	• Chip to chip/wafer
	  –> WLP
		• Chip in PCB
		• eWLP
		–> single/stacked layers

"...characterized by any combination of more than one active electronic component of different functionality plus optionally passives and other devices like MEMS or optical components assembled preferred into a single package that provides multiple functions associated with a system or subsystem."

Dr. Robert C. Pfahl, Jr. (iNEMI 2005)

Perhaps a Preferred Approach

A perspective from the end-user addressing new applications

- Resources for custom monolithic designs may not exist – limited R&D dollars

- Utilization of commercially available die reduces development costs and time

- Assembly at the package level eliminates incompatibility issues associated with working with multiple fabs

- Provide degrees of freedom for multiple sensing functionalities

Hetero- Advantages

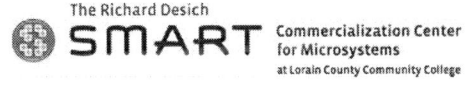

Integration flexibility, lower cost, faster time to market

- Standardized architectures and materials reduce process complexity

- Modular designs can be interchangeable

- Reuse of intellectual property shortens time to market

- Allows integration of multiple sensing functions

MEMS Integration Issues

Mary Ann Maher

SoftMEMS

SoftMEMS Introduction

SoftMEMS offers the broadest EDA integration platform and works with the best
of breed multi-physics simulators to enable the co-design of packaged
electronic circuitry with RF-MEMS, Microphones, Optical devices, Fluidics,
Biochips, Pressure Sensors, Accelerometers, Gyroscopes, etc...

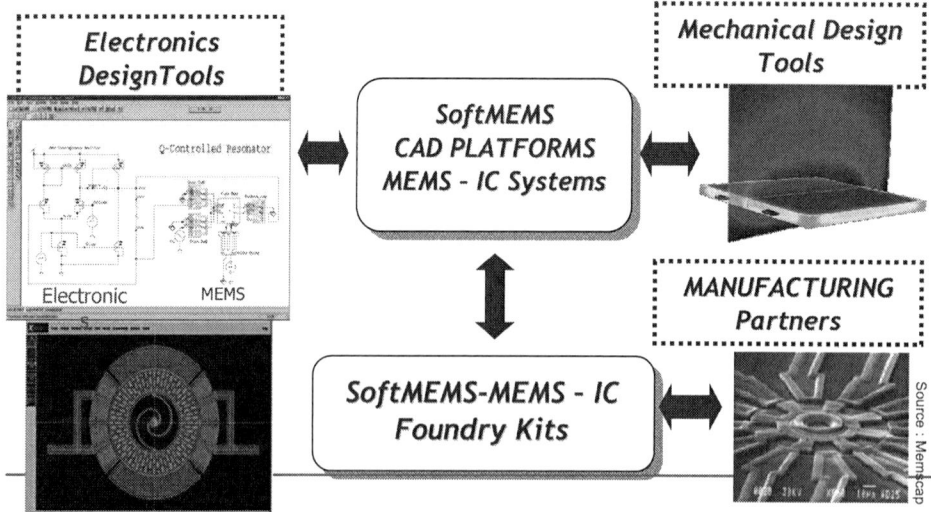

MEMS Integration Strategies

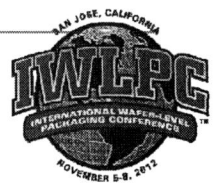

- Multi-chip systems-Hybrid integration
  - Die stacked
    - Typically MEMS die with wafer cap stacked with IC
  - Side by Side with wire bond
    - Multiple sensors- heterogeneous MEMS technologies
- Monolithic integration
  - MEMS on top, on bottom, within-"CMOS-MEMS"
  - MEMS may be capped
  - I/O interaction with MEMS needed
- Companies may have more than one strategy
  - That changes with time!
  - Product/Application dependent

Integrated MEMS

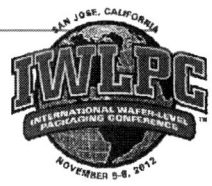

- Technology Challenges
  - MEMS post processing may be needed on IC wafer
  - MEMS and IC compatible process-temp, materials
  - Die level reliability issues- both MEMS and IC parts must be reliable
  - Mechanical properties of IC devices are not optimized
- Big Advantage
  - On-chip electronics- processing and compensation
- Co-Design Issues
  - MEMS "DRC"
  - Parisitics analysis on chip
  - Compatible design database for electronics and MEMS

Multi-Chip Systems

- ■ Technology Challenges
  - ❑ Limited I/Os needed for wirebonded solution
  - ❑ Wirebond-> moving to Interposer with TSV
  - ❑ More packaging levels compared to integrated system
- ■ Big Technology Advantage
  - ❑ Separately optimize fabrication processes
- ■ Co-Design Issues
  - ❑ Thermo-mechanical analysis of chip stack
  - ❑ Signal integrity

Conclusions

- ■ MEMS Integration-Decision based on many issues
  - ❑ Technology match- materials and process
  - ❑ Package cost vs. process cost
  - ❑ Die size/Device size match
  - ❑ Number of interconnects needed
  - ❑ Reliability issues
  - ❑ Sensing mechanism with electronics, capacitive, resistive etc.
- ■ Rise of IC fabs doing MEMS and fabless design houses will change the game and the strategy
- ■ MEMS integration ecosystem will enable more companies to take advantage of integration advances
- ■ Co-design ensures success with any MEMS integration strategy

Package Modeling of MEMS Devices

Manickam Thavarajah &
John Bloomsburgh
Fairchild Semiconductor

Diversified Packaging Needs

- MEMS based products (and their packaging requirements) are dramatically diversified:
  - By class of devices:
    - Mechanical devices (accels, gyros, pressure sensors, microphones, resonators, valves, etc.)
    - Optical devices (mirrors, spectrometers, gas chromatographs, displays, etc.)
    - Fluidic devices (reactors, pumps, filters, separators, etc.)
    - Bio/Nano devices (sensors, actuators)
  - By industry (same MEMS die may requires different packaging for different markets):
    - Military
    - Avionics
    - Process control
    - Automotive
    - Medical
    - Consumer
    - Etc.

Samples of MEMS Packaging

Military sensors. Kulite

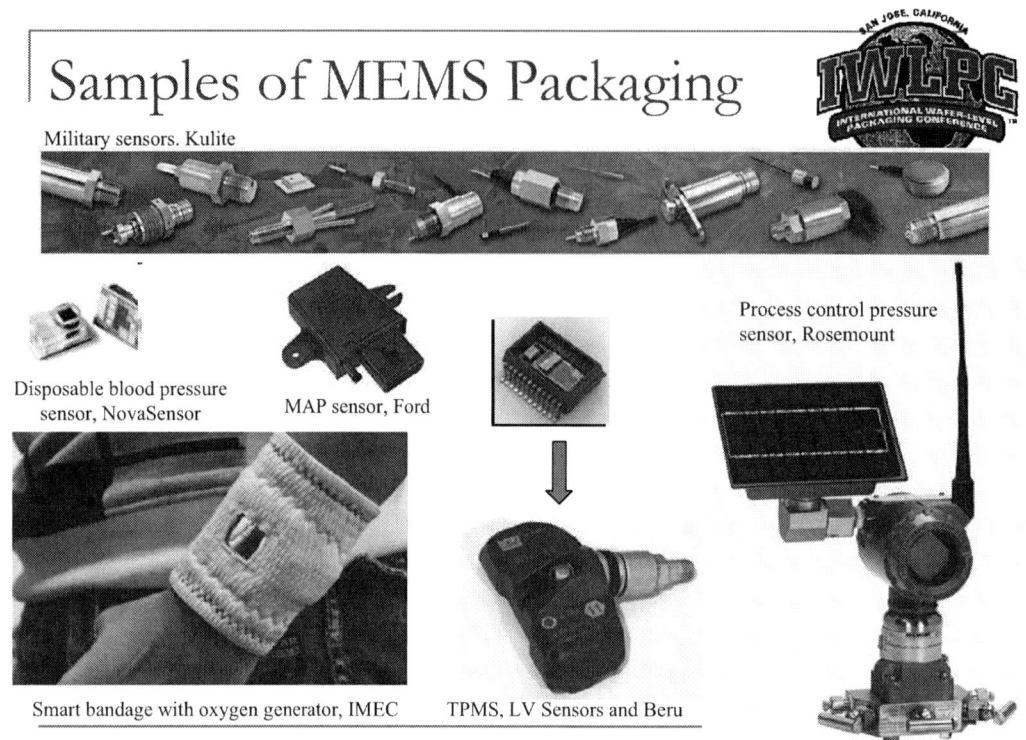

Disposable blood pressure sensor, NovaSensor

MAP sensor, Ford

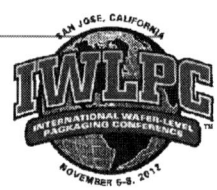

Process control pressure sensor, Rosemount

Smart bandage with oxygen generator, IMEC

TPMS, LV Sensors and Beru

Packaging Challenges for MEMS

- Just like IC packages, MEMS packages must be optimized for reliability, signal integrity, form factor, and thermal dissipation, but that's where the similarities end

- Primary packaging concern for most MEMS devices is performance degradation
- Many devices sense stress or displacement, often with Å resolution, which is easily degraded by package induced stress (e.g., overmolding or material TC mismatch)
- Moisture will have a big impact on signal connectivity with zF resolution

Packaging Challenges for MEMS

- MEMS devices often require a specialized interface with the outside world
  - Many devices interface directly with physical world
    - Die must be exposed to corrosive input gas or fluid media (e.g., pressure sensor, fluidic devices, etc.)

- Some MEMS operate at high temperatures (e.g., 1000°C for in-cylinder pressure sensor)

- Some devices are sensitive to processing vibration (e.g. during grind) and shock (from handling equipment such as pick and place)
  - Many devices are built from multiple wafers or dice, with challenging wirebonding requirements (e.g., wirebonding to vertical surface)

Package deformation under temperature change

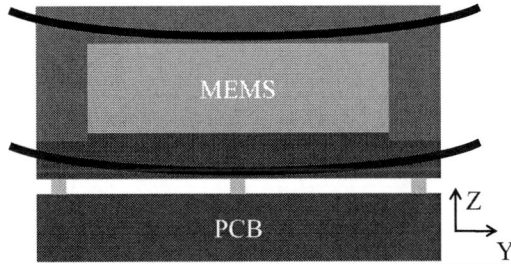

- Red = positive Z deformation

- When the temperature rises, the higher CTE materials at the bottom of the stack will expand more than the Silicon & plastic
  - This creates a "smiley face" bowing of the whole structure

Impact of package deformation on mirror

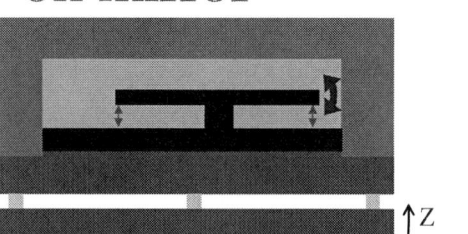

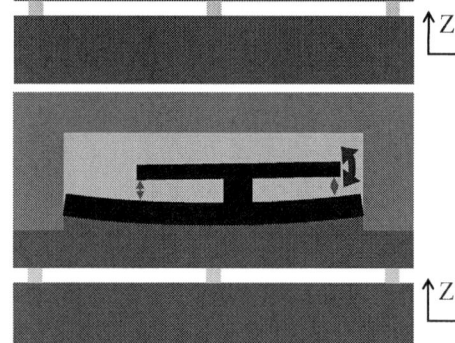

- If the mirror is offset to one side of the MEMS device, thermal stresses could induce bias problems

- Mismatches between symmetric sense components create errors that are difficult to detect in operation

- Die placement, mold compound variations, and other assembly effects can play a big role

---

Summary

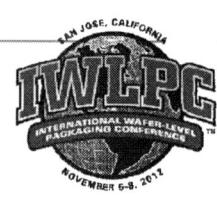

- MEMS packages have many of the same requirements as IC packages
  - Reliability, signal integrity, and small size/form factor

- Two main differences
  1. MEMS devices must interact with the outside world
     - Packaging must allow interfacial MEMS components access to the outside world while protecting everything else from the outside world
  2. MEMS are inherently Electro-***Mechanical*** components and the mechanical impact of package stresses will have a *direct* impact on MEMS performance
     - This impact must be studied and its effect minimized
     - FEA modeling is crucial
       - A good understanding of the MEMS operation **AND** the package materials, dimensions, and tolerances are crucial
       - Often, iteration between modeling and measurement is necessary to achieve full understanding of MEMS/package interactions
     - <u>Requires good partnership between MEMS designers and package designers</u>

- Minimizing the impact of the package on MEMS performance can only be achieved through synergistic package and MEMS design